# Using Computers

# Using Computers:
## The Human Factors of Information Systems

*Raymond S. Nickerson*

A Bradford Book
The MIT Press
Cambridge, Massachusetts
London, England

This book was set in Linotron 202 Baskerville by Achorn Graphic Services and printed and bound by Halliday Lithograph in the United States of America.

Library of Congress Cataloging-in-Publication Data

Nickerson, Raymond S.
  Using computers.

  "A Bradford book."
  1. Computers.   2. Electronic data processing.
  3. Human engineering.   I. Title.
  QA76.N497   1986        004'.01'9        85-24163
  ISBN 0-262-14040-3

To Doris

# Contents

# 4
## Anticipated Developments    54

# 5
## The Study of Person-Computer Interaction    74

# 6
## The Physical Interface    89

# 7
## The Cognitive Interface    112

# 8
## Software Tools    153

# 9
## Communication and Information Services    *171*

# 10
## Information Technology and Jobs    *200*

# 11
## Information Systems in the Office    *213*

# 12
## Designing Interactive Systems    *221*

# 13
## Some User Issues    *240*

# 14
## *Programming* 257

# 15
## Artificial Intelligence and Expert Systems    275

# 16
## Some Research Challenges    315

# 17
## Quality of Life: The Fundamental Issue    321

# 18
## *A Perspective* 348

# *Foreword*

This book has grown out of a long-standing interest in computers, how they work, what they can do, and what their existence means and will mean for society and for individuals. It may help put it in perspective if I note the events that have been most influential in sparking and sustaining that interest.

My first encounter with a computer was as an experimental psychologist with the Decision Sciences Laboratory of the U.S. Air Force Electronics Systems Division at Hanscom Field, in Bedford, Massachusetts. In the early 1960s the laboratory acquired a Digital Equipment Corporation PDP-1. This was the third or fourth PDP-1 to be installed, the first two or three of these machines having gone to the Massachusetts Institute of Technology, Bolt Beranek and Newman Inc., and possibly one other place. Prominent in my recollection of those days is the excitement that I shared with several colleagues—including Charles Brown, Donald Connolly, Carl Feehrer, Ugo Gagliardi, Ira Goldstein, and John R. Hayes—in using a computer to control experiments on perception, communication, decision making, learning, and related topics when this was a novel thing to do.

One especially keen memory involves a tiny but serendipitous excursion into computer graphics. In order to learn how to use the computer's oscilloscope display, I wrote a little program that would display any polynomial function over a specified range, perform a variety of specified operations on the function—multiply by a constant, add or delete a root, differentiate, integrate, change a coefficient—and immediately show the results. This was not a complicated program. The computer was used simply as a very fast clerk and draftsman. My only purpose in writing it was to get a little experience in working with a new display device. In fact, however, by interacting with this simple program, I believe I acquired an understanding of polynomial functions that was deeper, and certainly more graphic and dynamic, than what I had obtained from solving textbook problems in college math courses. My experience with this program, and with some related ones that permitted exploration of certain statistical constructs, left me with a firm belief in computer technology's potential for use in education and training.

The DSL experience sparked my interest in computer technology; subsequent affiliations with two other organizations have sustained and intensified it. The first was Tufts University and the second Bolt Beranek and Newman Inc. In 1965 Philip Sampson, Chairman of the Psychology Department at Tufts University and my thesis advisor, invited me to teach an introductory course on computers at Tufts. No such course had been offered on the campus before, and there was no established tradition for how one should be organized, so this was an opportunity to design something more or less from scratch. I taught this course for seven years, thoroughly enjoyed doing so, and stopped only when other job responsibilities made it impossible to continue. This experience forced me to look into some aspects of computer technology that I otherwise would not have studied; it also heightened my interest in computer applications and in the question of what implications the existence of this new and radically different type of machine has for the future.

In 1966 I joined the Information Sciences Division of Bolt Beranek and Newman Inc., which was directed by Jerome Elkind and John Swets. I am grateful to both of these people for giving me this opportunity, and to John Senders, a Principal Scientist at BBN at the time, for instigating this move. BBN has been my professional home for over eighteen years now, and it has been a good home indeed; I am indebted to many colleagues—far too numerous to attempt to mention by name—who have made BBN an intellectually exciting and congenial place to be. Many of these people are represented in the reference list at the end of the book.

One would have to be unusually insensitive to live among the computer-related research and development activities at BBN over the last two decades and not be provoked to think about them. The writing of this book has served to remind me of how extensive those activities have been. They have involved many aspects of computer and communication technology, including time-sharing, packet switching, network design and operation, electronic mail, artificial intelligence, and computer-assisted instruction. Of course, I am more likely to be aware of work going on at BBN than of similar work that is being done elsewhere. If I have failed to give commensurate recognition to other groups or individuals, it is owing to a lack of knowledge on my part, and not to any intent to slight.

The most immediate cause of this book was an opportunity in the spring of 1983 to chair a workshop on user–information system interaction sponsored by the National Research Council's Committee on Human Factors. The main objective of the workshop was to identify problems relating to the use of computer-based information systems that could be the foci of research sponsored by the Information Science and Technology Division of the National Science Foundation. To prepare for that workshop, I wrote a background paper, which turned out to be the beginnings of the book. When the idea of expanding the paper into a book was first entertained, the audience in mind was engineering psychologists, human-factors researchers, and people building computer-based systems for use by individuals not trained in computer technology. As the book evolved, the target audience broadened somewhat. While the main focus of the book still is on the human factors of computers and computer-based systems, I have tried to make it informative to anyone with an interest in such systems, how they are used, and what they mean.

I am grateful to my friend and colleague Richard Pew, who, as Chairman of the NRC Committee on Human Factors, asked me to chair the workshop; to the workshop participants—Sara Bly, Baruch Fischhoff, Henry Fuchs, Edmund Klemmer, Steven Leveen, R. Duncan Luce, Theodore Myer, Allen Newell, Howard Resnikoff, Brian Shackel, Daniel Westra, Robert Williges, and Patricia Wright—whose efforts assured the workshop's success; and to Robert Hennesey and Stanley Deutsch of the NRC staff, who, as Workshop Organizer and Study Director, provided invaluable help in arranging the workshop and getting out its report. I want also to thank Edward Weiss and Hal Bamford of the National Science Foundation for their sponsorship of the workshop, and my management, especially David Walden, President of BBN Laboratories, for their support of my participation in the workshop and the writing of the book. Although none of these people can be held accountable for the book's shortcomings, each of them shares some of the responsibility for the fact that it got written.

Several colleagues have been kind enough to read specific portions of the manuscript at various stages of its preparation and to provide very helpful comments on them. These include Madeleine Bates, John O'Hare, John Makhoul, Walter Reitman, and Natesa Sridharan. I have thoroughly enjoyed my interactions with Harry and Betty Stanton of MIT Press Bradford Books and am pleased to have this volume in their collection. Special thanks go to Joan Santoro and Diane Flahive for typing the manuscript and patiently enduring the many changes and revisions, and to Barbara Smith for helping find some elusive references.

My greatest debt of gratitude, by far, is to my wife, Doris, who has been a constant source of inspiration and support in more ways than I could possibly recount, not only during this project, but over many years. It is to her that, with much love, I dedicate this book.

# Using Computers

# 1

## *Introduction*

Species other than human beings use implements. A chimpanzee, for example, will employ a twig to extract insects from a porous tree trunk; it will use a leaf as a cup, a stick as a lever, a rock as a nutcracker. The differences between the most impressive examples of tool use by nonhuman species and what humankind has done in this regard, however, are sufficiently great to justify identification of our propensity to build and use tools as a distinctively human trait.

The tools that have been developed over the millennia constitute an impressive assortment indeed: tools for constructing objects from clay, wood, or metal; tools for weaving fabrics; tools for harvesting natural resources, such as timber and coal; tools for mending wounds and reconstructing diseased organs, for investigating worlds that are inaccessible to our unaided senses, for moving ourselves and cargos from place to place, for enabling us to communicate over long distances. Some of our tools, especially those of more complex design, we refer to as machines; but they are tools—things that serve as means to an end—nonetheless.

An often noted effect of the development of ever more versatile and sophisticated tools has been the corresponding decrease in the dependence on human muscle as a source of power. This, in turn, has changed the roles that people play in social units and has helped shape, in ways that are not always apparent, our attitudes toward ourselves. The idea of propelling ocean-going vessels by large crews of galley slaves chained to their oars is morally repugnant today. The temptation to be smug about our enlightened attitudes on such matters should be tempered, however, by a recognition that our moral judgment gets considerable reinforcement from the simple fact that as a means of propelling ships, human muscle is not economically competitive with the alternatives that technology has made available.

The history of the development of tools offers instances of profound alteration in human life—occasions when the development of a tool, by making it possible to do something that could not be done before or to do some familiar thing in a different way, has changed the course of history. The plow, the yoke, the wheel, the loom, the printing press, the steam engine, the airplane—each was the agent of such a transition.

Sometimes one cannot attribute radical change to the development of a single tool, but rather to a set of closely related developments that have had great impact over a short period of time. The point is illustrated by the history of farming in this country over the past 200 years. Whereas in the middle of the nineteenth century roughly 70 percent of the total U.S. labor force was devoted to farming, today about 3 percent grows enough food to feed the entire country, and to produce embarrassing surpluses as well. The shift from a dependence on human labor to the widespread use of machines took place gradually over several decades but got a big push with the development of such devices as Eli Whitney's cotton gin in 1793 and Cyrus McCormick's reaper in 1831.

While there can be little doubt that the tools that have been fashioned over the millennia have, on balance, produced enormous benefits for humankind, the story is not without its dark chapters. Many of the most ingenious of those tools have been implements of war, and the motivation for their development has been to provide the means to inflict death and destruction on some "other" subset of humanity. And many of the tools that were developed for more productive purposes have been employed by their owners to exploit their users. Even tools that are almost universally considered desirable and beneficial possessions can represent a threat of one sort or another: the automobile, perceived by many in our society as not only desirable but essential, is a case in point. While it provides us with unprecedented mobility as individuals, it is also directly responsible for about 50,000 highway deaths per year in the United States alone, besides being a major contributor to air pollution and the threatened depletion of fossil fuels. The automobile is convenient to use as an example of an important and valued tool that has some negative aspects, simply because it is so visible and the problems associated with it are so familiar; but one could illustrate the point with any number of other examples.

The question of how to design tools so as to ensure their usefulness to, and usability by, their intended users is one that toolmakers have addressed instinctively, if not explicitly, from the beginning of the toolmaking enterprise. A visit to a museum of hand tools suffices to impress one with the richness of human imagination and the sensitivity of toolmakers to the exquisitely subtle differences in the demands of superficially similar tasks.

Until fairly recently the problem of assuring a good match between tools and their users was left entirely to tool designers, who typically were also users of the tools they designed. But the tools that were developed became increasingly complex; and as the rate of increase in complexity accelerated during the middle of the twentieth century, the need arose for a new discipline devoted to the study of the interaction of people with tools, and particularly with those of sufficient complexity to be called machines. Many of the machines being developed were not designed by single individuals who fully understood their use and were themselves experienced users. The ways in which these machines coupled, or "interfaced," with their users became more complicated and the demands on the users were less well understood. Engineering psychology, human-factors engineering, or ergonomics, as the discipline is variously called, has been studying person-machine interaction and participating in the design of machines, especially of interfaces, for roughly four decades now. It has found much to do, and the impact the discipline has had on machine design has been substantial.

## *A New Tool*

At about the middle of this century a new type of machine appeared, one that was different in some fundamental ways from other machines we had become familiar with and had learned passably well how to design and use. We think of machines as assemblages of gears, levers, wheels, motors, and other hard components linked together so as to move in a coordinated fashion, when adequately fueled, in the performance of specific physical functions: lifting things, bending things, pushing things, pulling things. They are devices designed to change energy from one form to another, to manipulate forces,

and to accomplish work in the process. In the case of computing machines, however, energy transformation, force manipulation, and physical work are incidental, for the most part.
Computers are designed to transform information structures,
not energy. They manipulate symbols, not forces. And what
they do is more nearly analogous to thinking than to the performance of physical work.

The motivation for inventing new tools is usually a desire to
increase the efficiency with which familiar work is done. Sometimes, however, new machines have proved to have uses far
beyond those imagined by their developers. The designers of
electronic digital computing machines in the early 1940s were
primarily interested in computing projectile trajectories and
breaking communication codes (Goldstine 1972; McCorduck
1984): World War II was then at its peak. It is doubful if the
early developers of these machines, or anyone else for that
matter, had any notion of their potential range of applicability
or how ubiquitous they would shortly become.

I have quoted a story told by Lord Vivian Bowden before, to
make this point, and cannot resist doing so again. In 1950
Bowden was given the task of determining whether it would be
possible for a commercial firm to manufacture computing machines and sell them at a profit. The company in question was
Ferranti, which had just completed the first digital computer to
be built by a commercial firm in England.

> I went to see Professor Douglas Hartree, who had built the first dif-
> ferential analyzers in England and had more experience in using
> these very specialized computers than anyone else. He told me that, in
> his opinion, all the calculations that would ever be needed in this
> country could be done on the three digital computers which were
> then being built—one in Cambridge, one in Teddington, and one in
> Manchester. No one else, he said, would ever need machines of their
> own, or would be able to afford to buy them. He added that the
> machines were exceedingly difficult to use, and could not be trusted
> to anyone who was not a professional mathematician, and he advised
> Ferranti to get out of the business and abandon the idea of selling any
> more of them. (Bowden 1970, 43)

Professor Hartree's view appears to have been shared by
other people who thought about such things. Diebold points
out that "shortly after the computer was invented, a statement
was given wide circulation that all the computation in the coun-

try [United States] could be accomplished on a dozen—and later fifty—large-scale machines" (1969, 48). But if no one could foresee, when computers first appeared on the scene, how profoundly they would come to influence life, it took less than two decades for the scope of their potential impact to become clear. It is easy to find observations similar to the following two, made in the late 1960s:

The computer gives signs of becoming the contemporary counterpart of the steam engine that brought on the industrial revolution. The computer is an information machine. Information is a commodity no less intangible than energy; if anything, it is more pervasive in human affairs. The command of information made possible by the computer should also make it possible to reverse the trends toward mass-produced uniformity started by the industrial revolution. Taking advantage of this opportunity may present the most urgent engineering, social and political questions of the next generation. (McCarthy 1966, 65)

Today we are dealing with machines that can change society much more rapidly and profoundly than the machines that accompanied the industrial revolution of the late eighteenth and nineteenth centuries because they deal with the stuff of which society is made—information and its communication. (Diebold 1969, 4)

Computer technology has affected our lives in countless ways since these observations were made. The implications of the further development of this technology are impossible to foresee in detail with any certainty. However, many observers of the "computer revolution," as what we are currently witnessing is sometimes called, believe that its eventual effects will be at least as great as, and perhaps much greater than, those of the Industrial Revolution (Abelson 1982; Evans 1979; Toffler 1980). Some sociologists and futurists have asserted that the United States and other developed countries are in a state of transition, passing from an industrial society to a postindustrial society; they characterize the postindustrial society as an information society (Bell 1976, 1979; Evans 1979; Naisbitt 1984) and see the computer as the primary agent of this change.

In short, the computer is a new machine, a new tool, of enormous potential. It is perhaps the most awesome tool that has yet been developed. In three or four decades it has already transformed many aspects of life on this planet, and we are only beginning to learn how to exploit its capabilities. Like any powerful tool, it can be put to both effective and ineffective use and applied to both good and evil purposes. It is imperative that we learn to use it well and for humane ends.

### Information Systems

It is possible to define "information system" in such a way as to include DNA molecules and quasars. While such a definition could be useful in some contexts, it is too broad for the purposes of this book. Here we will think of an information system as any system whose main function is to "process" information for human use: to acquire it, organize it, move it from place to place, store it, and make it accessible to users. Our attention will be focused on information systems that make use of computer and communication technology in some significant way. Examples of the types of information systems that are of interest include electronic mail systems, word-processing systems, military command and control systems, computer-based information services or utilities, and personal computers that individuals use for their own purposes.

Many of the rapid societal changes we are experiencing stem directly from technological developments in methods for processing and disseminating information. In focusing on the plethora of recent developments that justify referring to ours as the information age, it is easy to overlook the fact that earlier advances in information technology—broadly defined—have also profoundly affected our lives. One of these, of course, was the invention of writing, which seems to have happened only about six or seven millennia ago. Another, which occurred a scant 500 years ago, was the development of printing technology—the invention of the printing press and the discovery of how to make relatively inexpensive paper. The invention of writing made it possible to accumulate knowledge, to store it, and to pass it on so that successive generations could build effectively on what had been inherited and learned. Printing technology democratized knowledge by making it accessible not just to a select few but to the masses; in doing so, it greatly accelerated the rate at which humanity's knowledge base grew. Much of the content of this book relates more or less directly to the possibility that information technology is poised for another quantum jump, equally far reaching in its effects. In the final chapter we return explicitly to this idea.

## *About This Book*

Computers and the types of information systems that computers make possible represent new challenges to those who wish to find ways to ensure that the machines we build are well suited to human use. Much of what has been learned about the design of person-machine interfaces over the past few decades is applicable in the case of computer-based systems. The existence of these systems raises some issues that have not been faced before, however. Never before have we dealt with machines that could talk, correct one's spelling, or diagnose a disease. Computer-based tools can do many of the cognitively demanding tasks that we used to consider achievable only by capabilities unique to human beings. They hold the promise, as many observers have pointed out, of extending our intellectual resources much as the Industrial Revolution extended our physical capabilities.

How are we to ensure that the computer-based tools developed over the next few decades will indeed be well matched to their users? How are we to minimize the human casualties of the further development of this technology? There were many casualties of the Industrial Revolution; can we manage the Information Revolution so that it does not have equally undesirable effects on some people? What can we do to increase the chances that the development and exploitation of this technology will contribute to the common good, to equity, to peace, to individual freedom and opportunity, to human dignity, and to the quality of life in general? These are the kinds of questions that have motivated this book.

My purposes are to provide an overview of where information technology is and where it appears to be headed, to review some of the human-factors research that has been done on computer-based systems to date, and to identify some of the issues and questions that are especially worthy of further research. The first few chapters following this introduction describe the development of information technology in the recent past, the uses of this technology in several fields, and some of the ways in which the technology is expected to develop in the foreseeable future. Subsequent chapters focus on the study of person-computer interaction, with special attention to the interface, both physical and cognitive. Consideration is then given to a variety of software tools and communication and information services that have been developed. Implications of information technology for employment, and especially for work that has traditionally been done in an office setting, are discussed. Various approaches to the design of interactive systems and some proposed guidelines are reviewed. Several user-related issues are then discussed. Some, but not much, attention is given to the activity of programming. Emphasis is placed on artificial intelligence and the current high level of interest in the development of expert systems. In the final chapters of the book an effort is made to identify some opportunities and challenges for research, and to speculate a bit on the potential that information technology holds for enhancing the quality of human life.

# 2

# *Backdrop*

## *Some Trends*

The major trends in computer technology over the past few decades are apparent: increasing speed, increasing reliability, decreasing size of components, and decreasing unit costs. How best to quantify these trends is less clear. One can find numerous reports of changes in various performance, size, and cost measures. These reports are not always easy to relate to each other, perhaps because terms are not always used the same way. In the aggregate, however, they provide a sense of the rapidity with which computer technology has been advancing. Here are some of them:

- The speed of semiconductor circuits has increased by about a factor of ten every decade over the past thirty years.
- Circuit reliability has increased by four to five orders of magnitude in 25 to 30 years (Branscomb 1982; Mayo 1977).
- Estimates of the rate of decrease in the cost of computing hardware, sometimes expressed per unit of computing resource (for instance, executed instruction) vary from 15 to 40 percent per year (Branscomb 1982; Dertouzos and Moses 1980; Knowles 1982). The rate of improvement has been faster for small general-purpose computers than for the largest machines (Branscomb 1982).
- The cost of primary memory has been dropping at comparable rates (Noyce 1977; Toong and Gupta 1982).
- The cost of a logic gate—the smallest part of a circuit whose output is determined by some Boolean combination of inputs— has gone from about $10 when gates were built with vacuum

tubes to a penny or less for those on integrated circuits (Mayo 1977).

- Over three decades, high-speed memories increased in size from 1,000 words to 1,000,000 words and in speed from 20 microsecond cycle times to 20 nanosecond cycle times (Pohm 1984).

- The power dissipation of random-access memory went from about 500 microwatts per bit in 1970 to about 4 microwatts per bit in 1980.

- The number of active-element groups (logic gates or memory cells) that can be placed on a single semiconductor chip went from less than ten in 1960 to nearly half a million in 1981 (Johnson 1981).

- The cost of logic elements has been brought down so sharply by large-scale integration that Sutherland and Mead (1977) speak of them as being essentially free. They note that the costs of the parts of computing machinery that are devoted to communication have become the dominant cost factors relative to those that are devoted to switching.

- According to one estimate, we now produce in one year computational capacity that would have taken a million years to produce at the production rates of 1960 (Preston 1983).

- The available computing power in the United States has been estimated to be growing at the rate of about 40 percent per year (Branscomb 1982).

- Communications costs have not dropped as rapidly as computing costs, but they are dropping. It has been predicted that a two-way-voice channel on a communications satellite may cost less than $10 per year before the end of the century (Clayton and Nisenoff 1976).

- As of 1979, the U.S. Bureau of the Census estimated there were 15 million computer terminals and electronic office machines in the United States.

- The estimated number of active-element groups in the average U.S. home went from about 10 in 1940 to about 100 in 1960 to a few thousand in 1980.

- In 1984 the number of personal computers in homes in the United States was estimated at six or seven million (Business Outlook 1984). Over one million personal computers were sold in the United States in 1982 (Borgatta 1983).

- Computer programming is among the fastest growing job categories in the world, if not the fastest.

In contrast with the automotive industry, which, as Begg
(1984) points out, has seen no major technological developments since the construction of the first Model T Fords, the
technology underlying the construction of computers has
changed radically over only a couple of decades:

Circuits were implemented with what seem now to be massive valves,
20 years ago germanium transistors soldered onto boards were used.
Now, the same functions can be performed by a die cut from a slice of
silicon crystal (a wafer) which has been etched by exotic acids, baked
with poisonous gases at high temperatures and bombarded with ions.
. . . There is less correspondence between the technology of the first
computers of the 1940s and 50s and those of today than there is
between that of the Model T Ford and that of the space shuttle. (Begg
1984, 9)

Toong and Gupta (1982) make the same point by contrasting
computer technology with the aircraft industry. If the latter,
they suggest, "had evolved as spectacularly as the computer
industry over the past 25 years, a Boeing 767 would cost $500
today and it would circle the globe in 20 minutes on five gallons
of fuel" (87).

## The Importance of Miniaturization

The basic operational components of all computers are "gates"
that accept inputs from one or more sources and produce outputs contingent on those inputs. In theory, it makes no difference whether the gates are built from mechanical switches,
electromechanical relays, vacuum tubes, transistors, or any of a
variety of other bistable devices. In practice, however, how they
are built makes a very great difference indeed. The operations
that computers perform are trivially simple; they manage to do
complex tasks by performing large numbers of simple operations in the right order and very quickly. The complexity of the
tasks that could be performed by a computer built with electromechanical relays or vacuum tubes is constrained by the
limited number of components that such a computer could
have. A machine with a few tens of thousands of such components presents unmanageable problems of heat dissipation and
maintenance.

The transistor was invented in 1948 by John Bardeen, Walter Brattain, and William Shockley, who later shared a Nobel Prize for their work. The invention of the transistor and the subsequent development of solid-state devices not only enhanced the development of very powerful computers but were essential to it. The dramatic decreases in the cost of computing resources have been made possible primarily by success in devising methods for miniaturizing circuits and packaging large numbers of them on individual silicon chips. And this success has been truly spectacular. As of 1977 an integrated circuit on a chip about one-quarter of an inch square could embrace more electronic elements than the most complex piece of electronic equipment that could be built in 1950 (Noyce 1977). Writing in 1983, Preston observed that "the cost of a single transistor or bit of memory is less than a hundredth of a cent, circuits operate for thousands of hours without failing, and a computer unit that far exceeds the capabilities of the ENIAC measures about a centimeter on a side, consumes a fraction of a watt, and can be bought for only a few dollars" (Preston 1983, 466; the ENIAC, one of the first digital computers, was built in the mid-1940s and contained about 17,000 vacuum tubes).

For the first decade or so following the introduction of the integrated circuit by both Texas Instruments and Fairchild Semiconductor in 1959, the number of transistors that could be placed on a single chip roughly doubled every year. It is obvious that such a rate of increase could not continue indefinitely; if it could, it would be less than 300 years until the number of elements on a chip equaled Eddington's estimate of the number of atoms in the universe, $10^{79}$. In fact, the rate of increase has already slowed noticeably: since 1973 the number of transistors per chip has been increasing by about a factor of four every three years instead of every two. Even that rate would extrapolate to over ten billion transistors per chip by the end of the century. Expectations are that the rate of increase will continue to slow over the next few years and that eventually greater packing densities will be precluded by certain fundamental physical limitations; the limits have not yet been determined, however, and one can find predictions of as many as a billion transistors per chip by the year 2000 (Meindl 1982; Robinson 1984c).

Perhaps the most spectacular result of this trend to date has been the development of the microprocessor or computer on a chip. Toong (1977) predicts that the advent of the microprocessor will probably touch more aspects of daily life than have been affected by all computer technology heretofore. Indeed, microprocessors are popping up everywhere one looks, not only in industrial devices but in household appliances, wearable or personal articles, and toys as well. In Toong's words, "the potential applications of microprocessor technology are so numerous that it is hard to visualize any aspect of contemporary life that will escape its impact" (160).

Storage technology is also advancing rapidly. Optical disks with very large capacity are beginning to be commercially available (Klimbie 1982). The packing density of these disks is limited only by the wavelength of light and has now approached that limit with bit representations of less than a micron. One multidisk system manufactured by Teknekron Controls has a capacity of $2.2 \times 10^{13}$ bits (Hawkridge 1983).

This number is reminiscent of Licklider's (1965) estimate of $10^{13}$ bits as the capacity that would be needed to hold all of solid science and technology. This estimate was based on work by Bourne (1961) and Senders (1963) from which Licklider derived $10^{15}$ bits as the total amount of information then contained in the world's libraries and 15 to 20 years as the doubling rate. (We should note in passing that these estimates refer to textually encoded information and do not take into account pictures and other nontext representations.) At the time of Licklider's estimates, the capacity of the largest random-access memory was somewhat under $10^7$ bits. Licklider noted that if the size of the largest random-access memory doubled every two years, "it would be possible to put all of the solid literature of a subfield of science or technology into a single computer memory in 1985" (1965, 17). (He assumed that a subfield of science or technology contained about one one-thousandth as much information as the whole, or $10^{10}$ bits.) The multidisk system of Teknekron Controls is not a completely random-access memory, so we should not conclude from the fact that it exists that Licklider's projection was too conservative. It is interesting, however, how very large quasi-random-access computer memories are becoming relative to the amount of information estimated to be stored as text in the libraries of the world.

The effort to increase the size of random-access memories

through integrated circuitry has intensified in recent years as the number of suppliers has increased: 64K random-access memory (RAM) chips, which are state-of-the-art, are being built by some 18 companies; 256K RAMs are beginning to be available commercially. A few companies, IBM among them, have built experimental one-million-bit RAMs (Robinson 1984b).

The reduction in costs associated with large-scale integration and miniaturization technology in general results from various factors: savings in the labor necessary to interconnect elements that are on the same chip; savings in maintenance costs, owing to the greater reliability of the interelement connections on an integrated circuit; and savings in the associated costs of power transformers, floor space, air conditioning, and so on.

Several other methods of storing information in very small areas have been developed. Examples include the use of charge-coupled devices (CCDs) and of magnetic bubbles. On a semiconductor memory chip one bit of information is represented by an electric charge that covers an area of about 100 square micrometers and consists of about 1.5 million free electrons and an equal number of holes (places that could be occupied by electrons but are not). In a charge-coupled device a bit is represented by only about 50,000 electrons and holes. This reduction in size is bought at the price of greater vulnerability to state changes induced by particles from ionizing radiation from a variety of uncontrolled sources (Zeigler and Lanford 1979). The solution to the "soft failure" problem, as it is called, is the use of redundancy—error-detecting and error-correcting codes—in the representation of information.

In a magnetic-bubble memory one bit of information is represented by the polarization of a tiny area of a thin film of magnetic material. The most common material at present is synthetic garnet, which permits the use of bubbles about 0.5 micrometer in diameter. Experimental work is being done with metallic glasses that may make it possible to use bubbles of only about 0.1 micrometer in diameter. The use of normally conducting vortexes in superconducting metallic glasses as the means of representing information could mean an increase in packing densities of another two orders of magnitude, since the vortexes measure only 0.005 to 0.01 micrometer (Chaudhari, Giessen, and Turnbull 1980).

## Computer Types and Generations

Numerous classificatory terms have been applied to computers. The terms *microcomputer, minicomputer, computer,* and *supercomputer* obviously represent some kind of progression, although the boundaries between the categories are not always clear. The same is true of the terms *first, second, third, fourth,* and *fifth generation* as applied to computing machines. The first-generation computers were built on vacuum-tube technology; those of the second, third, and fourth generations used transistors, integrated circuits, and very large scale integration, respectively. All of these machines shared a basic architectural feature: they were serial-processing machines in the sense that each relied on a single central processing unit (CPU) to do the work. Operations were performed one at a time by copying information from selected memory registers, bringing the information into the CPU, modifying it in some way, and, typically, storing it back in memory.

A key feature of fifth-generation machines, it is expected, will be parallel-processing architecture: several processors, possibly a very large number, will operate simultaneously. Parallel architectures are now economically feasible, inasmuch as the cost of producing processors is very small. How to exploit parallel architectures is a question for research, however, because little is known about partitioning complex problems in such a way that they can be approached in a parallel fashion. Moreover, putting many processors together in the same machine introduces new problems of interprocessor, or interprocess, communication. These too are problems about which relatively little is known.

For the first couple of decades after they appeared on the scene, computers were simply computers. All of them were too large, expensive, and difficult to maintain for single users to own and operate them. The first major distinction based on size was that between computers and minicomputers. This distinction gained wide currency with Digital Equipment Corporation's substantial success in marketing its PDP-8, which was introduced in the mid-1960s and remained the most widely used minicomputer for at least a decade. The last few years have witnessed a proliferation of computer manufacturers, new computer designs, and a great increase in the range of machine sizes.

Among the better known of the largest computers yet built are the Cray-1 and the Cyber-205. The Cray-1, introduced in 1976, was a successor to the Control Data Corporation's 6600 and 7600; all three machines were designed by Seymour Cray. The Cray is a vector computer, which means that it can perform the same operation on several variables simultaneously. It is also exceptionally fast when operating with scalars, having a 12-nanosecond cycle time. The Cyber-205, introduced in 1980, was a successor to CDC's STAR (String and Array Processor) and the one-of-a-kind Cyber-203. A detailed comparison of the Cray-1 with the Cyber-205 and with yet another supercomputer in the same class, the Burroughs Scientific Processor (BSP), may be found in Kozdrowicki and Theis (1984).

Two supercomputers manufactured in Japan, the Fujitsu FACOM-APU and the Hitachi M200HIAP, are believed to be within the same general performance range as the Cray-1 and the Cyber-205 (Buzbee, Ewald, and Worlton 1982). China also has a supercomputer, called the Galaxy, which may be slightly less powerful than the Cray-1. The Cray-X-MP, which became available in late 1983, is two to five times as powerful as the Cray-1. Other supercomputers introduced in 1983 are the Fujitsu VP-200 and the Hitachi S-810/20, both vector machines built in Japan. The Cray-2, scheduled to be available in 1985, is expected to be three to four times as powerful as the X-MP (Walsh 1983).

Comparing the speed of different supercomputers is not straightforward (Levine 1982). The number of executable instructions per second (IPS) is not really an adequate measure, because what can be accomplished by a single instruction differs greatly from machine to machine. The measure often used is megaflops (million floating-point operations per second). Even this measure is somewhat imprecise, however, because the number one obtains depends on the problem that is programmed and the specific way in which it is coded. Coding that is optimal for one machine may not be optimal for another. Moreover, there is a difference between peak rates and sustainable rates. It is possible to obtain rates for short bursts of computation that cannot be sustained over long periods in any practical program. This is because the possibility for concurrent operations is exploited more effectively by certain parts of a program, such as loops involving vectors or arrays, than by others. A common set of programs used to compare the speeds of supercomputers is known as the "Livermore Loops Benchmarks" (Riganati and Schneck 1984). These programs exercise a machine in a variety of ways, and it is possible for one machine to outperform another on some loops but underperform it on others. All supercomputers can attain speeds of a few tens of megaflops; some can do much better than 100 megaflops. Because supercomputers are, by definition, the fastest computers available at any specific time, we would expect the criteria for inclusion in this class to become increasingly demanding.

Recently the U.S. National Science Foundation selected four universities—Cornell, Illinois, Princeton, and the University of California at San Diego—as centers for work on supercomputers of the future. Each center will receive from 7 to 13 million dollars per year from the NSF and a roughly equal amount from state, industry, and local institutions. The establishment of these centers is in keeping with the NSF's decision to spend about $200 million over five years to improve academic computing facilities and to create a system of networked computing resources available to scientists throughout the academic community (Waldrop 1985).

At the other end of the size continuum are the personal computers, or PCs. The term is generally applied to machines that are inexpensive enough to be purchased by individuals for home or office use. Sometimes, but less frequently, it is applied to machines that may cost a few tens of thousands of dollars but can be used as personal tools by individual scientists or other professionals working on compute-intensive tasks. Machines in the latter category are also often called professional computers or workstations.

Toong and Gupta (1982) define a personal computer as a system that has all of the following characteristics:

- Price: less than $5,000.
- Secondary memory in the form of cassette tapes or disks.
- Primary memory capacity of 64 kilobytes or more.
- Ability to handle at least one high-level language such as BASIC, FORTRAN, or COBOL.
- Interactive dialog capability.
- Wide distribution through mass-marketing channels, primarily to people who have not worked with computers.
- General-purpose capability.

The emergence of such machines has been one of the most significant recent developments in computer technology. Reductions in the costs of logic and storage have meant that useful amounts of both could be made affordable to many individuals, and decreases in component sizes and power requirements have made it possible to package this logic and storage in devices that are portable and do not require special, temperature-controlled environments.

The personal computer market grew from essentially nothing in 1976 to more than $2 billion per year by 1981, and it is expected to reach $6.5 billion per year in 1985 (Toong and Gupta 1982). The market is divided into four segments: business, home, science, and education; business, the largest segment, has accounted for more than half of the sales. Business applications are expected to continue to provide the largest market for personal computers in the near future; the other markets are expected to grow steadily, however, and the home market could increase explosively at some point.

A term that has been used with increasing frequency over the past few years is *workstation*. The dividing lines between personal computers, professional computers, and workstations are not sharp. A workstation can be many things to many people; as the term is typically used, however, it connotes a station that provides the user with a variety of computer-based capabilities including word processing, electronic mail, information management, high-quality graphics, and a powerful set of program-building tools. A workstation is usually expected to have some relatively high-speed hard-copy output device (a printer) and a variety of input options (keyboard, light pen, mouse). A workstation is seen as a personal thing; it is not shared, although it may provide the user access to some shared facilities.

Schoichet (1981) compares several personal workstations (the Xerox 860, the AM Jacquard-J500, the Convergent Technologies Workstation, the Artelonics Series 1000, the Three Rivers PERQ, the Apollo Computational Node, and the AXXA System 90) with respect to a variety of features: character formation, graphics capability, interaction aids, CPU, whether it has a standard bus, RAM capacity, mass-storage capacity, whether it has a local-area network, programming languages available, and operating-system features. Other examples of personal workstations are the Xerox Star (Seybold 1981; Smith, Harslem, Irby, and Kimball 1982; Smith, Irby, Kimball, Verplank, and Harslem 1982) and the Apple Lisa (Ehardt 1983). Workstations with significant amounts of local processing power and both primary and secondary storage, tied together by local-area networks so as to permit communication among stations and the sharing of resources that are still too expensive to be located in individual stations, seem to provide the best of both time-sharing and dedicated stand-alone computing systems.

How much processing power and memory capacity and what kinds of support functions one wants at one's own workstation, and which resources one wants or needs to share, depend on one's application and one's budget. For the professional whose activities are relatively information-intense, a workstation should probably have enough processing power and memory capacity to support symbolic processing (text and symbol manipulation) and high-resolution color graphics. That probably means one megabit of memory. One also will surely want some type of hard-copy output device, such as an inexpensive printer. Shared resources would probably include a high-speed, high-quality printer, a hard disk, and very large capacity bulk storage. Of course these desiderata will expand to match what becomes technically and economically feasible as a result of further advances in the technology.

## Resource Sharing and Computer Networks

Since the earliest efforts to apply computers to practical problems, a goal of both manufacturers and users has been to find ways to help users share resources. A primary reason for this interest has been cost: the first computers were very expensive to build, and few people who had an interest in using them could afford to own one. The first response to this problem was the creation of computing "centers." Such centers typically housed one or more computers and the necessary ancillary equipment to provide computing services to some community of users—a research laboratory, a university, a corporation—and they were operated by a center staff. An end user delivered a program to the center, where it was entered into a queue to be run when time on the machine became available; the user obtained the results from the run hours or days later.

While the computing-center, or "closed-shop," approach did distribute the costs of a system over a community of users, it had serious drawbacks. Because the equipment was very expensive, it had to be kept busy if its costs were to be recovered. Users found themselves standing in line, as it were, waiting their turn on the machine. Getting a program to run satisfactorily often involved several cycles of writing code at one's desk, submitting that code to the computer center, waiting hours or even days to learn whether the program had run without problems (which it rarely did the first time), returning to the coding task to find the problems and fix them, and then resubmitting the revised code to the center for another try. Diagnostic information was provided by the system to facilitate the debugging task; but even with such information in hand, fixing a faulty or incomplete program could be a formidable challenge, and often several iterations of the entire process were necessary before the job was done.

A second approach to the resource-sharing objective, and one that addressed some of the shortcomings of the closed-shop arrangement, involved the development of software that would permit several users to use the same computer simultaneously and from remote locations. Time-sharing, as this approach was called, was first demonstrated at Bolt Beranek and Newman in 1962 (Bertoni and Castleman 1976). It made possible a qualitatively new kind of computer use. Users could interact with the computer from a distance, often over telephone lines, and could access the system at their convenience. There were problems, of course, stemming from the fact that the demands that a community of users placed on a system varied considerably over a 24-hour period. The quality of the service one received (in particular, the speed with which the system responded to one's inputs) typically was higher when the system was lightly loaded (3 A.M.) than when it was heavily loaded (midmorning, midafternoon). But these were manageable problems, and time-sharing greatly increased the number of people for whom the use of computing resources was economically feasible and at least somewhat convenient.

During the 1960s time-sharing facilities were established throughout the United States and in several countries in Europe. Each such facility had dozens or hundreds of users. People working on the same system not only shared hardware but in many cases found it easy and desirable to share programs and data files as well. The inevitable next step was to connect different time-sharing systems so that resources could be shared *across* systems as well as within them.

Much of the theoretical basis for the first efforts to develop computer networks may be found in a series of Rand Corporation memoranda, most of which were written by Paul Baran in 1964. Two published reports describing this work are Baran (1964) and Boehm and Mobley (1969). Small experimental networks were designed by computer scientists at the Computer Corporation of America in 1965 (Marrill and Roberts 1966) and at the National Physical Laboratory in Middlesex, England, shortly thereafter (Davies 1968; Davies and Barber 1973).

The development and application of networking technology was greatly stimulated by the Advanced Research Projects Agency of the U.S. Department of Defense (ARPA). An experimental four-node network was implemented under ARPA's auspices in 1969; known eventually as the ARPANET, it has been growing in size and complexity ever since, and now contains over a hundred "host" computers distributed over the United States, including Hawaii, and Europe. The network was designed, implemented, operated, and maintained by Bolt Beranek and Newman under contract to ARPA from 1969 to 1975 and thereafter under contract to the Defense Communications Agency, to whom responsibility for operation was transferred in 1975. It is a packet-switched, store-and-forward network (about which more later) that uses wideband (50-kilobit) circuits leased from AT&T and other carriers as the internode communication links (Heart 1975; McQuillan and Walden 1977; Roberts 1973).

The computers on the ARPANET are of various types made by several manufacturers. Communication among these computers is facilitated by interface message processors (IMPs) that act as translators between computers that speak different languages, as it were (Heart, Kahn, Ornstein, Crowther, and Walden 1970). Access to the ARPANET is obtained either through host computers that function as resource nodes on the network or through terminal devices such as terminal IMPs (TIPs), which make it possible to connect directly to a network without going through a host computer (Kahn 1972b; Ornstein, Heart, Crowther, Rising, Russell, and Mitchell 1972).

The ARPANET was conceived as an experimental network—a resource for studying store-and-forward data com-

munication technology, for conducting experiments, for
making measurements, and for developing models that would
be useful in this technology. The builders of the ARPANET
also wanted to learn how this technology might be used to
facilitate interaction and collaboration among people working
on common technical problems but located in different areas
(Heart, McKenzie, McQuillan, and Walden 1978; Kahn 1972a;
Roberts and Wessler 1970). The computers that were to be
interconnected were all time-sharing facilities, each already
providing computing resources to remotely located users. The
goal was to enable users on any of the computers in the network
to use programs available on any of the others. It was assumed
that this sharing of resources would decrease the duplication of
effort that occurs when each local facility must create all the
software and data files it wants to use.

For such a network to be truly useful, it had to be highly
reliable, it had to provide users with very fast access to the
desired information and resources, and it had to be less expensive than having the desired resources locally. The successful
implementation and operation of such a network also required
solution of several difficult technical problems involving topol-
ogy (network layout and interconnectivity), error control, and
intercomputer language translation (Frank, Kahn, and Kleinrock 1972; Heart, Kahn, and Kleinrock 1972; Heart, McKenzie, McQuillan, and Walden 1978; Kahn and Crowther 1972;
Walden 1972). A collection of early papers on the ARPANET
may be found in *AFIPS Conference* Proceedings, volumes 36
and 40. Other useful references regarding computer networks
include Chu (1974), Blanc and Cotton (1976), and Newell and
Sproull (1982).

The ARPANET experiment was sufficiently successful to
stimulate interest in networking among other agencies and private business. Both IBM and Digital Equipment Corporation
have developed networks to permit interconnections among
their own machines. IBM's network is called SNA, for System
Network Architecture, and DEC's is called DECNET. Several
commercial networks have been established to sell network services to businesses that do not own their own network facilities.
Among these are GTE's TELENET, CCA's TYMNET, and
Control Data's CYBERNET. There are also a variety of networks, some of them government funded or partially govern-

ment funded, to service special communities of users. These include CSNET (Computer Science Net), which links computer scientists at universities across the country; AGNET, which provides services of special interest to farmers, in addition to facilities of more general interest, such as electronic mail and teleconferencing; and DEAFNET, a network established for people with severe hearing impairments. DEAFNET provides not only conventional electronic mail capability but bulletin boards, news, and weather reports as well.

The problem of information security is much greater, of course, when a system is connected to a network that makes use of communication lines used by the general public. Military systems for which security is important typically do not use such lines but are "closed": they use only communications facilities that are dedicated to that particular use. Secure networks also employ end-to-end encryption and tightly controlled user-authentication schemes. The Defense Department plans to split the ARPANET into two networks: MILNET, a secure network for military hosts, and R&DNET, for the remaining ones.

Network technology has developed in a variety of directions. There now are local-area networks that connect machines in a given building or small geographical area, and wide-area networks that connect systems across a continent or around the world (Tanenbaum 1981). There are private networks, owned and operated by businesses or corporations, and public networks that sell network services to subscribers much as the telephone companies sell their services. Some network nodes are connected by coaxial cables, some by fiber-optic cables, some by radio transmission, some by satellite transmission. One of the better-known local-area networks, Ethernet—which is offered jointly by Xerox, DEC, and Intel—connects various devices and resources (personal workstations, file servers, disks, printers) with a coaxial cable, whose maximum length is about half a mile. Communication between networks has been accomplished through the development of internet "gateways."

Perhaps the two most important points about networks are that they greatly increase the opportunities for resource sharing and that they represent a new medium for communication. Host computers serve both as providers of resources to other network users and as users of resources located at other nodes on the net. In theory all the resources on a network are accessible to all its users. In practice it may not work out quite this way; nevertheless, access to a network typically means access to many widely distributed resources.

The extent to which computer networks are used for inter-person communication has come as something of a surprise. Originally designed to permit the transmission of data from point to point and to provide users with access to remote facilities that could be applied to computational problems, networks have turned out to be very useful for purposes of communication. The development of electronic mail gave networks a new and largely unanticipated function. Electronic mail was not planned; it got its start when programmers realized it would be possible to send and receive messages over an existing network and simply wrote the code that would permit them to do so.

The term "revolutionize" is greatly overworked today, but I can think of no more appropriate one to use in describing the implications of the emergence of computer networks. The establishment of worldwide computer networks has the potential of greatly increasing the accessibility that individuals have to other individuals, to major repositories of the accumulated knowledge of humankind, and to computing and information processing resources of every type. Such networks have the potential of making government be participatory in ways never possible before.

We may hope that in making it easier for people to communicate directly with other people around the world and to avail themselves of powerful information resources of various types, this technology will diminish the misunderstandings that stem from lack of communication and will facilitate the maintenance of world peace. One can hope, too, that it will prove to be an effective weapon against the common enemies of humankind everywhere—hunger, ignorance, and disease.

### The Merging of Computer and Communication Technologies

Modern electronic information systems involve two major technologies: computation and communication. Although, as we have noted, the costs of communication have not fallen as rapidly as the costs of computation over the past few decades, it is expected that as a consequence of the increasing use of digital transmission and other innovations the communication costs will drop more rapidly in the future than they have in the past. (Most of the world's communications systems are designed to transmit analog signals. Digital transmission has several advantages over analog transmission, however, and the expectation is that in time most systems will be fully digital and will accommodate data, voice, and imaging.)

One of the major reasons for the anticipated accelerated decrease in communication costs is the increasing use of computing resources in these systems. Indeed, the distinction between communication and computation technologies, while still an important and useful one, has become blurred in recent years. For many systems it is difficult to make the distinction at all. This merging has been extensive enough to prompt the coinage of new terms, such as "compunications" and "telematics" (Nora and Minc 1980). The following examples of developments (from Nickerson 1980) illustrate the point. Although the primary purpose of each is to facilitate communication, all of them make use of computing resources to accomplish that purpose.

- Packet-switched computer networks
- Packet radio
- Satellite communication stations
- Hybrid communication systems
- Electronic mail
- Computer-based telephone systems
- Computer-mediated teleconferencing
- Information utilities

Consider packet-switched computer networks. Packet switching is a technique for routing data transmissions that is designed to make efficient use of the channel capacity of the network as a whole. The method is most easily understood by way of contrast with the more conventional method of circuit switching. In a circuit-switched system, when a connection is desired between two nodes, one of the many possible routes for establishing the connection is selected by the switching circuitry, and that route is used until the transmission is finished. This is the method that has been used for many years by the telephone system; when someone in New York places a call to a number in Los Angeles, the circuitry selects a route that connects the calling and receiving phones, and that route is maintained, even during pauses when nothing is being transmitted, until the end of the conversation is signaled by one or the other party hanging up the phone.

A packet-switched network works differently. Here the message to be transmitted is divided into *packets,* each of which contains only a small fraction of the entire message. Each packet is given a *header* that contains, among other things, the address to which the message is being sent. Each header-packet combination is dispatched individually as the message is composed, and finds its own way to its destination. Packets from the same message may travel to the same destination by quite different routes; what route a packet takes on any particular leg of its journey will depend on what is available at the instant the packet is dispatched from its current way station. Because the packets take different routes, they may not arrive at the destination in the order in which they were sent; but no matter, they will be reordered and the message reconstituted by the computing machinery at the receiving end. The advantage of this way of doing things is that circuits are not tied up when there is nothing to transmit (as during pauses in phone conversations); the method makes more efficient use of the available bandwidth and thereby reduces the communication costs.

Computing resources are needed throughout a packet-switched network. Input nodes partition the message into packets, attach to each packet an appropriate header, and decide where to send the packet on the first leg of its journey. Intermediate nodes decode packet headers, select subsequent intermediate destinations, and send messages back to nodes from which packets have been received to acknowledge their receipt. Final-destination nodes order correctly the packets they receive and reconstitute the messages that were originally sent.

When one looks carefully at this process, it is difficult to say where communication ends and computation begins. A similar observation could be made with respect to all the examples in the list and numerous others as well. In contrast to commercial radio, which broadcasts over a preselected frequency, packet radio broadcasts each packet of data over whichever of a designated number of frequencies is available at the moment. This dynamic frequency allocation illustrates what is sometimes referred to as the "anarchy band" concept (Jackson 1980). Problems of contention and noise arise when packets from different sources are sent over the same frequency at the same time;

these are resolved, however, by retransmission until the sending radio receives confirmation from the receiving radio that the packet has been received. Since the broadcast of a given packet occupies a very brief period of time, contention is not a serious problem. Computing resources are used in a packet radio system for such functions as dynamic frequency allocation and contention resolution; they also help solve problems that arise when the radio transceivers are not stationary. Interest in developing packet radio technology runs high at the moment, fueled by the prospect of wireless access to computer networks and the possibility of mobile digital communication facilities.

The use of earth-orbiting satellites to transmit digital information over long distances has increased communication possibilities greatly. Because of their very broad bandwidths, satellite stations provide much versatility in transmitting data, voice, facsimile, or video. The early satellites operated in the 4- to 6-gigahertz (GHz) frequency range. Higher frequencies, up to at least 30 GHz, are expected to be used in future satellites. The higher frequencies permit the transmission of more information per unit time; however, they are also more subject to atmospheric disturbances than are the lower frequencies. One way to counteract these disturbances is to transmit the same signal from different locations on the assumption that disturbances encountered in one of the locations would be unlikely to occur at the same time in the other. Various types of error-detecting and error-correcting codes, all requiring the use of computers, have also been devised to compensate for noise.

Satellite transmission has also enhanced the usefulness of the telephone. In the two decades that have passed since the Communications Satellite Corporation (COMSAT) and the International Telecommunication Satellite Organization (INTELSAT) were formed in the early 1960s, the number of active international telephone circuits, worldwide, has grown by a factor of 400 and the cost of a telephone call has declined by 12,000 percent (Pollack and Weiss 1984). The sixth generation of INTELSAT satellites is scheduled to be in use by 1986.

Some computing has been involved in the operation of telephone systems at least since the introduction of direct dialing. Such computing was necessary to decode the dialed number and to effect the establishment of the desired circuit. The amount of computing has increased over the years with the introduction of such features as direct long-distance dialing, automated directory service, and, most recently, business phone systems that enable the user to forward incoming calls to other extensions, to define one- or two-digit codes for certain frequently used numbers, to put a party on hold while taking or placing a second call, to alternate between calls, to set up conference calls from one's personal phone, to transfer calls from one extension to another, to "camp on" a busy line so that the number will be rung as soon as the ongoing call is completed, and so on.

The merging of communication and computation technologies is one of the most significant aspects of modern information technology. It is this merging that is responsible, to no small degree, for the great increase in information accessibility that we are beginning to witness. Pool (1983) has described several recent trends in the flow of information through communication media and its absorption by potential recipients. The amount of information made available to the public has grown rapidly over the past few decades (Pool's observations are based on data spanning the period 1960–1977), but the amount we absorb has not kept pace:

More and more material exists, but limitations on time and energy are a controlling barrier to people's consumption of words . . . . this difference between trends in supply and consumption means that each item of information produced faces a more competitive market and a smaller audience on the average. People see or hear a decreasing proportion of the total information that is available to them. (609)

Pool notes that there is an interesting and quite regular relationship between the cost of transmitting a given number of words by means of a particular communication medium and the number of words made available through that medium per unit time. Not surprisingly, the lower the costs the greater the use of the medium. Between 1960 and 1977 there occurred a significant decrease in cost and increase in use for the electronic media, while the amounts remained relatively constant for the print media. Much of the growth in information flow has apparently been due to broadcasting. By 1977 about two-thirds (69 percent) of the words reaching Americans through measured media came by way of television or radio. That was up from 58 percent in 1960.

A particularly thought-provoking development is the reversal of the trend toward domination of the flow of information by a few mass media in the 1960s and early 1970s, in favor of a movement toward greater diversification. By the mid-1970s the mass audience seems to have begun to fragment as a result of an increase in the number of broadcast channels available—because of cable TV systems and added stations—and there has been a marked growth of individually addressed point-to-point media. "Computer networking," Pool notes, "is for the first time bringing the cost of a point-to-point medium, data communication, down to the range of costs characteristic of mass media" (611). These trends seem to contradict the familiar view that society is becoming more and more homogeneous. And they constitute a basis for hope that as information technology is further developed, it will provide increased opportunities for individualization and personal choice.

### The "Accelerative Thrust" in Information Technology

If one likes to think in terms of round members, one can mark the beginning of the second half of the twentieth century as the commencement of the Computer or Information-Technology Age. Since the appearance of the first commercially available computers, the UNIVAC-I in 1950 and the IBM-701 in 1953, the development of the computer industry itself, and the concomitant proliferation of uses for its products, have been truly remarkable. A little history may help put into perspective what we are currently witnessing.

When the first device was built for the purpose of assisting a person to compute is impossible to say. The abacus, or soroban, apparently discovered independently by the Greeks and the Chinese, is several thousands of years old. It is still widely and effectively used in the Eastern Hemisphere. Various other ingenious mechanical devices that assist in counting and calculating have been invented and used by different cultures over the centuries. Coming closer to our time, however, we can note a handful or people who invented things or put forth ideas that can be considered as directly paving the way for the modern digital computer.

The French philosopher-scientist Blaise Pascal built a mechanical calculator in the middle of the seventeenth century. To accomplish the carry, Pascal invented the technique that is still used in many mechanical counters, such as odometers. He concatenated a series of disks, each numbered from 0 to 9, in such a way that when a disk was moved from 9 to 0, a ratchet caused the disk to its left to advance one digit. Pascal's machine accomplished multiplication and division by repetitive addition or subtraction. About thirty years later, in 1675, Gottfried Leibniz—German philosopher, diplomat, scientist, mathematician—succeeded in building a machine that could multiply and divide directly.

Around 1745 another Frenchman, Joseph Jacquard, developed a technique by which holes punched in cards were used to control the selection of threads in the weaving of complicated patterns by machine. The "Jacquard loom" was perfected several years later and revolutionized the weaving industry. Another pioneer of the technique of coding information so that it could be represented by holes in cards was Herman Hollerith. His methods were employed in compiling the United States census of 1890. Hollerith's work led directly to the development of a large variety of punched-card equipment that later found use in many business data-processing systems.

One of the most celebrated names in the history of computers is that of Charles Babbage. A professor of mathematics at Cambridge University, he was one of those inventive individuals whose ideas are far ahead of the technology of their times. He planned and started to build two machines: the first he called the Difference Engine; the second and more sophisticated machine he referred to as the Analytical Engine. Babbage started his project around 1822 but never completed either machine. Some of his ideas, however, have stood the test of time and are reflected in the design of modern computers. It was he who first proposed storing a program within the machine.

One of the reasons Babbage was frustrated in his attempts to implement his ideas was that the technology necessary for building a practicable computing machine had not been developed. During the century that intervened before the next serious attempt to build a computer, this situation changed drastically. Electricity was harnessed and put to innumerable uses, communication engineering became a highly developed technology, information theory was established as an area of mathematics, and, what is perhaps as important as any other single factor, the need for high-speed calculating and information-manipulating devices became acute. One might argue that the general level of technical sophistication and a ready, albeit unwitting, market for information-processing devices made it inevitable that the electronic digital computer would be developed when it was.

Among the first of the twentieth-century computer pioneers was Howard Aiken, who in 1937 designed a machine called the Mark I. The machine was built by IBM and presented to Harvard University in 1944. It was not quite a digital computer in the modern sense of the word, but it was a large step in that direction. It was programmed by means of switches and plugboards.

During the remainder of the 1940s and the early 1950s the paced picked up considerably. In Great Britain a machine containing about 2,000 vacuum tubes was designed in secrecy and used very effectively, beginning in 1943, to crack the codes produced by the Nazi coding machine Enigma (Evans 1979; McCorduck 1979, 1984; Winterbotham 1974). Alan Turing, of later Turing-machine fame, participated in the design of this machine, which was referred to as Colossus.

At the Moore School of Engineering of the University of Pennsylvania, J. Presper Eckert and John Mauchly collaborated on the design of a machine that became known as the ENIAC (Electronic Numerical Integrator and Calculator). The purpose of the machine, which was completed in 1946, was to produce mathematical tables required for the firing of projectiles. Like the Mark I, ENIAC was programmed by means of switches and plugboards and was not a stored-program machine. It did, however, use vacuum tubes (about 17,000 of them) instead of mechanical gears and switches to accomplish its calculations.

While working on ENIAC, Eckert and Mauchly, with the help of others, designed a second macine, the EDVAC (Electronic Digital Variable Automatic Computer). A true stored-program computer, it incorporated some of the ideas of John von Neumann, the Princeton mathematician who is usually credited with the insight concerning the advantages of storing data and the program in the same store and in the same form. (Babbage's original design for his Analytical Engine distinguished two different storage components, one to hold data and the other to hold instructions.) The EDVAC was completed in 1952.

The EDSAC (Electronic Delay Storage Automatic Calculator), another of the earliest stored-program computers, was built in the early 1950s at the University of Cambridge, England, under the direction of Maurice Wilkes; it too incorporated some of von Neumann's ideas. At about the same time a computer was designed and built at the Institute for Advanced Studies at Princeton, whose director was J. Robert Oppenheimer, by a team that included John von Neumann, Herman Goldstine, and Julian Bigelow. The reader is referred to Goldstine's book (1972) for a full account of this project and for a general history of the development of computer technology as well.

Following these developments and the introduction of the first commercial machines in the early 1950s, the technology progressed at an ever accelerating and nearly explosive rate. We may summarize the history of the development of computing machines by distinguishing several epochs: (1) hundreds of millennia during which little, if any, computing was done; (2) a few thousand years of using such devices as the abacus or soroban; (3) about three hundred years of development of germinal ideas and the electronic technology on which modern computers are based; (4) the development of several experimental electronic calculating machines during the latter part of the first half of this century; (5) the first use of commercially available computing machines in the 1950s; and (6) rapid acceleration of computer and communication technologies and applications during the last two decades.

The growth curve implied by this summary is a familiar one, an instance of what Alvin Toffler calls the "accelerative thrust" that characterizes our times:

Whether we examine distances traveled, altitudes reached, minerals mined, or explosive power harnessed, the same accelerative trend is obvious. The pattern, here and in a thousand other statistical series, is absolutely clear and unmistakable. Millennia or centuries go by, and then, in our own times, a sudden bursting of the limits, a fantastic spurt forward. (1970, 26)

Norbert Wiener makes a similar point:

The period during which the main conditions of life for the vast majority of men have been subject to repeated and revolutionary changes does not even begin until the time of the Renaissance and the great voyages, and does not assume anything like the accelerated pace of the present day until well into the nineteenth century. . . . There is no use in looking anywhere in history for parallels to the successful inventions of the steam engine, the steamboat, the locomotive, the modern smelting of metals, the telegraph, and the transoceanic cable, the introduction of electric power, dynamite and the modern high-explosive missile, the airplane, the electric valve, and the atomic bomb. (1950, 32)

The words were written before computers had become a commercial reality. It is doubtful whether even Wiener, who understood the potential of computer technology and cybernetics probably as well as anyone of his time, clearly anticipated how rapidly this technology would develop in the next few decades.

• • • • • • • • • • • • • • • • • • • • • • • • • • • • • • • • • • • • • • • • • • • • • • • •

The electronic digital computer has been in existence for only four decades, but it has had a profound influence on our lives in that short time. The technology itself has developed extremely rapidly and in unanticipated ways. The emergence of computer networks, the development of the microprocessor or computer on a chip, and the merging of computer and communication technologies are but a few of the more recent chapters in this story.

Unlike almost everything else, the cost of a unit of computing resource has steadily declined. This has been possible because of the success of computer scientists and engineers in developing techniques for storing, processing, and transmitting information that require less and less material to build and power to operate. The advance of the technology has been propelled also by the extraordinary usefulness of the machines that have been built. The demand for computing resources has exceeded original expectations by many orders of magnitude.

It has often been pointed out that truly exponential curves occur only in mathematics, that no natural process can grow exponentially indefinitely. Be that as it may, the accelerative thrust of twentieth-century technology is surely a real phenomenon of inestimable significance. What the growth curves will look like centuries or even decades hence, we can only surmise; it is a relatively safe wager, however, that the computer will be one of the key factors in shaping the future not only of technology but of civilization itself.

# 3

# *Uses and Users of Information Systems*

It would be quite impossible to list here all of the ways computers are being used today. In science, in engineering, in medicine, in business, in education, in the arts—in essentially every field of human endeavor—the computer has proved to be an astonishingly useful device. Computer technology touches the lives of each of us in countless ways every day. We use products that have been designed, manufactured, and distributed with the help of computers; we draw checks on banks that depend on computers to carry out the associated account transactions; we buy food and other goods at stores that use computers to record point-of-sale transactions and to control inventories; we read magazines that have been produced with computer-controlled techniques. We are undoubtedly far more dependent on computers than we realize. If the plug on every computer in the country were suddenly pulled, we would become quickly and painfully aware of this dependency: planes would not fly, payrolls would not be produced, telephones would not work, many household appliances would not function, electric power systems would fail, presses would stop rolling. The list of consequences could be extended indefinitely.

The astoundingly rapid proliferation of computer uses can be attributed to two facts: (1) many things can be done more conveniently and less expensively with the help of computers than without it, and (2) in an increasingly complex society there are some functions that, without computers, probably could not be performed at all. In this chapter we will consider just a few examples of the many uses to which this technology is being put.

## Farming

Though not usually thought of as a high-technology industry, farming is far more automated than most aspects of manufacturing. The automation of farm work, coupled with the development and use of fertilizers and pesticides and the application of genetic engineering, has made it possible for only about three percent of the U.S. labor force to produce more than enough food for the entire population, whereas at one time nearly three-quarters of the labor force was engaged in farming.

It is easy for us to forget how rapidly this change took place. Rasmussen (1982) points out that most of the tools employed by farmers at the time of the American Revolution differed little from those in use 2,000 years before. In the ensuing 200 years, however, farming became the most highly mechanized industry in the world. Mechanization is only one of the factors that have made a nearly 20-fold rise in productivity possible over about 120 years: the other changes that have occurred are the consequences of increased knowledge about agriculture, achieved through controlled research.

A spectacular illustration of automation in farming has been visible to anyone who has flown over the Midwest in the summertime in recent years. The numerous circles one sees below are crop fields that are watered by systems of pipes, as much as a quarter of a mile in length, that sweep the disk-shaped areas as they pivot about the center of each field. Each pipe is supported by a series of mobile towers regularly spaced along its length. Center-pivot irrigation has been in use for almost 30 years, so its development was not dependent on the availability of computers. It illustrates quite dramatically a major way in which automation has affected farming, however, and information technology is now being used to increase further the efficiency of the technique. Programs have been developed, for example, to specify when to irrigate and how much water to use, based on calculations of the amount of water lost to a given field through evaporation and transpiration, the amount of water available, the type of crop and its stage of growth, and weather information (Splinter 1976). The use of computer-based models relating crop growth to numerous determining factors assures a more nearly optimal utilization of limited water supplies.

Many of the other technological developments in farming predate computers, but farmers have begun to use computers and will undoubtedly use them more and more in managing their increasingly complex and mechanized operations (Holt 1985). These uses include the continuous planning of fertilization and irrigation to match soils, crops, and weather conditions; the mixing of feeds for livestock; financial planning and management; livestock health maintenance; herd improvement; and production monitoring.

Two ways in which computers are currently being used by dairy farmers are noted by Anderson (1983). The most widespread use is through subscription to computer-based data services that process information provided by subscribers and issue periodical reports on such things as milk production, livestock health, and feed. Second, dairy farmers are beginning to involve computers directly in their day-to-day operations: adjusting feed to the milk production of individual cows; monitoring costs and yields, again on individual cows; monitoring health; and improving the herd through selective breeding. Because profit margins are often very narrow in farming, and subject to large swings as a function of many variables, the ability to adjust quickly to changing conditions can mean the difference between prosperity and financial disaster. As one dairy farmer puts it, "In the past thirty years it was farm machinery that gave a big advantage, but in the next twenty years, it will be information that makes or breaks a farmer" (Anderson 1983, 35).

## Manufacturing

Factory automation is not a new concept. The A. O. Smith Corporation had a fully automated factory for producing automobile frames in the early 1920s, long before computers were part of the industrial scene (Bollinger 1983), and there are numerous other examples of the automation of manufacturing processes that predate computers. However, the appearance of computers in the factory extended greatly the range of operations that could be automated.

The conventional model of a factory has been one in which there is a flow of material through a system. Raw materials and prefabricated or partially fabricated parts enter the system at one end and finished products emerge at the other. The process by which the raw materials and parts get transformed into finished multipart products includes a variety of activities such as cutting, milling, polishing, painting, and assembling. Many of these processes are now performed under computer control. The use of computers in process control is a more time-sensitive application than many others, because in this case the computer is monitoring, controlling, and otherwise interacting with events in the real world. To do this successfully, it must be able to determine when those events occur and to initiate control actions within the time constraints imposed by the process.

Although perhaps less a part of our stereotype of a factory, no less important to its operation is the flow of information— inventories, parts lists, specifications, instructions for particular operations, schedules, purchase orders, sales orders, and so on. Several isolable but interrelated information systems are involved. Production schedules must be consistent with sales projections, which are based at least in part on received orders. Parts inventories and purchases must be consistent with production schedules. One of the goals of any manufacturing operation must be to match production as closely as possible to demand, in terms of both quantity and timing. Large inventories, of either parts or finished products, are expensive. On the other hand, the inability to produce in sufficient quantity to meet the demands of the marketplace means losing opportunities to competitors.

Ideally, a manufacturer wants to minimize the time that any part or product sits idly in inventory, but also to assure sufficient stock to meet demand. Realization of both objectives requires accurate forecasts about how quickly finished products can be absorbed by the marketplace and careful planning and scheduling of the production process so that parts are acquired or produced only very shortly before they are needed. If either the information that is used for scheduling or the scheduling tool itself is faulty, the consequences can be costly. Among other things this means that planning tools probably must be capable of being adapted and adjusted on the basis of experience so that they will become more accurate and efficient through use.

The more automated the manufacturing process becomes, the more important are the specifics of the information systems that control the automated tools. Numerically controlled machine tools have been used for some time now to perform various manufacturing operations. These tools are sometimes linked via computer networks with other similar tools working on other aspects of the process so that a product component may pass from one tool to another in the production and assembly process. Several numerically controlled machine tools may be driven by a single computer that in turn gets its directions from another computer higher up in the hierarchy. With such a system one can think of the entire manufacturing process, or at least that part of it that is automated, as being directed by a set of hierarchically organized information structures—namely, the programs that operate the machine tools and the computers that control them.

Gunn (1982, 116) argues that the greatest opportunity for improving productivity in manufacturing lies not so much in the development of ways of making the individual operations more efficient as in the organizing and managing of the complete manufacturing enterprise: "The complexity of the modern factory is daunting: in some plants thousands of parts must be kept in stock for hundreds of products. Indeed, the complexity of the operations has sometimes led to a situation resembling grid lock on the factory floor: it is not uncommon for a metal part to spend 95 percent of the time required for its manufacture waiting in line for processing." The computer's most important contribution to factory productivity, Gunn suggests, will be its capacity to link design, management, and manufacturing into a network of commonly available information.

The roles of human beings in the manufacturing process have been changing, and are continuing to change, drastically. More and more of the jobs performed at all levels of an operation will be information-handling jobs and will require interfacing with an information system in some way.

### Retailing

The introduction of information systems in retail stores, including grocery stores, has been quite visible to all of us. One major function of these systems is to capture information about purchases at the point of sale or the checkout counter. This information helps managers to maintain accurate and up-to-date records of what is being sold each day, to plan wholesale buying, to control inventories, to allocate storage space, to plan pricing changes, to speed up billing, and to provide consumers with detailed information about their purchases.

Such systems would seem not to pose much of a training problem for sales clerks. Issues of job satisfaction may arise, however, as phases of sales work become increasingly automated. There are other issues, too, relating to aspects of these systems that are not visible to consumers. The information system of which the sales terminal is a part serves as the central nervous system of the entire retail operation, and, as in the case of manufacturing, its most important implications are less for the efficiency of individual activities than for the coordination of the enterprise as a whole. From this perspective the users of such systems are not only the sales clerks but the managers, the buyers, the accountants, the space planners, and anyone else who needs to know what is and what is not selling and how this depends on such variables as season, day of the week, time of the month, advertising, product layout and display, and pricing. In a highly competitive business the organization with the most timely and accurate information is in the best position to modify to its advantage those variables it controls and to adapt its operation to those it does not.

## Defense

Information systems are centrally important for all levels of military planning and operations. Without accurate and timely information, effective command and control at any level are impossible. Military documents these days typically refer to command, control, communications, and intelligence ($C^3I$) as one integrated entity. Elements of $C^3I$ include "fixed, surface-mobile, and airborne command and control centers; surface, airborne, and satellite systems for warning, surveillance, reconnaissance, weather data imagery, and intelligence; surface, air, and space-based communications that carry the information to its many destinations" (Zraket 1984, 1306).

When one thinks about military applications of computer systems, one probably thinks first of such things as fire control, intelligence analysis, and weather forecasting. Some of the most important and widespread uses of computers within the military, however, are very similar to those found in the commercial sector. A case in point is information management, which includes the maintenance of files and the preparation, editing, and transmission of messages of all sorts.

The U.S. Defense Department is presently working on modernizing the information system that supports the operations of its World-Wide Military Command and Control System. A recent report by the Assistant Secretary of Defense for Communications, Command, Control, and Intelligence (1981) identifies, as among the major problems, the following two: inadequate on-line software development and data management tools, and user-computer interface deficiencies. By way of correcting the latter, the system's designers hope to provide a user-oriented automated message-handling capability that will include "supporting functions such as the composition, coordination, and transmission of messages developed by users and the automated receipt, distribution, and accounting of messages received by the users. Other functions such as maintenance of historical message files, on-line preparation of private user-oriented files, and gathering of statistics will be included" (Assistant Secretary of Defense 1981, 41). Because the primary mode of use will be an interactive one, the design of workstations will be a focus for research.

Several recurring themes are seen in Defense Department planning documents for computer equipment and systems. These include a trend toward distributed operating systems and data bases (machines interconnected via networks and networks interconnected via internet gateways); efforts to standardize software (for instance, adoption of Ada as the standard Defense Department language) to facilitate its transfer from one machine to another and to increase programmer productivity; a growing reliance on certain basic software tools, such as electronic mail systems and information management systems more generally; increased interactive use of computers by people at all levels within the command structure; efforts to design "friendly" and natural interfaces and user aids; and safeguards against human error (Assistant Secretary of Defense 1981; Bartee, Buneman, Gardner, and Marcus 1979; Center for Tactical Computer Systems 1980; U.S. Army Materiel and Readiness Command 1980; Dertouzos 1980; Jones 1980).

The need for human-factors research is particularly acute here; though faced with a shortage of skilled personnel and a limited time for training because of the high turnover rate, the military services are engaged in developing ever more complex computer-based systems—and, of course, the consequences of failure due to human error could be grave indeed. Each of the services maintains an active program of research and development on various aspects of information technology and, in particular, on issues relating to user-system interaction. The focal agency for the Defense Department's efforts, however, is the Advanced Research Projects Agency. A major new ARPA project in this area, known as the Strategic Computing Program, is discussed in chapter 4.

## General Management

A distinction has sometimes been made between people who use computer systems because their jobs require it and those who do so (or do not do so) by choice. This is an important distinction. People whose jobs require the use of a computer system will use it—more or less effectively—independently of how well or poorly it was designed. People who have a choice will often simply decline to use a system—even one that could do them some good—if using it, or learning to do so, is too much trouble.

General managers and executives in most businesses are among the people who have a choice, and although the data are sparse, it seems to be the case that relatively few of them so far have made extensive use of the numerous systems purportedly designed with them in mind. There are several possible explanations. Perhaps the systems really do not provide the functionality executives need or want. Perhaps they provide the functionality, but the difficulty of operating them, or learning to do so, is not considered worth the effort (efficient use of many systems requires an ability to type, which not all executives have, or aspire to have). Perhaps "pride of rank," as the reluctance to engage in activities that are not typically performed by people in high-level positions is sometimes called, has been a factor.

Whatever the case, there is some evidence that high-level managers, including chief operating executives, are beginning to overcome their reluctance and to make personal use of computer-based information systems. Rockart and Treacy (1982, 83) attribute this trend to three factors: "User-oriented terminal facilities are now available at an acceptable price; executives are better informed of the availability and capabilities of these new technologies; and, predictably, today's volatile competitive conditions heighten the desire among top executives for ever more timely information and analysis." The authors observe that the patterns of use of computer systems by executives represent variations on a few basic themes. These themes serve to define a new kind of information system that they refer to as an "executive information support," or EIS, system. EIS systems share a central purpose (provision of information for planning and control), a common core of data (data on important business variables, such as general-ledger accounting variables; budgeted, actual, and projected sales; and so forth), two principal uses (to provide access to business data and to perform personalized analyses of these data), and a support organization (coaches and assistants to help managers use the EIS effectively).

It seems reasonable to assume that use of computer-based information systems by executives will increase in the future. The systems will undoubtedly be made more useful as their designers acquire a better understanding of what kinds of resources various executives need to do their jobs. Interface designs will improve. Pride of rank, to the extent that that is now an issue, will probably become a non-issue, or perhaps appear in a quite different form: a personal computer, or a terminal to a central information system, in one's office may become a status symbol and an emblem of managerial sophistication.

Solid evidence that electronic devices, and in particular computer terminals, are becoming firmly entrenched in offices is provided by the fact that designers of office furniture are beginning to work on the problem of accommodating such devices in both functional and aesthetically appealing ways. Astute furniture designers will no doubt produce some designs that have an appropriately executive flair.

### Education and Training

Among the most rapidly growing communities of computer users are primary- and secondary-school students. The number of personal computers in U.S. schools approximately tripled from the fall of 1980 (31,000) to the spring of 1982 (96,000) (National Center for Education Statistics 1980, 1982). As of 1984 the number was estimated to be about 300,000. This trend seems certain to continue.

The widely publicized report *A Nation at Risk*, issued by the

National Commission on Excellence in Education (Gardner
et al. 1983), identifies five "basics" on which four-year high
school programs should focus: English, mathematics, science,
social studies, and computer science. Specifically, its authors
recommend that all high school students be required to take
four years of English, three years of mathematics, three years
of science, three years of social studies, and one-half year of
computer science. "Computers and computer-controlled
equipment," they note, "are penetrating every aspect of our
lives—homes, factories, and offices"; and technology is radically transforming many occupations. The objectives of instruction in computer science are summarized as follows: "The
teaching of *computer science* in high school should equip students
to (a) understand the computer as an information, computation, and communication device; (b) use the computer in the
study of the other Basics and for personal and work-related
purposes; and (c) understand the world of computers, electronics, and related technologies" (26).

While not including it among its set of basic academic com-
petencies (reading, writing, speaking and listening, mathematics, reasoning, and studying), the College Board (1983)
recognizes computer competency as "an emerging need" and
notes that students entering college would profit from a basic
knowledge of how computers work, some ability to use them,
an awareness of the various ways in which they are used, and
some understanding of the issues—social, economic, ethical—
associated with their use.

The National Science Board Commission on Precollege Education in Mathematics, Science, and Technology has taken the
position that "computers can effectively be integrated into virtually all teaching and learning areas" (1983, 92) and has urged
that computers be considered not merely as mathematical tools
but as facilitators of learning. The commission identified several educational objectives in the area of computing skills, some
of them appropriate for all students and some for students who
show special interest in computing or plan to pursue further
study in scientific and technical fields. Those in the first category were as follows:

- Basic knowledge of how computers work and of common computer
terminology, including a general understanding of the various applications of computers.

- Experience in using the computer as a tool, which should include experiences in the use of standard applications software such as word processing systems, and filing systems.
- Familiarity with one high-level computer language and ability to use that language as a means of interacting with the device.
- Ability to use a computer language to do problem-solving tasks in the context of normal academic experiences and at a level which reflects the individual student's level of ability.
- General understanding of the problems and issues confronting both individuals and society as a whole in the use of computers, including the social and economic effects of computers, the history and development of computing, and the ethics involved in computer automation. (100)

Evidences of the growing interest in computers in the classroom abound. In scanning the March 31, 1983, issue of *MTA Today*—the news journal of the Massachusetts Teachers Association—I estimated that about one-third of its space was devoted to the use of computers in the classroom. Recently the California Assembly voted to make computer proficiency a requirement for high school graduation (*Boston Globe*, May 28, 1983, 23). Clarkson College in Potsdam, New York, is reported to be issuing a Zenith Z-100 computer to every entering freshman (Metz 1983). Several other colleges have announced the intention of requiring that every student own a personal computer, presumably on the assumption that continuous access to a machine will be necessary in order to get one's school work done. In 1983 Brown University equipped a classroom with 55 high-performance workstations with graphics capability, connected by a high-speed network; the room is used for courses in computer science, differential equations, differential geometry, and neurosciences (van Dam 1984). As van Dam puts it, "Now an instructor can introduce a new topic by talking his way through an animated sequence of images viewed by all the students and letting them work independently on the same 'interactive' movie" (158).

A computer-simulated laboratory permits a student to experiment not only with models of objects, relationships, and systems that exist in the real world but also with imaginary worlds that do not or could not exist. The computer provides the opportunity for studying processes that, unlike most of the phenomena described in elementary physics books, cannot be described mathematically. Wolfram (1984) has suggested that the computer is not only making possible the study of phenomena far more complex than those that could be studied by more traditional means, but also prompting a new way of thinking. One aspect of this involves viewing scientific laws as algorithms.

Carnegie-Mellon University has announced plans to develop jointly with IBM an integrated computer network for the University community. The network would connect the personal computers owned by faculty, staff, and students to each other, to large centralized computers, and to the library. "The ultimate aim is to have a personal computer for each student, for all administration and staff members who need one, and for each faculty member who wants one" (Cyert 1983, 569). The total number of computers expected to be in this network is about 8,000. In announcing the plan, Richard Cyert, president of Carnegie-Mellon, said, "I believe that this system will have consequences that one day will be looked upon as a revolution in higher education" (Cyert 1983, 569).

Project Athena, a university-wide program "to integrate modern computer and communications capabilities into all phases of the educational process," was initiated at the Massachusetts Institute of Technology in the spring of 1983. The initial resources for the project included grants of equipment, software, services, and funds from DEC and IBM totaling about $50 million; MIT plans to raise an additional $20 million to supplement that support. The aim is not to facilitate learning about computer and communication technology per se, although that is likely to be one consequence of the experiment, but rather to find ways to apply computational resources to the teaching of all academic subjects in such a way as to facilitate learning.

Athena resources are allocated by two committees with faculty and student members from across the institute, which review project proposals and grant funding and computing resources three times a year. Plans call for the publication of a summary of Athena projects about every six months. Approved projects as of May 1984 included the development of an electronic journal, computer models for public decision making, computer-enhanced education, space-flight dynamics, graphical modeling of molecular structures, software for foreign language instruction, computers in writing, and computational biology.

Perhaps even more than in general education, the potential

of computers in specialized training has been of interest to researchers for some time. Computer-driven simulators have been used to teach such skills as airplane piloting, nuclear power plant control room operation, and surface ship and submarine piloting. Because full-scale life-size simulators are very expensive, considerable attention has also been given to the possibility of developing relatively inexpensive miniature simulations of these control situations with sufficiently realistic computer-controlled graphics to make it possible to train at least subsets of the skills required in these complex tasks.

Training has been recognized as a high-priority problem within the U.S. Department of Defense. A report from the Defense Science Board 1981 Summer Study Panel on Operational Readiness with Higher Performance Systems (Office of the Undersecretary of Defense for Research and Engineering 1982) suggested that training may be the most important element in the operation and maintenance of weapon systems. For fiscal year 1984, individual training at service schools was estimated to cost about $13.4 billion and to consume about 20 percent of the total labor allocated to the services (Assistant Secretary of Defense for Manpower, Revenue Affairs, and Logistics 1983). Also indicative of the Defense Department's awareness of the importance of training is the Report of the Defense Science Board 1982 Summer Study Panel on Training and Training Technology, issued by the Office of the Undersecretary of Defense for Research and Engineering in November 1982. Among the specific recommendations of this report, one of special interest in the present context is the recommendation to accelerate the use of computer-based instructional methods and/or embedded training systems. The report notes that advanced training technology involving microcomputers, interactive video, or arcade-like games is well suited for situations in which training time, space, and equipment are limited.

Computer-aided instruction has long been viewed as potentially capable of alleviating training problems considerably (Baker and Knerr 1981; Orlansky and String 1979). To date there is little evidence that its impact has been great; but many researchers believe that until recently neither the hardware nor the software necessary for exploiting computer-assisted instruction on a large scale was widely available. That situation appears to be changing rapidly.

Especially good candidates for computer-based training are the operators of computer-based information systems. Realization that such systems have the potential of training their own users has given rise to the idea of "embedded training." An embedded training system is one that is contained within the system the user is being trained to use. Thus, for example, a data-management system might contain within it some tutorial software designed to give the novice user instruction in the system's use. The embedded-training approach has been widely endorsed, and one now more or less expects any new computer-based system of any complexity to have some ability to train its own users. Few if any systems have yet been developed that can rely on embedded training to the point of making more traditional training methods and documentation completely unnecessary, but progress in that direction is being made. The design of effective embedded-training techniques will be a continuing challenge to system developers for some time to come.

## Research

Physicists have come to depend on computers to satisfy the computational demands of experiments with high-energy particles; biochemists use them to help determine the nucleic acid and amino acid sequences that compose DNA and protein molecules; and space exploration would be out of the question without the involvement of computers. It is not surprising that computers are heavily used in such obviously compute-intensive scientific enterprises as these; they have also proved to be very useful tools in areas of investigation that have not traditionally been classified as compute-intensive. Experimental psychology is a case in point.

Computers have been used in psychology research laboratories since the very early 1960s. In what may have been the first published experiment run under computer control (Swets, Millman, Fletcher, and Green 1962; see also Licklider 1961; Swets, Green, and Winter 1961; Swets, Harris, McElroy, and Rudloe 1964, 1966), the subject's task was to learn to identify nonverbal sounds that varied in several dimensions: frequency, amplitude, interruption rate, duty cycle, and duration. The point of the experiment was to investigate the effectiveness of certain principles that were often used in programmed instruction, such as continual interrogation and overt response, immediate feedback or knowledge of results, learner-controlled pacing of the lesson, and presentation of successive items conditional upon previous performance.

The computer generated the stimuli, presented them to the subjects, recorded the subjects' responses, provided immediate feedback regarding the correctness or incorrectness of those responses (when the experimental condition called for that), and summarized the results from each experimental session. The pioneering character of this study was soon recognized by several investigators interested in possible uses of digital computers in psychological research (Green 1963; White 1962; Mayzner and Dolan 1978).

A second noteworthy feature of these early computer-controlled experiments is that several subjects were handled simultaneously by the same computer, each of whom could proceed through the session at his own pace. This was made possible by the use of a "sequence break" system that had been designed for the PDP-1 computer by, primarily, Edward Fredkin of Bolt Beranek and Newman and Benjamin Gurley of the Digital Equipment Corporation (Licklider 1962). Thus the procedure involved a primitive form of time-sharing, and it is, so far as I know, the first example of this type of computer use on record.

Through the first half of the 1960s the fact that a psychological experiment had been run under computer control was worth a few sentences in the method section of a report. This use of computers soon became so common, however, that it came to merit only a casual mention, if any at all. The usefulness of computers for processing data was obvious from the beginning. What took slightly longer to become apparent was that they could be used to advantage to generate and present stimuli, measure and record responses, and perform other tasks as well. Moreover, computers made it possible to use experimental paradigms that were difficult, if not impossible, to implement without them. And because changing the computer's program in effect changed the computer from one machine into another, a single computer could take the place of memory drums, tachistoscopes, timers, relays, and numerous other laboratory devices. Accounts of the early applications of computers to psychological research may be found in Borko

(1962), Green (1963), Mayzner and Dolan (1978), and Uttal (1967).

Researchers in the humanities are also finding ways to use computers to advantage (Raben 1985). A Museum Computer Network has been formed to give scholars and students access to information and visual representations of artifacts (Vance 1969). Classicists have compiled a computerized database of Greek texts called the Thesaurus Linguica Grecae, which includes works from about 750 B.C. through A.D. 600. The database contains about 52 million words, which is estimated to be more than 80 percent of the extant corpus of ancient Greek literature. The software that provides access to this database is called Ibycus (Glazebrook 1984); it can search the entire New Testament for a specified letter string in eleven seconds.

Computer-based analyses of text have been used to characterize the writing styles of specific authors and the differences among them. Sometimes these analyses have been used in an effort to resolve questions of disputed authorship. Mosteller and Wallace (1964), for example, analyzed some of the *Federalist Papers* for the purpose of determining whether they were written by James Madison or Alexander Hamilton. The analysis showed that function words (prepositions, articles, conjunctions) served as better discriminators than did content words (nouns, adjectives, verbs), because the latter were too context-dependent. By applying Bayesian decision techniques and other methods, Mosteller and Wallace inferred that Madison was the author of the papers in question. Several other examples of computer use in the humanities and fine arts are given by Sedelow (1970).

An example of how networks can facilitate research by making the results of ongoing work generally and rapidly available to other researchers is the emerging system of information exchange and research tools for molecular biologists. In 1982 the National Institutes of Health awarded a contract to Bolt Beranek and Newman to establish a data bank known as Gen-Bank, which is to be a central repository for DNA sequences as they become identified. In 1984 the NIH awarded a contract to IntelliGenetics to establish a national computer resource, to be called BIONET, that would give researchers in molecular biology ready access to the DNA sequences stored in GenBank or to the amino acid or protein sequences stored in other similar central repositories. BIONET is also to provide software tools for searching for and comparing sequences (Lewin 1984). Such facilities are expected to reduce duplication of effort and to accelerate research in molecular biology considerably. The existence of a network dedicated to serving this user community and of extensive accessible databases that can be shared is expected, also, to facilitate communication and collaboration among the scientists doing the research.

## Users of Information Systems

Who are the users of information systems? More importantly, who are they likely to be in the future? As the preceding comments suggest, farmers will be users, as will manufacturers, merchants, and schoolchildren. Law enforcement officers also will use information systems in investigating crimes; truckers will use them to connect with cargos awaiting shipment, so as to minimize the distance they travel without paying loads; physicians will use them in diagnosing diseases. Indeed just about everybody, old and young, skilled and unskilled, will make use of information systems in one way or another.

One might argue that since the average person has gotten along quite nicely without much in the way of computing resources in the past, it should not be assumed that he will want or need to use such resources in the future. On the other hand, the average person got along quite nicely without electricity, automobiles, telephones, radio, or television until quite recently. He got along without readily accessible books until a few hundred years ago and without written language until about six or seven thousand years ago. To say that the vast majority of humankind has managed without computing resources in the past, and has not felt deprived on that account, tells us essentially nothing about what the demands for such resources are likely to be in the future.

This does not imply that all systems should be designed to accommodate everybody. People will use computers for different purposes and different systems will be developed to meet specific needs. It makes sense to ask about any particular system, what are the characteristics of its intended users. It is interesting to consider, in this regard, whether users might be broadly classified in useful ways. There have been several attempts at such classification (Cuff 1980; Ramsey and Atwood

1979). One basic distinction is that between computer professionals and all other users; Shackel (1981) summarizes the results of several efforts to develop a taxonomy of users who are not computer professionals and concludes that there is so far no satisfactory way to classify them. Not only do users differ considerably from each other, but any given user changes over time as a consequence of learning.

Eason, Damodaran, and Stewart (1975) have proposed a top-level classification scheme that distinguishes among three types of users: clerical staff, managers, and specialists. Users in these groups are expected to differ with respect to the degree of flexibility and adaptability they will demand of a system, the amount of effort they will be willing to make in learning how to use a system, the degree to which they will be willing to modify their own standard operating procedures, and the amount of information they will need regarding details of system operation. The same writers also note the importance of distinguishing various media and modes of interaction. Problems often arise when systems designed with one type of user or use in mind are used by other users or in unanticipated ways.

Several writers have suggested that managers are less likely than technical specialists to be willing to invest much time in learning to use computer-based tools (Damodaran 1981; Eason 1981; Stewart 1981). Damodaran notes that managers often have someone else use the computer in their behalf. We have noted earlier, however, that the situation with respect to manager-users may be changing and that the reluctance of people in this category to use computer-based systems personally may be declining (Rockart and Treacy 1982). We may expect that as more computer-based tools are developed that speak directly to managers' needs, use of these systems by managers will grow.

Stewart (1981) defines a specialist as one who works primarily within a particular discipline and has a body of specialized knowledge and techniques appropriate to his area. According to Stewart, specialists are likely to use computers for a variety of purposes, ranging from the performance of highly structured tasks involving data input for predefined outputs, to the writing of new applications programs. Specialists may be willing to invest considerable time and effort in learning how to use computer-based tools, provided those tools really help them to perform their tasks.

Clerical workers are less likely than either managers or specialists to have the option of not learning to use a computer system. If their job requires that they learn it, they will do so. How long it takes them to learn it and how effectively they use it will depend strongly, however, on the same types of factors that will determine whether managers and specialists use a system in the first place.

Schoichet (1981) distinguishes four types of personal workstations on the basis of user categories: operational, professional, managerial, and the equipment manufacturers themselves. Users of operational stations include clerks, secretaries, and administrative assistants. Workstations of greatest interest to this group focus on word processing, data entry, filing, and other administrative functions. The professional user group includes engineers, purchasing agents, bank loan officers, and brokers. Because a large percentage of office paperwork is done on behalf of this group, Schoichet suggests that the greatest opportunities for productivity increases lie in integrating the various systems that support their work. The workstations that address this market segment typically provide high-resolution bit-map screens, disks, and high-bandwidth local-area network interfaces.

Schoichet notes that what would constitute an ideal workstation for the managerial user is not yet clear. Although a great deal of paperwork is done for managers, their primary job is working with people. They will use the system less frequently than operational or professional users and will require an "extremely easy-to-use interface." Their use will be more for purposes of communication than computation; hence, electronic mail is seen as a particularly important capability for the systems they will use. The capabilities of systems targeted for the equipment manufacturers will be intermediate between those for operational users and those for professional users.

Another basis for classifying users has been level of skill or expertise. In spite of many efforts, however, no one has yet devised a straightforward way to partition users according to skill level that does not involve a gross oversimplification of the facts. Consider, for example, the terms *casual, computer-naive, intermittent, nondedicated, nonprofessional,* and *inexperienced,* all of which have been used to characterize people for whom computer use does not constitute a way of life. While some of these terms are used interchangeably, they are not synonymous. They refer to quite different factors—including seriousness of use, knowledgeability, frequency of use, purpose of use, and level of training or experience. Moreover, any one of these terms may be used in somewhat different ways by different writers. Cuff (1980) lists several characteristics of a casual user: poor retention of detail, propensity for error, need for a safety net, limited typing ability, limited initial training, reluctance to use documentation, and intolerance of structural formality. One should not assume, however, that other people using the term *casual* have precisely this list of characteristics in mind.

In short, describing "the user" of an information system is not a trivially easy task. Given the very great variability within any user group, providing such a description may not even be a meaningful objective. User taxonomies are probably at best stimulants to thinking about user needs, and not to be taken seriously beyond that. Without question, however, users differ in a variety of important ways, including purpose of use, frequency of use, knowledge of computers in general, knowledge of the particular system being used, types of programs used, and degree of choice in use. It is probably also true, though, that certain principles of interface design apply broadly across both systems and user groups. The articulation of these principles is a major challenge to human-factors research.

# 4

## Anticipated Developments

Given the rapidity with which information technology has advanced over the last thirty years, and the apparent absence of any insuperable obstacles to its continued rapid advance, we can hope to identify only a small fraction of its many future developments and implications. The following list is offered as a speculative sampling of what is likely to happen in the next few decades.

- Continued reductions in the size, power requirements, and cost of computing and information-storage elements, accompanied by further gains in speed and reliability.
- Submicron features on VLSI chips, and one million or more active-element groups per chip.
- Emergence of new computer architectures featuring high degrees of parallelism.
- Write-once, read-many-times optical disks capable of storing trillions of bits.
- Increase in the amount of computing resources and computing activity per capita. Phipps (1982) has estimated that the per capita consumption of electronic circuits in the United States, which was about 10,000 in 1980, will be about 2 million by 1990.
- Increasing availability of computing resources to the average person. Some experts believe that the cost of memory and processing power will be so small within a few decades as to be inconsequential. The notion seems to be that we should assume there will be ample computing resources to satisfy everyone's needs at costs that are within everyone's means (Giuliano 1982).

- A blurring of the distinction between computers and other devices as microprocessors become standard components in major vehicles, factory and office equipment, household appliances, and personal articles.

- A blurring also of the distinctions among word-processing systems, personal computers, professional computers, smart terminals, and workstations. Already manufacturers are offering systems with the ability to emulate other systems, so that, say, a personal computer can mimic a terminal with which the user may be familiar.

- Proliferating applications of new communication facilities: fiber optics, microwave broadcasting, satellite broadcasting.

- Proliferation of office networks linking document preparation, communications, and information management functions.

- Feasibility of electronically conducted "real time" polls and referenda. It has been predicted that 80 to 90 percent of the homes in the United States will have two-way television by 2000 (Dunn 1979).

- Much use of synthetic speech as an output medium.

- Limited but useful continuous-speech recognition capability.

- Growing emphasis on the importance of software, and in particular of "intelligent" software that will provide users with cognitive access to computing resources.

- Availability of a great deal of special-purpose software, some developed by professional programmers and some by amateurs to meet their own needs and interests. This software will include a wide variety of resources—programs to manage personal or family finances, to plan menus, to lay out a family garden, to facilitate home repair, to plan vacations, and to help in many other ways.

- Widespread use of microprocessors in monitoring the consumption of energy in various contexts (vehicles, home and office heating, refrigeration) and adjusting the control parameters of the monitored systems so as to maximize efficiency of energy use.

- Greater use of computer modeling and computer-generated animation for educational purposes—letting students observe processes (chemical reactions, physiological processes, interactions of celestial systems) as well as read descriptions and mathematical representations.

- Greater accessibility of people to other people via portable terminals providing linkage to computer networks.
- Proliferation of computer-mediated information services: electronic news services, want ads, job posting.
- Electronic accessibility of very large data bases, both general- and special-purpose.
- Shopping from the home with the help of video-disk-based electronic catalogs.
- Payment of bills from home computer terminals. Exactly how this will happen remains to be seen; perhaps one will send a message to one's bank to transfer $x$ dollars from a checking account to a Sears Roebuck charge account. Not the least of the challenges here is the development of foolproof procedures for user authentication.
- Application of "expert systems"—a few of which are already beginning to be used—to an ever-widening range of problems, including medical diagnosis, weather forecasting, oil exploration, investing.
- Larger range of options in home entertainment: video program storage, dial-up movies, sophisticated computer-based games.
- Proliferation of "intelligent" toys and games. (Already one can purchase for less than $100 a chess-playing machine that can beat most amateurs.)
- Significant changes in the jobs people perform. Computers will replace people on some jobs but will also create some new jobs. Whether they will create more or fewer jobs than they abolish remains to be seen.

Any list of this sort is bound to miss the mark: the most profound and far-reaching effects of information technology will probably be among those that no one anticipates. To borrow an observation of Linstone's (1975, 60): "In looking to the future we are in much the same position as the Greek temple builder trying to predict the future of building construction without a knowledge of steel, cement, or even the arch."

### New Computer Components

The quest of computer designers has been, in a word, speed. The way one gets a computer to do interesting and complicated things is to get it to do a very large number of simple things in a short period of time. If one thinks of a computer as a collection of interconnected bistable elements—which is an accurate if oversimplified characterization—one may say that speed is achieved in two ways: (1) by decreasing the time required to switch individual elements from one of their states to the other (opening or closing a gate, changing a memory element from one state to the other) and (2) by decreasing the time required to communicate the states of elements from one part of the machine to another.

In the first two or three decades of the computer age, attention was focused on the problem of decreasing switching times, inasmuch as this component was large relative to the time required to get electrical signals from place to place: the time required to open or close a mechanical relay or to turn a vacuum tube on or off is many orders of magnitude greater than the time it takes to transmit a signal over the distances involved in even the largest of the early machines. So the problem on which work was focused in order to increase the speeds of computers was that of developing basic components with faster switching times. Great progress has been made on this problem, and today machines are built of components that can change state in the range of hundreds or tens of nanoseconds. The limit has not been reached in switching times, however, and work continues on several types of faster components, including Josephson junctions and transphasors.

The Josephson junction, named for its inventor Brian Josephson, is still an experimental device, but many researchers believe it will eventually replace the transistor as the basic switching element of processors. The Josephson junction is made from two superconductors separated by a thin layer of insulating material. Even in the absence of a voltage across the junction, a current flows from one superconductor to the other by a process called quantum-mechanical tunneling. The application of voltage stops the current, so, like a transistor, the junction can act as a two-state switch. Unlike the transistor, however, the junction has only two terminals (the transistor has three), so that as a switching device it requires a different type of logic than does the transistor.

The Josephson junction has three advantages over the transistor: speed, power dissipation, and size. It can switch in about 10 picoseconds, or 1/100 the time required by a transistor; because it is a superconducting device, it dissipates about 1/1,000 the power that a transistor does; and it can be made several times smaller in linear dimensions. Unfortunately, Josephson junctions operate at the temperature of liquid helium, 4.2° Kelvin, so their packaging poses some difficult engineering problems. The expectation is that these problems can be solved and that the Josephson junction can become a practical building block for superspeed computers (Matisoo 1980).

Efforts are being made to develop a practical optical transistor, sometimes called a transphasor. Its operation is most easily understood by analogy with the familiar electronic transistor. An electronic transistor has three components: an emitter, a collector, and a base sandwiched between them. The device can be made to function as a gate because a small current applied to the base can cause a much larger current to flow from the emitter to the collector; in other words, application of a small current to the base effectively changes the transistor from an insulator to a conductor, or from an open switch to a closed one.

An optical transistor, or transphasor, is also composed of three elements. In this case a certain type of crystal is sandwiched between two partially reflecting mirrors. By judicious selection of the thickness of the crystal and the wavelength of a beam of coherent (not necessarily visible) light focused on it, the transphasor can also be made to function as a gate or switch: the presence or absence of a second (weak) coherent light source, or probe beam, can determine whether the device transmits nearly all or nearly none of the (relatively strong) constant beam. This arrangement is possible because the refractive index of certain crystals (such as indium antimonide) changes nonlinearly as a consequence of a change in the intensity of incident light; in particular, the crystal's refractive index increases sharply at some point with a small increase in incident intensity. The transphasor exploits this property in such a way that it is possible to effect a very large change in the amount of light transmitted by the constant beam as a consequence of a very small change in the intensity of the probe beam (Abraham, Seaton, and Smith 1983). In addition to being able to change state about a thousand times faster than an electronic transistor, the transphasor also has the advantage that it can act as a switch for several parallel beams simultaneously. Further, any given beam can be made to adopt any of several steady states rather than only two. Interest in building computers with optical components is high and the possibilities are being aggressively researched (Glass 1984).

As the switching times of components have gotten smaller and smaller, the other factor limiting computing speed—the time required to transmit signals from point to point—has become increasingly significant as a determinant of the overall speed of the machine. The signal transmission speed achievable in an electrical circuit is about 15 centimeters per nanosecond; so there is little room for improvement on this dimension, inasmuch as the maximum speed of any signal—physicists tell us— is the speed of light, or about 30 centimeters per nanosecond. Even a computer that used optical signals rather than electrical ones could not get signals from one place to another more than twice as fast as current technology permits.

This being the case, the only way to reduce signal transmission times so that they are commensurate with switching speeds is to reduce the distance over which the signals must travel. This is one among several reasons why miniaturization technology is such an important factor in the evolution of computer designs. When an entire processor can fit on less than a square centimeter, the time it takes to get a signal from any place to any other place within the processor cannot be very long. Switching times are now so fast, however, that even the time required to move a signal from point to point on a microchip is of some concern to the designer. Furthermore, the interconnections between components can account for as much as half the area on today's microcircuits, the remainder being shared by the active elements and the spaces that separate them; and as the number of active elements on the chip increases, the fraction of the chip devoted to interconnections is expected also to increase. Thus the task of the circuit designer today is quite different from that of a few short years ago. He must concern himself not only with the conventional problems of signal flow but also with the problem of organizing things in such a way as to minimize the distances that signals have to travel, both because of the space that interconnections take and because of the need to minimize signal travel time.

There is still a considerable way to go in increasing the speed of elementary components and in reducing their size and power requirements, before fundamental physical limits are reached. We are not quite sure what those limits are. With respect to power requirements, for example, there is a debate among physicists about whether it is, in principle, possible to build computational devices that operate on quantum-mechanical principles and dissipate arbitrarily small amounts of energy in the performance of a computation (Robinson 1984a). Feynman (1984) has proposed one design for such a machine.

At any rate, as the technology approaches the limits set by physical laws, the practical engineering problems that must be solved in order to get closer still grow more formidable. Another approach toward increasing the computer power that can be brought to bear on a particular problem at any given time is that of organizing the basic components in new ways—designing new system architectures, particularly architectures that permit many things to happen at once.

### New Architectures

The architecture of computers and of computer systems has been determined to no small degree by economic considerations. The more costly components have been produced in relatively small quantity and shared in one way or another. Processing components have cost much more than memory components, so most machines have had a single processor but lots of memory of various types. Also, the cost of computing has been greater than the cost of communication, so techniques were devised that permitted many users to communicate with a single machine and share its resources.

The steady decline in the cost of processing power is promoting two developments, one at the level of individual machines and the other at the level of machine complexes. For individual machines there is increasing interest in architectures featuring many processors operating in parallel (Levine 1982; Norrie 1984). It may well be that in time the distinction between processor and memory will become blurred and we will have active memories, built out of processors.

This trend in machine design is likely to affect the way people think about complex problems. Most of the problem solving that has been studied by psychologists in the laboratory is probably best considered to be serial problem solving; at least, that part of the process that is readily available to observation and introspection seems to be serial. The availability of machines that can effect parallel approaches will undoubtedly prompt the development and study of such approaches.

At the level of computer complexes the changing economics are fostering parallelism in the form of distributed systems and distributed databases. Instead of concentrating resources at a central location and providing access to those resources over communication lines linked to remote terminals, the new architectures will distribute the computing power and the data among many small but powerful machines.

The machines will be interconnected by data communication networks and the networks will be interconnected by internet gateways. Such a decentralization of computing resources seems inevitable, given the amounts of computing power and storage that can now be incorporated in (physically) small and relatively inexpensive machines. Moreover, distributed systems have the important advantage of being relatively robust. A multiply interconnected network will continue to function in spite of the disabling of a subset of its nodes—provided, of course, that none of the disabled nodes is the unique source of some function essential to the network's operation.

Tarasoff (1978) has suggested that the anticipated lower costs of data communications, brought about in part by satellite systems, may tend to make distributed data-processing approaches less attractive. If one of the major advantages of distributed data processing is that it minimizes the need to transmit large amounts of data over long distances for processing at a large central facility, then a drastic reduction in transmission costs might neutralize this advantage. This does not mean that distributed systems would no longer be desirable, however, but only that one of the current reasons for implementing them would disappear. If it should turn out that, in time, neither the centralized nor the distributed approach is clearly superior from a financial point of view, then which to use will be decided on other grounds. Perhaps many of the systems of the future will blend the two approaches. Considerable computing power and storage capacity will be distributed throughout a network, with most of the workstations having very substantial amounts of both; yet there will also be some remotely accessible centralized facilities that contain extensive libraries of software tools and information and can deliver unusual amounts of computer power for special tasks that require them.

The distinction between serial and parallel processing machines is, like most others, an oversimplification. All computers being built today perform some operations in parallel. The degree of parallelism can vary, however, over a considerable range, and concurrent operation is achieved through a variety of designs. Most large mainframes provide for the overlapping execution of individual instructions and parts of instructions. This is not usually what one has in mind when one talks of parallel architecture, however. Although there are many different architectures that are conventionally referred to as parallel, they fall into roughly two classes—SIMD and MIMD. In an SIMD (Single Instruction Multiple Data) machine, a single stream of instructions is applied to multiple streams of data. Such machines are often referred to as vector machines. In an MIMD (Multiple Instructions Multiple Data) machine, multiple streams of instructions are applied to multiple streams of data. These machines are generally referred to as multiprocessor machines (Browne 1984).

The first major effort to develop a multiprocessor computer began in the mid-1960s and eventuated in the ILLIAC-IV. Conceived by computer scientists (mainly at the University of Illinois), funded by the Defense Department's Advanced Research Projects Agency, and built by commercial computer companies including Burroughs, Texas Instruments, and Fairchild Semiconductor, the machine was originally designed to contain 256 processors. In part because the developmental costs proved to be far greater than anticipated, plans were changed while the machine was in progress, and the number of processors was reduced from 256 to 64. Uncharitable critics of the ILLIAC project point out that the completed machine was one-fourth the originally intended size and, at $30 million plus, four times as expensive as originally estimated.

After completion around 1972, the machine was installed at NASA's Ames Research Center in California, where it remained in operation for about ten years. The largest computer constructed up to that time, it was also the first to use semiconductor memories, and experience in building it reduced considerably the enthusiasm for experimentation with thin film as a primary-memory medium (Falk 1976). The ILLIAC-IV is often referred to as one of the first-generation supercomputers. Others of about the same vintage are Control Data's Star 100 and Texas Instrument's ASC. While only one ILLIAC-IV was built, by the late 1970s there were a small number of installations of both the Star 100 and the ASC. None of these machines is being produced today.

The ILLIAC-IV was a collection of 64 semi-independent processors whose activities were coordinated by a single master control unit. Associated with each processor was a memory unit, with a storage capacity of 131,072 bits (2,048 words of 64 bits each). The effective memory-access time, including the time required to coordinate the activities of the different memories, was about 350 nanoseconds. Thus the machine could execute instructions at the rate of nearly 3 million per second per processor, or nearly 200 million per second when all processors were simultaneously active.

In addition to the primary memories associated with each processor, the ILLIAC-IV had two auxiliary memories: a disk and a thin-film drum with a laser-beam writing mechanism, with capacities of a billion and a trillion bits respectively. Access time of the disk averaged 20 milliseconds; but once accessed, data could be transferred from the disk to a primary memory at the rate of half-a-billion bits per second. The data transfer rate for the high-capacity drum was about 4 million bits per second.

The intent was to apply the computing power of this machine to the solution of large-scale computational problems that could not be solved by more conventional serial-processing machines. In fact, while it was successfully applied to certain types of problems (especially to the solution of the Navier-Stokes equations used in aerodynamics problems), the ILLIAC-IV did not prove widely useful in solving complex problems, to the surprise and disappointment of its developers. The problem is not that there exist no tasks that require very large amounts of computation; the problem appears to be that we are not yet very proficient at designing algorithms for these tasks that effectively exploit a multiprocessor capability.

A familiar example of an application that requires very large amounts of computing capacity is worldwide weather forecasting. First, variables such as temperature, water vapor, cloud cover, air velocity, and pressure are measured simultaneously at many different locations around the world. These measurements serve as inputs to mathematical models, which predict how the conditions at each point on a three-dimensional grid blanketing the earth will change during a short time interval, say ten minutes. The new values of the variables predicted by the model then replace the original values and are used to run the model through the next cycle. The process is repeated enough times to cover the desired forecast period.

The number of calculations required per unit time depends upon the number of variables the model tracks, the spatial resolution of the measurement grid, the duration of the time step, and the details of the computations used to predict the value of a variable of the $n$th time period given the state of the model at the $n-1$th time period. For a 7-variable worldwide model with a grid of cubes 200 kilometers square and 10 layers deep that is being updated every 10 minutes and using 500 arithmetic operations to update a single variable, the number of operations required is roughly one-half billion for every update of the model. (Because one wants to be able to run these models much faster than in real time, this should not be taken as an indication of the processing speed required.) Increasing the resolution of the grid by a factor of 10 on each dimension, thus giving a resolution of 20 kilometers, would increase the computational requirements by three orders of magnitude, bringing them to 500 billion operations per update (Ingersoll 1983).

Other problems that require large amounts of computing power and therefore seem to be good candidates for the application of parallel-architecture machines include image analysis and computer vision (C. M. Brown 1984; Rosenfeld 1969), the design of very large scale integrated circuits (Antognetta, Pederson, and DeMan 1981), seismic exploration, fusion energy research, astrophysics and bioengineering (Fernback 1984), the modeling of plate tectonics, and the modeling of the behavior of objects or gases near a black hole (C. M. Brown 1984; Smarr 1985). A recent study on supercomputers sponsored by the Institute of Electrical and Electronics Engineers identified materials research, structural mechanics, chemical engineering, atmospheric sciences, physical oceanography, and theoretical physics as fields in which supercomputers could be particularly useful (Walsh 1984).

Perhaps a major reason why the ILLIAC-IV was not applied effectively to a larger assortment of problems is that it was not readily accessible to a sufficiently large community of users. As parallel-architecture computers become more commonplace, their availability will provide the impetus for many people to think about how to structure problem-solving approaches that exploit the capabilities of these machines.

The supercomputers mentioned in chapter 2, which are sometimes referred to as second-generation supercomputers, all have some degree of parallelism, although in most cases they are vector, or SIMD, machines, not true multiprocessors. These machines are beginning to be applied effectively to a variety of problem areas—there were an estimated 120 installations worldwide by early 1984, and the demand for them seems to be growing.

Several experimental multiprocessor machines are being developed. Waldrop (1984b) mentions the following:

- The Cosmic Cube, developed at the California Institute of Technology, with 64 processors, each claimed to be equivalent to an IBM personal computer.

- The Dado, developed at Columbia University especially for use by expert systems and knowledge-based programs; plans call for a 1,023-processor machine.

- Non-von (for non–von Neumann), also being developed at Columbia University: "the ultimate goal of non-von is one million processors, each with a tiny sliver of memory" (Waldrop 1984b, 609); the immediate goal is a 128-processor version.

- The Connection Machine, being designed at MIT and Thinking Machines, Inc. (Hillis 1981a); the ultimate goal is a million-processor machine.

The Butterfly computer is a multiprocessor (MIMD) machine designed and built with DARPA support by Bolt Beranek and Newman and currently being used to conduct research on computer vision, parallel computer architecture, electrical circuit simulation, finite element analysis, mutiplexing and demultiplexing of digitized voice, and a variety of communication network functions (Allik, Crowther, Goodhue, Moore, and Thomas 1985; Beeler 1985; Brown, Ellis, Feldman, LeBlanc, and Peterson 1984–1985; Crowther, Goodhue, Starr, Thomas, Milliken, and Blackadar undated; Deutsch and Newton 1984). The design permits the tight coupling of from 2 to 256 processors through a high-speed switch. Each processor is a Motorola MC68000 and all are identical except for input-output handling capabilities. Each processor has 1 megabyte of main memory (with the option of expanding to 4 megabytes) local to itself, but the switch that connects the processors provides each with easy access to the memories associated with all of the others. Thus one can think of the multiprocessor as having one main memory distributed over the entire configuration and readily accessible from any processor. The time required to complete a memory reference is 4 microseconds. The bandwidth through each path of the Butterfly switch is 32 megabits per second, so the total throughput potential of an $n$-processor machine is $32n$ megabits per second.

The design of the Butterfly makes it possible to distribute tasks to processors without regard to the location of the data associated with those tasks. The processors are thought of as a pool of identical workers. Tasks are distributed to these processors on a first-come first-served basis. The Butterfly's operating system, which is called Chrysalis, manages the assignment of multiple tasks to multiple processors by maintaining a "dual queue." When there are more tasks to be performed than there are processors available to perform them, a task queue is maintained; when there are more processors than tasks, a processor queue is formed. Because the processors are all identical, there can be an excess of processors or an excess of tasks, but not both simultaneously. So in reality only one queue is needed, and at any given moment it will hold either tasks or processors or possibly be empty.

While the development of parallel-architecture machines promises to increase the computing power that can be applied to complex problems by several orders of magnitude per unit time, learning how to use this capacity effectively may be more difficult than is realized. As Lenat (1984) points out, most hard problems have search trees that grow exponentially, so that "even a million-fold increase in computing power will not change the fact that most problems cannot be solved by brute force, but only through the judicious application of knowledge to limit the search" (212). One cannot assume that if it takes $n$ units of time to solve a problem with a sequential single-processor machine, the same problem can be solved in $n/m$ units with $m$ processors working in parallel. As the number of processors grows, so also does the complexity of the problem of coordination and interprocessor communication. Although considerable effort is now being devoted to learning how to use parallel-architecture machines, the constraining factor for the immediate future is likely to be a lack of knowledge of how to make the most effective use of their potential.

## Anticipated Developments in Communication

The hybrid communication systems being developed or planned, which combine earth-orbiting satellites with terrestrial radio and wire (or optical fiber) links, will make it easier to move very large amounts of information from place to place. The Xerox Corporation's Xerox Telecommunications Network (XTEN), for instance, uses satellites for long-distance transmission, terrestrial microwave for short-distance hops, and wire to connect terminals to local nodes; other planned hybrid networks include the Advanced Communications Service of AT&T and the Satellite Business System, a joint venture of IBM, Comsat, and Aetna Insurance. Satellite communication links are expected to stimulate the data-processing and office-automation industries by reducing information transfer costs. New markets for such services as electronic mail and digital facsimile are expected to emerge, and the costs will be low enough to make such services attractive. Predicasts (1984) has predicted that the number of domestic satellites will triple (from 22 to 69) between 1983 and 1995 and that, because of technological advances, the capacity represented by these systems will increase at an even greater rate. Revenues generated by satellite communication are expected to go from about $700 million to about $5.6 billion during the same period.

With cable and direct-broadcast satellite television delivering a wide choice of programs, the home viewer, equipped with a cassette video recorder or other storage device, will be able to customize his televiewing. Advances in communication technology will also offer the possibility of two-way communication, which has obvious implications for education, as well as for social and political processes.

## Some Major Research and Development Efforts

Recognizing the economic advantages to be gained from the further development of information technology—and the consequences of being left behind by others who do a better job of it—governments and business alike have undertaken a number of long-term projects to advance this field. Japan has launched several national efforts, most notably the National Superspeed Computer Project and the Fifth-Generation Computer Project, aimed at making it a world leader in information technology by the end of the decade. Problem areas that are seen by the Japanese as requiring significantly augmented computing resources include "nuclear fusion, image analysis for the Earth Resources Satellite, meteorological forecasts, electrical power systems (power flow, stability, optimization), structural and thermal analysis, and very large scale integrated circuit design and simulation" (Buzbee, Ewald, and Worlton 1982, 1189; Feigenbaum and McCorduck 1983; Moto-Oka 1982, 1983).

The goal of the National Superspeed Computer Project is to build a machine faster by three orders of magnitude than existing supercomputers—a parallel-architecture machine with an execution rate of about 10 billion floating-point operations per second (BFLOPS), 1 billion bytes of memory, and a memory transfer rate of 1.5 billion bytes per second. The project involves a partnership between Japan's Ministry of International Trade and Industry (MITI) and six major Japanese computer companies: Fujitsu, Hitachi, Mitsubishi, Nippon Electric, Oki, and Toshiba; total funding over about seven years, beginning in 1982, is believed to be about $200 million. When the project is completed in 1989, all of the participating firms will be allowed to make use of the results of the effort in their own product developments. Both Josephson junction devices and gallium arsenide technology are receiving much attention in this project (Buzbee, Ewald, and Worlton 1982; Walsh 1983).

The Fifth-Generation Computer Project, which also receives its support in part from MITI and in part from industry, was started in 1982 and is expected to continue for ten years. Initial plans called for spending about a billion dollars, half from the government and half from private industry, over that period. Goals include increasing productivity in low-productivity areas, meeting international competition and facilitating international cooperation, assisting in the saving of energy and resources, and coping with an aging society. This project focuses on issues of usability and applications, and it calls for the development not only of new architectures but also of knowledge-based expert systems, natural-language understanding, speech input-output, reading capability, and a variety of other aspects of artificial intelligence (Buzbee, Ewald, and Worlton 1982; Japan

Information-Processing Development Center 1981; Sunn 1983). The Japanese have stressed the social implications of information technology, and in particular of artificial intelligence and expert systems (Moto-Oka 1982).

Japan's apparent commitment to becoming the world leader in the development and exploitation of information technology has spurred other countries and consortia on to launch similar projects. Great Britain has announced a five-year, $300-million program, which involves government, university, and industrial scientists and is known as the Alvey Project, after John Alvey, the chairman of a committee whose work led to the plan. The committee recommended that efforts be concentrated on four areas: software engineering, man-machine communication, artificial intelligence, and very large scale integration. Supervised by a directorate within the Department of Industry, the project will also involve the Department of Education and Science, the Science and Engineering Research Council, and the Ministry of Defense (Dickson 1983).

The European Strategic Program for Research and Information Technology (ESPRIT) is a planned collaboration by the European Economic Community (EEC) and private industry to assure European participation in research in this rapidly moving field. Costing $1.3 billion over five years, the projected work will cover five areas: "(i) advanced microelectronics, aimed at designing, manufacturing, and testing very high speed and very large scale integrated circuits; (ii) software technology, embracing what is described as 'the management practices for information technology as well as the scientific knowledge underlying them'; (iii) advanced information processing, including the exploitation of VLSI; (iv) office systems; and (v) computer integrated manufacturing" (Dickson 1984a, 28). An interesting feature of the ESPRIT collaboration is the requirement that each funded project involve researchers from at least two EEC countries (Dickson 1984a). Another European cooperative program is the recently approved plan to establish a center for research on artificial intelligence and expert systems in Munich; the center will be sponsored by three computer manufacturers, Siemans of West Germany, International Computers Limited of England, and Bull of France, with an operating budget of about $7.5 million a year (Dickson 1984a, b).

In the United States a group of companies, including Advanced Micro Devices, Allied BMC Industries, Control Data,

DEC, Eastman Kodak, Harris, Honeywell, Martin Marietta Aerospace, Mostek, Motorola, National Semiconductor, NCR, RCA, Rockwell, and Sperry, formed a research consortium in 1983 under the name Microelectronics and Computer Technology Corporation (MCC). Its headquarters are in Austin, Texas. Policy and direction are provided by a board of directors and a technology advisory board, each of which contains a representative from each of the participating companies. MCC plans to spend about a billion dollars during its first ten years; among the subjects for research are human-factors technology and artificial intelligence. This venture is financed entirely by private funds, with each shareholder company contributing from $150,000 to $250,000 initially. In return for their support the shareholder companies have privileged access to the results of MCC research and development for a specific period of time.

Also in the United States, the Defense Department's Advanced Research Projects Agency has proposed to spend about $600 million on a strategic computing program over five years beginning in 1984. Because ARPA has been a major force in the development of information technology (including time-sharing, packet-switched networking, and artificial intelligence) in the United States to date and is leading the effort toward further development, its Strategic Computing Plan is worth noting in some detail. The plan spells out ARPA's intention to apply machine intelligence technology to problems of national defense. It calls for the development both of new computer architectures featuring a high degree of parallelism and of new software techniques to exploit these architectures. Heavy emphasis is placed on the use of "expert systems" that not only "mimic the thinking and reasoning process of humans" but are "equipped with sensory and communication modules enabling them to hear, talk, see, and act on information and data they develop or receive" (Defense Advanced Research Projects Agency 1983, preface, p. 1). Intended applications include autonomous vehicles, "expert associates," and large-scale battle-management systems.

The autonomous vehicles that are envisioned are to be equipped with sensing systems, possibly including visual systems patterned after the architecture of the human retina. Arrays of thousands of image pixels, each associated with a few switches and communicating with neighboring cells, are considered possibilities. The intent is to "provide special vision subsystems that have rates as high as 1 trillion von Neumann equivalent operations per second" (DARPA 1983, preface, p. 4) by the late 1980s. (A von Neumann equivalent operation is the equivalent of a single operation in a single-processor machine.)

Autonomous systems, as the term is used in the ARPA plan, are true robotic devices: "They are able to sense and interpret their environment, to plan and reason using sensed and other data, to initiate actions to be taken, and to communicate with humans or other systems" (DARPA 1983, chapter 5, p. 21). An autonomous land vehicle of the type that is projected would be able to negotiate a 50-kilometer cross-country trip using digitally stored terrain and environmental data as well as information acquired via its visual sensing system. It is estimated that the expert navigation system would require a computational capability of 10 billion to 100 billion von Neumann equivalent instructions per second.

The "expert associates" described in the ARPA plan would be able to interact with human beings via speech and could be trained and personalized for specific situations and individuals. A pilot's associate, for instance, could perform such functions as lower-level instrument monitoring, control, and diagnosis. The development of these devices will involve dealing with a variety of person-machine interface issues as well as with questions of function allocation, manual backup for system failure, and user acceptance.

The battle-management system envisioned in the ARPA plan "would be capable of comprehending uncertain data to produce forecasts of likely events, drawing on previous human and machine experience to generate potential courses of action, evaluating these options and explaining the supporting rationale for the evaluation to the decision maker, developing a plan for implementing the option selected by the decision maker, disseminating this plan to those concerned, and reporting progress to the decision maker during the execution phase" (DARPA 1983, chapter 5, p. 27). It too is expected to interact with users by means of speech and natural language. The application of machine-intelligence technology to large-scale battle-management problems will confront issues of communication, information integration, and distributed control and decision making.

The ARPA plan also calls for "an extensive infrastructure of computers, computer networks, rapid system prototyping services, and silicon foundries to support these technology explorations" (DARPA 1983, preface, pp. 4–5). The idea is that such an infrastructure will facilitate both the development of experimental systems and the sharing of results and ideas among the community of researchers working on these problems.

The intelligent functional capabilities identified in the ARPA plan are those that are considered especially critical for the military applications addressed by the program. Yet because these capabilities include natural language, vision, speech, navigation, planning, and reasoning, clearly any progress that is made toward providing computers with them will contribute to a technology base that has many other applications. The plan stresses the importance of transferring to industry the technology that is developed within the program.

Implementation of the plan will require progress on several fronts. It will depend, for example, on our learning to program highly parallel machines effectively. How difficult it will be to program a machine with hundreds, perhaps thousands, of processors remains to be seen. If the intent is to develop systems that will mimic the thinking and reasoning processes of humans, we need also to understand these processes better than we do now. We especially need to understand better what it is that human beings do particularly well and what they do poorly. We can hope that the reasoning capabilities programmed into intelligent systems of the future will surpass those of human beings in certain respects. At least they should be free of the more common reasoning deficiencies that people exhibit and that have been widely documented (Nisbett and Ross 1980; Tversky and Kahneman 1974; Wason and Johnson-Laird 1972). There are other aspects of human thinking, however, that may well turn out to be both critical to the performance of certain tasks and impossible to simulate. While some progress on the problem of codifying expertise has been made, relatively little is known about what constitutes expertise and how to represent it in a machine. Even less is known about how to mechanize common sense—a topic we will return to in chapter 15.

• • • • • • • • • • • • • • • • • • • • • • • • • • • • • • • • • • • • • • • • • • • • • • •

In looking to the near future of information technology, we can be confident that many of the well-established trends of the recent past will continue. These include decreases in the size, power requirements, and costs of the basic elements from which computers are made, accompanied by increases in their speed and reliability. These trends will be maintained both by further refinements in existing production techniques and by the development of qualitatively new techniques. Much effort during the immediate future will be put into the design of new architectures. Attention will be focused both on the development of multiprocessor machines and on the implementation of distributed operating systems and distributed databases. Much energy will continue to go into the design of powerful low-cost machines for home and personal use.

A greater fraction of resources than in the past will be devoted to the development of software to exploit more effectively the wide range of machines that will exist. Much attention will be given to the development of programs that embody nontrivial amounts of intelligence or expertise. However, there are many unexplored ways in which unintelligent and nonexpert programs can be used to great advantage, and we can expect more work on software of this type.

Major advances in information technology in general, as well as progress toward specific objectives, are more or less assured during the next few years by the considerable resources that are being committed to these ends in various parts of the world. Predictions as to what the advances will be are bound to be imperfect, however. Some of the announced goals will prove more difficult to attain than was originally thought, while some of the most significant developments will not have been anticipated at all. We should, of course, do our best to anticipate the future and to shape it, but having done so we should be surprised not to be surprised.

# 5

# *The Study of Person-Computer Interaction*

Person-computer interaction has been a focus of attention from researchers since the development of time-sharing in the early 1960s. One of the first and most influential papers on the subject was Licklider's (1960) forward-looking discussion of the prospects of a symbiotic relationship evolving between computer users and their machines. Several other broad-ranging discussions of person-computer interaction have been published between then and now (Kemeny 1972; Licklider 1965; Martin 1973; Meadow 1970; Miller 1969).

Among the earliest papers to call attention to human-factors issues in the design of computers was one by Bridgewater (1954). Other discussions that have emphasized these issues include those of Karlin and Alexander (1962); Nickerson (1969); Shackel (1969); DeGreene (1970); Parsons (1970); Shackel and Shipley (1970); Bennett (1972); Dunn (1976); and Miller and Thomas (1976). General reviews of research on person-computer interaction include Davis (1966); Licklider (1968); Nicholson, Wiggins, and Silver (1972); Frederiksen (1975); Rouse (1975); and Kriloff (1976). A good source of review material is the *Annual Review of Information Science and Technology,* published by Knowledge Industry Publications, White Plains, New York.

A very useful bibliography of work on human factors in computer systems has been produced by Ramsey, Atwood, and Kirshbaum (1978). It lists 564 references, providing an abstract and evaluative comments for each, and its subject index simplifies the task of identifying the relevant work in a given subarea of the field. Another large bibliography—on computer graphics, interactive techniques, and image processing—has been compiled by Pooch (1976).

## *Types of Person-Computer Interaction*

Most of the tasks for which people use computers interactively are highly structured, with the computer playing a relatively straightforward, predetermined role. However, from the time that attention was first given to person-computer interaction, there has been interest in the possibility of involving computers in the more creative aspects of complex tasks. Licklider (1960), for example, noted the importance of exploring ways of involving computers in problem formulation and planning. Citrenbaum (1972; Kleine and Citrenbaum 1972) has focused on planning as an activity that is difficult to automate but one that might be facilitated by interactive techniques. A few other writers have also emphasized the possibility of using interactive systems to help formulate as well as solve problems (Cushman 1972; Hormann 1971a, b; 1972; Newell 1965; Sackman 1972). It seems safe to say, however, that in spite of this long-standing interest in getting computers to assist in the more creative aspects of various tasks, most of the interactive uses to date have involved the more mundane activities.

Nevertheless, the ways in which people interact with computers have changed drastically since the latter first appeared on the scene in the 1950s. As we have already noted, the size and costliness of the first computers dictated a certain style of use. Programming was done off-line, and only after doing their best to anticipate and eliminate all the "bugs" did programmers have the results of their labor punched onto cards and presented to the computer center for compilation and a trial run. The queue of programs waiting to be run was typically long, because the economics of the situation made it necessary to amortize the great cost of a machine over many users, which meant making sure the machine was seldom idle. The correct way to visualize this situation is to have a large machine at the center of the image and lines of users on the periphery waiting to get a moment of its time.

The invention of time-sharing represented a major step forward by providing users with remote access to computing resources and the ability to interact with the computer more or less continuously in real time. In visualizing this mode of operation, it is still appropriate, however, to have the machine in the center of the image. What is different from the earlier situation is that the machine now shifts its attention from user to user so rapidly that (ideally) each user may feel as though he is getting its full attention, and the users are not obliged to come to where the machine is in order to use it. As in the earlier case, economics dictated that the machine be kept very busy, and again this was done by letting enough users connect to the system to ensure long (albeit invisible) queues. Overloading often degraded system performance to an intolerable level.

With the introduction of personal computers (PCs)—in particular, PCs that can double as terminals to a network of computing resources—the image should change. Now it seems appropriate to picture the user occupying the central position, with the PC and various other resources around him. Economics do not demand that the PC be kept busy all the time. And although some of the centralized resources that one might wish to tap are still likely to be very costly, the amortization of their costs over many users does not necessarily require the type of queuing that can disrupt an interactive session. The storage capacity of the PC, including its local secondary memory, makes it unnecessary to transfer information continuously from a centralized data base to the user: data transfers can now be made in large blocks over wideband paths and hence need not disrupt a work session—at least not frequently.

Thus we see that over the short period of time that these machines have been around, several modes of interaction can be identified, each of which was typical for a few years: hands-on use of large (by early standards) machines by specialists, sharing of centralized facilities through batch-processing or closed-shop operations, on-line interactive use of time-shared facilities through remotely located terminals, independent use of individually owned PCs, the use of PCs both as stand-alone computers and as access ports to resource-sharing networks. The last type of interaction, which is an especially versatile one, will be available to more and more people in the near future. The future is also sure to produce other forms of interaction that we have not yet seen.

We have much greater latitude and freedom now than we had a decade ago in designing work environments that are well suited to human beings who happen to be users of computer systems. Some of the constraints that were imposed by the economic realities of the past have become less severe or have disappeared. As the economic impediments to the design of maximally useful and usable systems diminish, the limits of our imaginations become the more important constraints.

Any use of a computer requires an interaction in some sense, because it involves a person and a machine performing in complementary ways. The term "interaction" has generally been restricted, however, to situations in which the coupling between person and machine is fairly tight and their exchanges take place on a time scale that is more or less characteristic of human conversation. The term was not usually applied to the way in which users and computers related in a batch-processing environment, and, indeed, it became widely used only after the development of time-sharing.

Interactive problem solving in a practical situation in which the computer and the user each play a critical role is nicely illustrated by an air traffic control application described by Whitfield (1976) and Whitfield and Stammers (1976). One of the computer's primary functions is to detect potential collisions by plotting the positions and projected trajectories of aircraft in a given airspace. When the computer detects a potential collision or near collision, it signals the controller, who may then instruct the computer to project the flight paths, given specified modifications of their parameters. The controller then selects a set of parameters that will get the aircraft off the collision course.

For a growing number of people the computer is becoming a truly multipurpose tool. Such users may engage in several different activities—for instance, programming, debugging, text editing, file searching, and mail processing—in a single work session, and they often do not finish one task before switching to another (Bannon, Cypher, Greenspan, and Monty 1983). Interaction command protocols, or command histories, reveal frequent skipping from task to task or nesting of tasks one within another.

Bannon, Cypher, Greenspan, and Monty suggest that one way of thinking about a work session is to treat each task as a separate work space and to characterize the user's pattern of activity as moving from one work space to another. Each work space may be thought of as an environment designed to facilitate the achievement of a particular goal or set of closely related goals, and equipped with tools and data relevant to those goals. This concept of work-session organization leads the authors to suggest several guidelines for interface design, including the following:

- The load on working memory required to switch from one task to another should be kept acceptably small.
- It should be convenient to suspend and resume tasks.
- The history of a user's activity within a work space (the record of the command sequence) should be available for redoing the same task or a similar one.
- Users should be able to organize their work spaces to reflect functional groupings—for instance, hierarchical groups of work spaces maintained in various states of completion for possible reactivation.

### Approaches to the Study of Person-Computer Interaction

Computers are qualitatively different from other machines with which we are familiar. It should therefore come as no surprise that methods found useful in the study of other person-machine systems are not as effective when applied to computer-based systems. One problem is that much of what is going on at the person-machine interface is not observable, because it is going on either inside the machine or inside the user's head.

Three approaches that have been tried are modeling and simulation, observation, and experimentation. Mathematical models and simulations have often been developed for purposes of identifying system bottlenecks and predicting the effects of modifications in either hardware or software components. The task of simulating or modeling a system increases, of course, with the complexity of that system; but since design errors are likely to be especially costly in a complex system, the simulation approach is considered worth taking (Proctor 1963). System models have been very useful in helping to configure systems, plan expansions, balance resources, establish pricing policies, and level loads. They tell us little, however, about the nature of the interaction between a system and an individual user.

Human performance models have been developed that do emphasize the interaction between people and machines, but for the most part the machines on which the developers of these models have focused have not been computer systems.

An extensive review of performance models that have been applied to a wide range of person-machine systems has been prepared by Pew, Baron, Feehrer, and Miller (1977).

Many of the early studies of the interaction of users with time-sharing systems were observational in the sense that they involved attempting to obtain measurements on real systems and real users without controlling the independent variables as one would in a formal experiment. Most of this work was not theory driven, and one of the problems was to identify those measures that would yield useful data. The following is a partial list of the rather broad assortment of measurements that were made: frequency of work sessions; duration of work sessions; diurnal pattern of work session starts; user interarrival times; frequency of use of individual commands; number of commands executed per unit time; number of input-output lines per session; time required for execution of benchmark programs; system response time; user response time; interaction cycle time; task turnaround time; output-time/compute-time ratio; program size; connection accessibility; statement interpretation rate; overhead (proportion of time spent running an executive program); console time per session; CPU time per session; console time versus time of start; CPU time versus time of start; computer time per task; number of user input lines per session; rate of user requests; input rate (characters per second); output rate; output/input ratio (Boies and Gould 1974; Bryan 1967; Carbonell, Elkind, and Nickerson 1968; Grignetti and Miller 1970; Grignetti, Miller, Nickerson, and Pew 1971; Haralambopoulos and Nagy 1977; Jutila and Baram 1971; Rodriguez 1977; Sackman and Citrenbaum 1972; Scherr 1965; Shaw 1965).

Some of these measures were of interest primarily for their potential usefulness in identifying system bottlenecks and matching system resources with user demands or in developing price-differential incentives to help level the load on a system over the hours of the day. Others were obtained in the hope of gaining some insights into the interactive process and of describing various aspects of that process in quantitative terms. Some investigators have studied users interacting with time-sharing systems with a view not only to producing statistical descriptions of use patterns but also to determining how the quality of the user's work or satisfaction with the system relates to the parameters of the use patterns.

The studies cited have yielded a considerable amount of data. It is reasonable to ask what conclusions might be drawn from them. The answer is, not many. One difficulty is that in the absence of a theoretical framework it is not clear what most of these measures mean. One might, of course, look to such measures to get some insights into how a theoretical framework might be structured. If one could find some invariants, one might hope that one had obtained some clues about stable aspects of person-computer interaction. As it happens, probably the most obvious characteristic of the numbers is the very great variability among measures of the same type. Ratios of 5 or 10 to 1 have not been uncommon (Sackman 1970b; Thomas 1977).

A question that received considerable attention during the late 1960s was that of the relative merits of time-sharing and batch-processing approaches to computer use (Adams and Cohen 1969; Gold 1969; Grant 1966; Schatzoff, Tsao, & Wiig 1967). Perhaps the first reported attempt to compare experimentally the effectiveness of time-sharing and batch processing was that of Grant (1966; Grant and Sackman 1967), who studied the performance of programmers coding and debugging programs. Gold (1969) studied nonprogrammer professionals who were using the computer to help solve problems. Both studies showed better performance with the time-sharing systems. Sackman (1968; 1970a, b) reviewed several early studies that compared batch and time-sharing use. Ten such studies were summarized by Sackman and Citrenbaum (1972) and the results compared with respect to person hours, computer time, costs, and user preference. Users tended to prefer the on-line or time-sharing mode, and it required fewer person hours than did the batch mode to accomplish a given task.

Studies comparing time-sharing and batch approaches have not escaped criticism. Lampson (1967), for example, criticized the Grant and Sackman (1967) report for presenting many statistics without carefully considering the appropriateness of the measures taken or the meanings of the numbers obtained. Lampson's points undoubtedly apply with equal force to many subsequent studies of the same genre.

One observational method employed by system designers is what might be called the post-mortem approach. In this case one identifies the faults and limitations of existing systems or tries to determine why systems have not been used by people to whom they could, presumably, be useful (Alter 1977; Franklin and Dean 1974; MacDonald 1965; Nickerson 1981). Unfortunately, it is easy to find fault with existing systems but quite another thing to build better systems. Perhaps, however, such criticisms will serve to call attention to certain types of problems and thus reduce somewhat the probability of their recurrence.

Having users keep logs or diaries, in which they record certain activities or events soon after they occur, is another way of acquiring data about system use. A problem with this approach is the researcher's lack of control over the completeness or truthfulness with which the users keep their records (Epstein 1981). Recently researchers at IBM have developed an observational method that involves having a second computer time and record the keyboard activity while a user is using the system of interest. Later the recorded interaction is played back for observation and analysis (Neal and Simons 1983). The developers of this approach consider one of its advantages to be its nonintrusiveness on the user's work.

An early focus of experimental work was the question of how the user's performance of some task or attitude toward the computer related to the system's response time (Boehm, Seven, and Watson 1971; Morefield, Wiesen, Grossberg, and Yntema 1969; Miller 1965, 1968). This interest has continued, and system response time is still a topic of speculation and some experimentation (Bergman, Brinkman, and Koelega 1981; Butler 1983; Goodman and Spence 1981).

Miller (1965) suggested that the delay in system response that a user finds tolerable depends on the nature of the interaction. He proposed a taxonomy of user inquiries and estimated what would be an acceptable delay for each of the types of inquiries in the taxonomy. Acceptable delays, according to Miller's analysis, vary from a small fraction of a second to several minutes, depending on the specifics of the situation.

Boehm, Seven, and Watson (1971) tested the idea that if access to the computer were restricted for a fixed period of time ("lockout period") following a computer output, users might be induced to concentrate more on their problem-solving strategy than on tactics and their performance might thereby be improved. The users were given a map showing streets, freeways, and accident frequencies at different intersections; their task was to decide on the locations of three emergency hospitals and to specify decision rules regarding the use of freeways and secondary roads for hospital access in such a way as to minimize the average waiting time per emergency for the area. The computer was able to give the user information on request regarding the effectiveness of any assignment of hospital locations and decision rules. So the user's role was to suggest solutions, and the computer's was to quantify their implications. The independent variable of interest was the time the user was prohibited from typing in new input after the computer had provided a response to a proposed solution. This varied from 0 to 8 minutes. The main result obtained using lockout periods of different durations was very large individual differences within groups. But users objected to a brief lockout even when their performance was not impaired by it.

One of the more ambitious studies of effects of system response time on interactive problem solving was conducted by Morefield, Wiesen, Grossberg, and Yntema (1969; Grossberg, Wiesen, and Yntema 1976). These investigators, using themselves as subjects, explored the effects of system response time on the use of the Lincoln Reckoner on a variety of problems. The Lincoln Reckoner was a system that had been designed for use by scientists who were not necessarily programmers. It provided them with tools for doing numerical computations on arrays of data. The system's "nominal" delay was varied from 1 to 100 seconds. (The actual delay was plus or minus 10 percent of the nominal delay.) The subjects used the system to solve several preselected problems that required significant amounts of computation. Again the main effects of artificially increasing the response time varied with the difficulty of the problem on which subjects were working.

Although system response time has been of great interest to investigators from the very first efforts to study person-computer interaction, the evidence is not strong that this variable is a very significant determinant of the quality of one's performance when using an interactive system. Of course, computers have steadily grown faster, and machines that yielded the response times of 10 or 15 years ago would not survive in today's market. So the assertion that response time has not been shown to be a major determinant of user performance should be interpreted in light of the fact that today's machines are very fast, and their response times very short, by yesterday's standards. This is not to suggest that response time is no longer an issue. It may be, however, that it has less effect on user performance than on users' attitudes. Noticeable delays are annoying, and that seems to be true almost independently of the length of the delays. What one wants when one is sitting at a computer terminal is instantaneous response—always. Users recognize that that is an unrealistic desire: they know that while the responses to some inputs can be instantaneous, those to others cannot. They acquire expectations as to how long it should take to do what. And while even expected delays may be a source of some annoyance, those that are unexpected or perceived as unnecessary are especially frustrating. It seems unlikely that this situation will change. As computers get faster, more of the operations that now cause delays will appear to be performed instantaneously, but there will always be uses that tax the computational resources that are being tapped. Dissatisfaction with the state of the art is likely to be a fact of life indefinitely. No matter how fast and responsive computer-based systems become, there will always be the challenge to make them faster still, and to enlarge the set of things that can be done more or less instantaneously.

## The Possibility of an Experimental Science

The three approaches that have been used in the study of person-computer interaction—simulation or modeling, observation, and controlled experimentation—will undoubtedly all continue to be used; each has something to recommend it, and the three complement each other. Of special interest to the experimentalist, however, is the design of controlled experiments whose results will be genuinely useful to, and used by, the designers of future systems.

In the past there have been various impediments to experimentation on person-computer interaction. A few years ago Nickerson and Pew (1977) noted as problems associated with doing research on person-computer interaction "the cost of computer systems and operations, which tends to make them relatively inaccessible to researchers, the difficulty of gaining adequate control of experimental situations, the tremendous variability in most performance measures that have been taken, the lack of generality of findings, and the rapid development of computer technology (which can make results obsolete before they are reported)" (258). Experiments with real time-sharing systems have been difficult to perform and their results inconclusive, in part because these systems have been expensive and few experimenters have been in a position to manipulate their parameters and operational features solely for purposes of experimentation. When experiments have been done, they have often produced results that were either not readily interpretable or without great practical significance. In general, interpretability and practical significance of experimental results have tended to be inversely related: in order to obtain the control over a situation that would permit one to get unambiguous results, it has been necessary to choose system-specific or situation-specific questions that are of limited interest.

It may now be possible to develop research methodologies that will define a bona fide experimental information science. The economic considerations certainly have changed and continue to change dramatically. No longer is it necessary to tie up a very expensive facility in order to conduct experiments on user-computer interaction. Moreover, numerous experiments have been done during the past couple of decades, and there is undoubtedly something of value about methodology to be learned from these efforts—from both the more successful and the less successful ones. Developing more effective experimental methods must be considered one of the primary challenges for this field.

Computer technology is moving so fast that experimenters risk seeing the results of their research become obsolete before they can be applied. How can one ensure that one's work will not be outdated by the time it is completed? One answer is that if research is to produce results that are of more than passing interest, it must be addressed to issues that are system independent and to the discovery of general principles that are applicable across a broad range of equipment. But how is one to do research that is not system specific? Clearly, we are interested in real systems and how users interact with them. Moreover, it is primarily by using real systems that one acquires an understanding of the technology's capabilities and limitations—and its potential.

Fortunately, the insights gained by using a specific system are not necessarily without relevance beyond that system. Fortunately too, the physical properties of the system—excepting the interface—are probably of little concern to the vast majority of users of information systems. The important issues from the user's point of view relate to functionality and convenience of use. Users are unlikely to want to know, for example, whether the information that they are accessing is stored in a magnetic bubble memory, on an optical disk, or in a hologram; what they care about is how accessible the information is and whether the system can deliver it in a useful form.

Card, Moran, and Newell (1980b, 1983) have taken text editing as a paradigm of human-computer interaction and have studied it extensively. They give the following reasons for this choice:

(1) The interaction is commonly rapid: a user completes several transactions a minute for sustained periods. (2) The interaction is intimate: A text editor, like all well-designed tools, becomes an unconscious extension of its user, a device to operate *with* rather than operate *on*. (3)Text editors are probably the single most heavily used programs: There is currently a massive effort to introduce text-editing systems into offices and clerical operations. Even in a systems programming environment, one study (Boies 1974) found that 75 percent of the system commands issued were text-editor commands; and (4) computer text editors are similar to, and can therefore be representative of, other systems for human-computer interaction: Like most other systems, they have a discrete command language and provide ways to input, modify, and search for data. The physical details of their interfaces are not particularly unique. Because of these similarities, progress in understanding user interaction with text editors should help us to understand interaction with other systems as well. (1983, 101)

Moran (1983) has presented a framework for relating the way tasks are conceptualized independently of computer systems to the way they must be formulated in order to be performed by a specific system. Called ETIT, for External-Task-to-Internal-Task analysis, the framework consists of three parts: an external task space, an internal task space, and a mapping from the former space to the latter. The problem now becomes that of representing a specific editing task in terms appropriate to the external task space and then mapping this representation onto a representation using the command language of a specific editor. In some cases mapping an external task function onto an internal task function requires nothing more than a change of function name (if that); for example, a *move* function in the one domain may translate into a *cut* in the other. In other cases an external task function may have to be represented by a sequence of internal task functions plus a proposed method for making the relationship between external and internal representations explicit. If the proposed model for external representation is sufficiently general, then showing how it maps onto different internal representations (different editors) may help make clear the similarities and differences between the editors and facilitate the transfer of knowledge about one of them to the task of learning how to use the other. To use Moran's terms, one may compare editors by comparing their ETIT mapping rules. This framework is a promising one. More are needed, however, for guiding research and shaping a coherent presentation of research results.

### Assigning Functions to People and to Computers

The problem of allocating functions between people and machines predates computers, of course. The principle that people should do those things that they can do better than machines, and machines should be given tasks that they can perform better than people, seems an eminently sensible one, and most discussions of function allocation have accepted some version of it (Balzar and Shirley 1968; Chapanis 1965; Edwards 1962; Hormann 1965; Jordan 1963; Rieger and Greenstein 1982). Fitts (1962) has argued that asking whether a person or a machine is *better* at a particular task can oversimplify the problem somewhat, because the "best" allocation, in some sense, is not always the preferred one; one does not typically choose silver as an electric conductor, for example, even though it has less resistivity than copper. According to Fitts, "the central issue in choosing components for a complex system is usually not so much which component will do a *better* job, as which component will do an *adequate* job for less money, less weight, less power, or with a smaller probability of failure and less need for maintenance" (35). One might object that Fitts gives "better" too narrow a connotation and that a broader one would include considerations of cost and efficiency.

With the development of increasingly sophisticated systems, the appropriateness of the better-able principle probably should be questioned on other grounds. To be sure, we may assume that at least for the foreseeable future there will be things that people can do and machines cannot. And certainly there are things that machines can do and people cannot. In these cases the principle applies and the question of allocating functions presents no problem. However, the number of tasks that can be performed by either people or machines is growing. Consider the partitioning of tasks proposed by Licklider and Clark in 1962 (p. 114):

a. To select goals and criteria—human;

b. to formulate questions and hypotheses—human;

c. to select approaches—human;

d. to detect relevance—human;

e. to recognize patterns and objects—human;

f. to handle unforeseen and low-probability exigencies—human;

g. to store large quantities of information—human and computer; with high precision—computer;

h. to retrieve information rapidly—human and computer; with high precision—computer;

i. to calculate rapidly and accurately—computer;

j. to build up progressively a repertoire of procedures without suffering loss due to interference or lack of use—computer.

Several of the tasks that are assigned to humans in this partitioning can be performed today, at least in limited ways, by computers. Undoubtedly, computers will get better at those tasks as time goes on. The question is, when a task can be performed equally well by a person and a machine, how does one decide who or what should perform it? Weizenbaum (1976) urges that we keep before us the distinction between questions of feasibility and questions of appropriateness. It might be possible, for instance, to build a computer system capable of conducting a psychiatric interview; whether it would be appropriate to delegate such a function to a machine is another matter.

One of the challenges for research is to identify problems for which the abilities of people and those of computers complement each other particularly well. The traveling salesman problem, for example—to find the shortest route connecting all of a large number of cities—is too complex for people to solve without the help of computers and not sufficiently well understood to permit the writing of a non-exhaustive-search program that would solve it automatically. Such problems lend themselves to an interactive approach, in which the person suggests possible solutions and the computer checks them out. This approach is not guaranteed to yield the shortest possible route, but it will fairly readily identify one that is among the shortest. The interactive approach has in fact been explored for the related problem of finding routes for transit systems (Rapp 1972).

• • • • • • • • • • • • • • • • • • • • • • • • • • • • • • • • • • • • • • • • •

Computers and information systems come in many shapes and sizes, as do the people who interact with them. Consequently, the study of person-computer interaction is bound to be a many-faceted undertaking. Computer-based systems are so pervasive that the idea of person-computer interaction is almost as encompassing as that of person-machine interaction. There are, perhaps, some things that can be said that apply to the general topic, but for the most part, experimental research stands a better chance of producing interpretable and useful results if it is focused on specific aspects of the topic.

The trick here, as in any experimental science, is to identify problems that are manageable on the one hand and nontrivial on the other. Sometimes researchers in this area have acquired control over variables at the expense of trivializing the problem being investigated. Unambiguous results may be attained, but they are of very little interest to anyone. At the other extreme, investigators sometimes have addressed issues and questions of obviously great interest but have had so little control over the variables affecting the outcome that results are not readily interpretable. It is to be expected that as the field matures, research methods and paradigms will be developed that are more effective and efficient than those that have been used in the past, and we will get better at asking questions that are both answerable by experimental means and worth asking.

A particular problem associated with human-factors research relating to computer-based systems is the rapidity with which the field is advancing. The question is, how to do research that will yield results that are not obsolete before they are published. If work is to be of more than passing interest, it must be addressed to generic and system-independent problems.

The issue of function allocation is more complicated than the simple    let-computers-and-people-each-do-what-they-do-best principle would suggest. It will almost certainly get more complicated in the future, as the set of functions that can be performed either by people or by computers continues to grow.

# 6

## *The Physical Interface*

The dictionary definition of *interface* that best applies to person-machine interaction is "the place at which independent systems meet and act on or communicate with each other" (*Webster's New Collegiate*). In the context of a discussion of person-computer interaction, the term refers to the media by which information is passed between the person and the computer. On the computer's side are such things as keyboards, light pens, mice, printers, plotters, and video displays. On the person's side are, primarily, eyes, ears, and fingers.

Licklider (1965) rejected the term *interface* in favor of *inter-medium*. Interface, in his view, with its connotation of a surface or plane of separation, does not do justice to the complexity of the issues involved. The important areas for concern, he felt, lie on both sides of the literal interface and include factors that are not visible at the point of contact. I agree with Licklider's observation but will use the word interface nevertheless because it has become part of the accepted terminology.

While effective interaction requires that information flow both from the person to the computer and from the computer to the person, curiously it is conventional when talking about the interface to use terminology that focuses on the computer. Thus when we use the terms *input* and *output* in this context, we almost invariably mean input to the computer and output from the computer. The challenge for human-factors people, however, is to assure that output from the computer constitutes suitable input for the person and, conversely, that input to the computer is something that is convenient for the human to put out.

The interface may be described in terms of the devices and the sensory-motor system that effect the information exchange, or in terms of the forms that the information takes for purposes of exchange. For convenience, we can refer to the one as the physical interface and to the other as the cognitive interface. The former will be the focus of this chapter and the latter of the next.

### Manual Input Devices

Several studies have compared the effectiveness of different input devices (keyboard, light pen, stylus and tablet, joystick, track ball, mouse, touch pad, touch screen) in performing specific tasks (Albert 1982; Card, English, and Burr 1978; Edwards 1984; English, Engelbart, and Berman 1967; English, Engelbart, and Huddart 1965; Goodwin 1975; Heglin, Saben, and Driver 1972; Hillis 1981b; Morrill, Goodwin, and Smith 1968; Seibel 1972). These studies have produced much useful information, but there have not yet emerged either definitive design criteria for devices, excepting keyboards, or guidelines for determining the relative merits of the different devices for specified types of applications. (On keyboards, see Alden, Daniels, and Kanarick 1972; Dolotta 1970; Duncan and Ferguson 1974; Ferguson and Duncan 1974; Hanes 1975; Hirsch 1981; Hornsby 1981; Klemmer 1971; Remington and Rogers 1969; Seibel 1972.)

The vast majority of the keyboards for terminals and microcomputers have the standard QWERTY layout. In spite of numerous attempts to demonstrate that this is not the best arrangement, and a few experiments showing that specific others—including chord schemes—may have some advantages (Kroemer 1972; Martin 1972), the QWERTY scheme is not likely to be discarded in the near future. Any change that would require relearning by everyone who has acquired typing skill would be strongly resisted.

Other aspects of keyboard design do vary considerably from keyboard to keyboard: sizes of keys, spacing between keys, sculpting of the individual keys and the entire keyboard, dynamics of the keys, positioning of special function keys. Anecdotal evidence suggests that heavy users of these devices (such as secretaries on word-processing systems) develop strong preferences with respect to keyboard design that vary somewhat from person to person.

Membrane keyboards have some clear advantage over conventional full-travel keyboards in that they are lighter and more flexible, and production and maintenance costs are lower. There has been some concern, however, that such keyboards might be unacceptable to skilled typists and might yield higher error rates, because the feedback the user obtains from the movement of the keys has been lost. Loeb (1983) found that the performance of non–touch typists was about the same on both types of keyboards. Touch typists performed much better initially with a conventional keyboard than with a membrane keyboard; the difference decreased markedly, however, during three hours of practice with the latter. Given that the touch typists had had many hours per week of experience with the conventional keyboards, the results from this study raise the possibility that with comparable amounts of experience with both arrangements one might be able to use the membrane keyboard as effectively as the more conventional one.

Efforts have been made to use the touch-tone telephone as an input device to an information retrieval system, with synthesized speech as the output mode (Witten and Madams 1977). The telephone does not compare favorably with full-alphabet keyboards as a general-purpose manual input device, but it is quite adequate for certain specific applications, such as querying specialized databases. Its great advantage lies in the fact that so many people have phones.

Among the most useful input devices that have been developed are the light pen, the stylus and tablet, and the mouse. The light pen, which is used in conjunction with a cathode-ray tube (CRT) as an input device, is a light-sensitive cell mounted in the tip of a pen-shaped instrument. The term "light pen," or worse, "light gun," is misleading. One is apt to infer that the instrument emits light as a pen emits ink, or a gun bullets. The analogy is erroneous, however. The light pen is a sensing device: it does not emit light, it detects it. Moreover, it signals the computer *when* it has detected light, not *where*.

The principle of operation is simple. One points the pen at some region of the display surface. If the CRT's electron beam is positioned on the spot at which the pen is pointing, the pen sends a pulse to the computer, indicating that it has just detected light. When the computer receives a pulse from the light pen, the program that is controlling the computer's activity can determine where the electron beam is positioned at the moment. The event that triggers the pen is the positioning of the electron beam on the spot where the pen is pointing. To the eye it might appear that many spots on the display are simultaneously and continuously intensified; however, this results from the speed with which the beam can be moved from one spot to another, the persistence of the phosphor on the scope face, and the hysteresis of the eye. The sensitivity range of the photo cell in the light pen is adjusted so that the cell will be triggered only if it happens to be pointing at a spot at the same instant that the beam is focused on that spot. One of the problems associated with the design and manufacture of light pens is that of ensuring that they will not be triggered by stray light.

A distiction is sometimes made between *pointing* and *drawing* as two ways of using the light pen. In the former case the pen is used to point to some part of a computer-generated display; in the latter it is used to input freehand drawing. The principle of operation is the same in both cases, and from the programmer's point of view both are accurately described as pointing. Drawing involves the use of a "pen-tracking" prodecure in which a "target" is made to move in response to a movement of the pen over some part of the target's area. The following procedure illustrates how pen tracking might be done. As a target we might use a three-by-three square of pixels illuminated on the display surface (see figure 1). The pen-tracking program would be written (1) to determine when the pen is pointing at some part of the target and (2) to move the target, whenever the pen is pointing at some part of it other than the center square, so that the center square *will* be where the pen is pointing. Suppose the pen is pointing at the target's center square, as shown in figure 1a, and the user begins to draw a line. As soon as the pen has moved far enough that it points to some square other than the center one, as in figure 1b, the program will move the target so that whatever square is being pointed at becomes the center one, as in figure 1c. Thus, the program tracks the movement of the pen. The coordinates of the successive positions of the center of the target will represent the track of the pen, and by storing these away, the program can keep a record and maintain a display of what the user has drawn (figure 1d).

We should note that the user must initiate his pen movement where the target happens to be. If the target is positioned in the upper-right-hand corner of the display, and he wants to draw something in the lower-left-hand corner, he must first point at the target where it is and move it to where he wants it to be. But

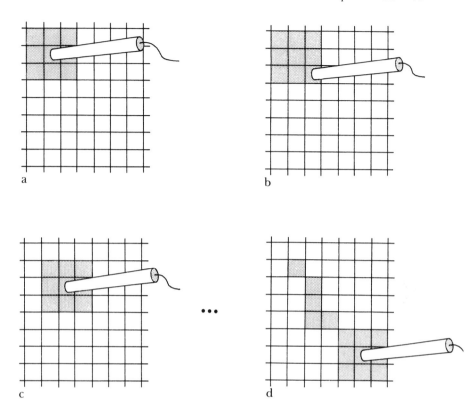

**Figure 1**

A light pen "draws" on the CRT by means of a pen-tracking procedure:
(a) pen is positioned on target; (b) user moves pen relative to target; (c)
program moves target so pen is again pointing to its center; (d) track of
pen's movement is displayed on scope if desired. (In these representations
the size of the target is enlarged relative to the pen size, for clarity.)

if he does not wish that movement to be interpreted as part of his drawing, he must have some way of communicating that fact to the machine. Often a switch or trigger is mounted on the pen for this purpose, one position indicating that the movement is to be considered part of the drawing, the other indicating that it is not.

Convenient as the light pen is, there are several problems associated with its use. The pen can be moved only at a moderate rate; if it is moved too quickly, the tracking program may lose it. Precision is not high: the area over which the photo cell is sensitive may extend to several adjacent dot positions, and this problem may be aggravated by parallax when the pen is not held perpendicular to the display surface. An advantage is that in pointing or drawing with a light pen one is doing something very similar to what one is used to doing with a conventional pen or pencil.

An alternative to the light pen as an input device to be used in conjunction with a CRT display is a tablet and stylus. The tablet is an etched circuit, usually embedded in epoxy, providing a square or rectangular writing surface, about 10 to about 20 inches on a side. One may think of the surface as a coordinate grid of conducting wires. The stylus contains an electrostatic probe that, when in contact with the tablet, will pick up a time sequence of pulses whose pattern represents the $x, y$ coordinates of the point touched. Sometimes the stylus also contains a control-signal switch that is closed by pressure on the stylus tip. In this case the stylus will be active only when sufficient pressure is applied. The coordinates of the tip position are delivered to a register, the contents of which can be read directly into the computer.

Usually the control program displays on the scope a point corresponding to the current position of the stylus on the tablet. Thus the user draws on one surface, but what he is drawing appears on another. One might wonder whether users would find this arrangement awkward. In fact, they adapt to it very readily.

The mode of operation of the tablet and stylus is fundamentally different from that of the light pen. In the case of the tablet and stylus, it is not necessary to maintain a target for pen tracking. The system is capable of signaling the computer not only *when* the stylus has touched the tablet but *where*, thus making fewer demands on the computer's time than when a light pen is used. Moreover, the programmer is relieved of the problem of developing pen-tracking programs.

In addition to these advantages, there are several other reasons why a tablet-stylus system might be preferred over a light pen. The tablet-stylus is considerably faster: it can deliver more *x, y* coordinates per unit time. It permits greater precision, and problems of parallax and triggering from ambient light do not exist. Tracing through paper is possible, and some tablets are translucent, permitting rear projection of the material to be traced. Tablets are normally mounted horizontally, so that they provide a somewhat more natural working surface. (Of course, there is no reason why CRTs could not also be mounted horizontally—although they usually are not.) Because the user is writing on one surface while observing the display on another, his hand does not obscure the material being displayed. The display itself need not be within arm's reach, thus permitting flexibility in the design of the workstation.

The mouse is qualitatively different from both the light pen and the stylus, but it serves the same functions of pointing and drawing. It does not resemble a pen or pencil either structurally or operationally. It was designed specifically for use with graphically oriented interactive systems and has some advantages over both the light pen and the stylus.

The mouse is a small flat-surfaced plastic device containing a ball mounted in such a way that it protrudes slightly from the casing's undersurface and is free to turn in all directions. When the ball rotates, a signal is sent to the computer indicating its direction of rotation. How this information is used depends on the program that is controlling the computer when it is received. The most convenient way to make the ball turn is to move the mouse across any flat surface; when this is done, the ball rotates in the direction in which the mouse is moved.

Typically, the mouse is used when one is interacting with a computer by means of a graphics-oriented program. By moving the mouse across a flat surface, the user may cause a cursor to move across the face of the display in (apparently) the same direction as the mouse. Moving the mouse when it is not in contact with a surface does not cause the cursor to move (because it does not cause the ball to rotate). This is convenient, because it means that one can move the cursor as far as one wishes in a given direction by a succession of short moves of the mouse over the same surface area.

A common use for the mouse is the selection of menu options. One makes the cursor point at the desired option by first rolling the mouse over a surface and then indicating that one has reached the desired option by pressing a switch conveniently located on the mouse's back. The device is also used for freehand drawing. In this case it is necessary to signal the computer (perhaps by holding down a switch) that the intent is for the cursor to leave a trace as it is moved around on the display. Display-generation software packages provide a variety of drawing aids to facilitate the composition of graphs, charts, schematic diagrams, and other graphics.

Because the mouse does not resemble a pen or pencil, or other familiar instrument, its use generally requires a bit more experimentation by beginners than does the use of a light pen or stylus. It is quickly mastered by most users, however, and is proving to be a very popular device.

Given the wide acceptance of the mouse, it was probably inevitable that someone would develop a device that could legitimately be called a CAT. This device—the capacitance-activated transducer—permits the user to move a cursor on the screen by moving a finger on the transducer pad. The motion of the finger across the surface of the CAT causes the cursor on the screen to move in the same direction. Increasing the amount of skin contact, either by pressing the finger harder or by using more than one finger simultaneously, will increase the speed at which the cursor moves.

### Visual Displays and Computer Graphics

For a long time the primary medium for storing and distributing information has been the printed page. While this has served us remarkably well, and will probably continue to do so for some time to come, it is a highly static medium and is therefore limited as a vehicle for conveying information about dynamic concepts, processes, and relationships. A computer-driven visual display is not (necessarily) static and opens up a new set of possibilities for presenting information in drastically different ways. Our understanding of the structure and behavior of the very small (molecules, for instance) and the very large (such as galactic systems) can be enhanced by the construction of digital models and their visual representation by means of computer graphics. Such systems permit the inspection of structures from various perspectives and the observation of behavior on different time scales (Langridge, Ferrin, Kuntz, and Connolly 1981). Although the possibilities have hardly begun to be explored, the potential of graphics as a powerful, "natural," and broadband means of conveying information to a user has been widely recognized (Atkinson, Dalvi, Drawneek, Fellgett, Hovland, Tring, Walker, and Whitfield 1970; Boehm, Seven, and Watson 1971, 1975; Joyce and Cianciolo 1967; Prince 1971).

Computer graphics is also seen as having great potential for application in business (Cortes 1983). As of the summer of 1983, *Computer Decisions* listed over 150 vendors of business graphics systems (Klein 1983). One questionnaire study that obtained responses from 57 companies or institutions revealed that 85 percent of the respondents used computer graphics for the preparation of presentations for management. The second and third most frequently reported uses were financial applications (57 percent) and marketing applications (52 percent). Schure (1983) suggests that "in the 1980s and 1990s sophisticated computer graphics power will be a corporate requirement, as fundamental to corporations' economic survival as reading, writing, calculations, and verbal skills are essential to present-day communications channels."

Some hint of what might be done is provided by occasional examples on television of animation, process simulation, slow-motion and time-compressed photography, dynamic changes of spatial scale (zooming, panning), and the mixing of photography and computer-generated graphics. Van Dam (1984) has suggested that graphics is on its way to becoming the standard form of communications with computers.

Until recently the widespread use of graphics has been inhibited by the relatively high cost of graphics terminals. That is changing. The projected large market for such devices is fueling a race to develop more versatile displays, and the competitive pressures of the marketplace assure that prices will continue to go down. While the display terminals that are currently available differ in many respects, probably the single factor that most affects price is display resolution. Very high resolution is not a requirement for many uses of displays, however, and the graininess that is associated with low resolution can be reduced with anti-aliasing techniques; color graphics terminals that are quite adequate for many purposes are now available for a few hundred dollars, and the cost will undoubtedly continue to drop.

Display technology is evolving rapidly. New and more versatile display terminals are being introduced to the market regularly. Increasing resolution, finer intensity control, and greater color selection are apparent trends in this evolution. New techniques that have emerged include plasma panels, light-emitting diodes and liquid crystal displays. Work is also proceeding on the development of three-dimensional displays (Getty 1982). Software for exploiting the display hardware is also being developed, and quite realistic graphics are now possible.

An aspect of display technology that deserves highlighting is the videodisk. Although this device has many uses that are independent of computers, it represents the potential for substantially enlarging a computer-based system's range of visual output. Not all videodisks work on the same principles. One of the major technologies uses a capacitance-stylus system for recording and playback. These disks are grooved, and readout is effected by a stylus that mechanically tracks the groove much as does the needle of a record player. An alternative approach records information on an ungrooved disk by burning tiny pits along spiral tracks on the disk face. In this case the encoded information is read from the patterns of reflection (or nonreflection) obtained by a pulsed laser beam focused on the disk surface. Such a disk typically can record 54,000 tracks, each of which is the equivalent of a single television frame. In still another common method, information is represented by transparent "holes" in a translucent surface and is read by the detection of light transmitted through the disk. Because the pits or holes need not be of uniform size, information can be stored in other than binary form. As of 1982 it was possible to store as many as 14 billion bits on each side of a videodisk. The possibility of "jukebox" arrangements, providing access to as many as 1,000 disks, raises prospects of enormous amounts— by today's standards—of relatively low-cost storage (Goldstein 1982).

Optical laser systems have an advantage over the capacitance-stylus system in that they permit random accessing of frames. Most videodisk systems also permit slow-motion or variable-speed viewing, as well as selection and freezing of individual frames. Although one thinks of videodisks mainly as devices for storing pictures, they can be used to store anything that can be encoded digitally. By coupling a videodisk to a computer, one can make the frame selection contingent on either programmed or interactive variables. One can also mix the kinds of video stimuli that are readily stored on videodisks with computer-generated graphics.

The market for videodisks is expected to grow explosively during the next few years. Competition among the major suppliers (including MCA, N. V. Phillips, RCA, and Matsushita) will be fierce. According to some estimates, annual sales could exceed $10 billion by 1990 (*High Technology* 1980).

Much is known about visual display design. There are guidelines for brightness, contrast, ambient illumination, resolution, flicker, character and symbol design, character organization, clutter, cursor design, and many other aspects of displays. Sources of human-engineering information for visual display design include Morgan, Cook, Chapanis, and Lund (1963); Woodson and Conover (1964); Meister and Sullivan (1969); Christ (1975); and McCormick and Sanders (1982). Sources of information about computer-driven displays and the human-factors aspects of their design include Barmack and Sinaiko (1966); Licklider (1967); Cropper and Evans (1968); Gould (1968); Burnette (1972); Grether and Baker (1972); Machover (1972); Martin (1973); Foley and Wallace (1974); Danchak (1976); Shurtleff (1980); and Galitz (1981). IBM has recently produced a booklet on the human factors of workstations, which discusses many issues relating to the design of visual displays and also considers the questions of radiation safety, keyboard design, and lighting (Rupp 1984). For a review of input devices for interactive graphics systems, see Ritchie and Turner (1975); and for a large bibliography on computer graphics, interactive techniques, and image processing, see Pooch (1976).

Stewart (1981) has studied several hundred video display terminals (VDTs) in work situations and has catalogued the relative frequencies of such difficulties as poor character spacing, poor character shape, and poor image quality. He considered also problems of heat, lighting, and ventilation and found such problems in a large percentage of the cases studied. Matula (1981) has produced a bibliography of studies of visual problems (eyestrain and visual fatigue) associated with the use of

VDTs. Other recent sources on the visual effects of working with VDTs are Oestberg (1976); Hunting, Laubli, and Grandjean (1981); Laubli, Hunting, and Grandjean (1981); Mourant, Lakshman, and Chantadisai (1981); Dainoff, Happ, and Crane (1981); Smith, Cohen, Stammerjohn, and Happ (1981); and Brown, Dismukes, and Rinalducci (1982).

Much of the work on VDTs as sources of stress or health problems has been criticized on methodological grounds. Among the undesirable effects sometimes attributed to long-term use of VDTs are mood disturbances (anxiety, depression, anger), stress symptoms, and indications of job dissatisfaction and boredom. Sauter, Gottlieb, Jones, Dodson, and Rohrer (1983) took survey data and objective measurements on 248 VDT users and 85 corresponding nonusers and found little, if any, evidence of detrimental effects of VDT use, with the possible exception of eyestrain. Some working conditions were judged to be less favorable, however, among the VDT users than among the nonusers. For example, users rated their workplace environment to be less pleasant and their chairs less comfortable than did nonusers. Users did not rate the lighting quality of their workplaces lower than did the nonusers. It is hard to know what to make of these results—the cause-effect relationship are not very clear.

A few investigators have tried to discover whether the visual processing of text is more or less difficult when the text is presented on a VDT than when it is presented conventionally on paper (Waern and Rollenhagen 1983; Wright and Lickorish 1983). Wright and Lickorish had subjects proofread text in which errors had been planted (misspellings, missing or repeated words, line interpositions), either on paper or on video displays. The proofreading was considerably faster and more accurate when the text was presented on paper. There is also some evidence that people tend to read text more slowly from video terminals than from the printed page (Muter, Latremouille, Treurniet, and Beam 1982). In interpreting these results, one must bear in mind that the subjects (presumably) had not had extensive previous experience reading text from visual terminals.

Some of the human-factors problems associated with the use of computer graphics are familiar problems but take on a somewhat different character in this context. Consider, for example, the problem of selecting a set of colors for a graphics terminal.

Assume a terminal that displays color by combining three primaries with adjustable intensities at each pixel. A display that allowed $2^8$, or 512, intensities on each of the three primaries would have the capability of producing $2^{24}$, or roughly 17 million, colors. Suppose now that only eight of the colors can be used by a given program but that those eight can be any of the 17 million possibilities. The human-factors problem here is to select the eight possibilities in such a way as to satisfy desired criteria, such as intercolor discriminability, consistency with established color-coding schemes, and so on.

The development of computer-based display technology also adds some new dimensions to the question of how best to code and represent information for human consumption. The linking of a computer to a high-quality display opens up the possibility of coding information dynamically in many different ways, including differential intensification, blinking, color changing, spotlighting, selective magnification, partial overlaying, insetting, and animation. Graphics terminals allow the user to interact with text in very different ways than do hardcopy devices. For example, one can indentify portions of a text by pointing or encircling. Given adequate symbol recognition capability, the use of hand-drawn proofreader symbols is possible also (Coleman 1969). Computer-based displays provide the capability to move selectively over a very large "virtual" display by zooming, panning, and lateral and longitudinal slewing. In general, they give us more flexibility than we yet know how to exploit.

One way to think of a display is as a window into the information that is stored in the computer's memory. With an appropriate command language one should be able to get the computer to display selectively various subsets of this information, at various specified levels of detail. At any time, the user should have the option of dropping to a level of greater detail, exploding a specified component of a display so as to get more detailed information with respect to it, or, conversely, of suppressing detail and backing off to get a more panoramic view.

Most currently available computer terminal displays have small display areas as compared to the visual scene with which an office worker may be accustomed to working—that is, an office or even a desk top that can display numerous documents simultaneously. Nickerson, Myer, Miller, and Pew (1981) distinguish between "hard" and "soft" approaches toward overcoming this limitation. The hard approach is that of increasing the display's physical size and resolution. Large solid-state flat-panel displays represent one example of this approach. The soft approach is that of improving the versatility of the small display, providing it with a considerable amount of local processing power and with the sophisticated software that can make the display serve as a window or as multiple windows opening onto a work surface.

Multiple windowing capability is a particularly versatile display tool that was developed as part of the Interlisp-D programming environment at Xerox's Palo Alto Research Park (Sheil 1983a, b). Windowing is now being offered as a feature on personal computers such as Apple's Lisa and MacIntosh. Examples of personal computer software systems that provide multiple window capability are those of VisiCorp and Micro-Soft. Both systems provide multiple windowing, but they do so in rather different ways. VisiCorp's system (VisiON) permits the partial overlaying of windows, whereas the MicroSoft system divides the available screen up among the windows, without permitting overlap, through a process called "automatic tiling" (Hirsh 1984).

It is apparent that there are a great many potential applications of computer graphics. For present purposes it will suffice to mention one by way of illustrating some of the human-factors issues that may arise. Imagine a dynamic computer-controlled map, able to display subsets of information on demand—expressways, toll roads, primary roads, secondary roads, restricted-access roads, rivers, dams, bridges, service facilities, recreation areas, biking or hiking trails, hospitals, and numerous other features. Imagine also a system that could suggest routes from A to B to match a traveler's objectives (minimize travel time, minimize costs, avoid traffic, pass through specified intermediate points, maximize points of interest, and so forth).

Or—to indulge imagination a bit further—suppose one could store in a very large computer database digital representations of photographs with a resolution of, say, a few meters, covering a large geographical area such as the continental United States. Imagine that one could use a graphics terminal as a window through which to view this entire area or selected parts thereof. Thus one might look at the entire country as it would appear to an astronaut from a height of a few thousand miles. One might then choose to zoom in on the state of California and see the equivalent of a Landsat photograph from a few hundred miles. One might then zoom in still further to look at the Bay area. One might ask the computer to locate certain points of interest by having them blink on the display or by causing arrows and labels to appear. Suppose one now wished to focus on New England. One might pan the display out from the Bay area so that once again it covered the entire United States and then zoom back in on the New England area. And so on. From this imaginary system it is but a short step, conceptually, to one that would permit not only the viewing of an area from above but also travel through it—for instance, simulated driving through its streets via display terminals. Some effort in this direction has already been made with the use of videodisks.

While realization of this fantasy is beyond the current state of the art, the consideration of it raises some questions about how best to match display technology to human capabilities and limitations. For example, when viewing a segment of a large display, how important is it that one be aware of that segment's location in the larger context? In viewing different segments of an underlying display, how important is it to maintain continuity? That is, can one work as effectively with a series of snapshots as one can by moving from one segment to another by means of slewing or panning and zooming operations? These questions already have some practical importance vis-à-vis displays of integrated circuits, which are maps of sorts, that permit one to zoom in on arbitrarily small segments of the total circuit area. One means for providing the contextual information, in this case, is an inset that gives a dynamic scale-model representation of the displayed area mapped onto a block representing the entire "virtual display surface."

Figure 2 illustrates the idea. An entire circuit, containing a few tens of thousands of active-element groups, is shown in 2a. A small portion of the circuit, specifically the portion enclosed in the small white box in 2a, is shown, greatly enlarged, in 2b. Working only with displays of this type, representing various portions of the circuit, the designer could easily get lost, especially inasmuch as there may be many small regions of the circuit that, when magnified, look very much alike. The use of an inset map to help the designer keep his place is illustrated in

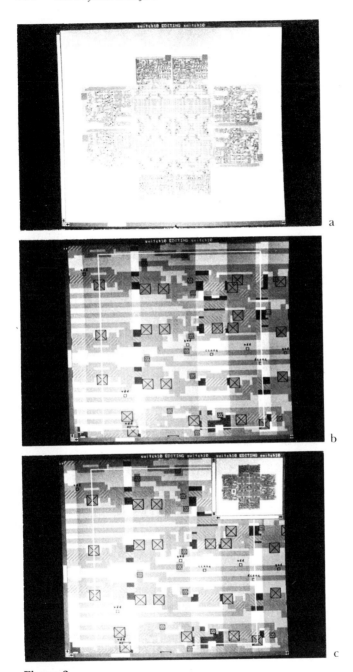

a

b

c

**Figure 2**

A segment (2b) of a circuit (2a) is displayed; an inset maps the displayed area onto the circuit as a whole (2c). The circuit is the switch chip for the BBN Butterfly multiprocessor (see p. 65). The pictures were made by Ray Tomlinson and Dean Spoffard.

2c. Here a minified representation of the entire circuit, with the region represented by the enlargement marked, is shown in the inset.

The Defense Mapping Agency has been making increasing use of digitial representation of its enormous amounts of cartographic data (Severance and Granato 1983). Hardcopy products (conventional paper maps) now account for less than 50 percent of the Agency's support to the military services, and the emphasis on digital products is expected to keep growing. Cartographic data are used in two ways, according to Severance and Granato: (1) to produce hardcopy maps and (2) to facilitate assessment and analysis of the battlefield. The picture they present is that of a commander "fully cognizant of the entire battlefield, having at his touch tremendous volumes of intelligence data." Such data would be used to help predict where the enemy can move, how fast, and over what routes. An important element in the design of systems that use large amounts of digital data, they point out, is a sensitivity to the user's own information-processing capabilities and limitations. One of the major concerns of designers of military displays for representing tactical situations is that of clutter: there is a tendency to put as much relevant information as possible on the screen at the same time. The problem of clutter is not unique to displays that are used for military purposes. A variety of ways to compensate for clutter by "highlighting" salient aspects of a display have been proposed (Knapp, Moses, and Gellman 1982; Hemingway, Kubala, and Chastain 1979). These include the use of color, intensity modulation, flashing, and iconic or pictorial encoding.

The general goal in using displays as decision aids in any context is to provide the users with just enough information to support their decision-making activities but not to inundate them with more than they can effectively use. The research question is: How much is just enough? Presumably a partial answer is that it depends on how the information is organized. But there too is an area for research.

### Input-Output Bandwidth

Computers can accept information at much higher rates than people can input it, and they can output it at much greater rates than people can assimilate it. On the other hand, computers are not yet good at accepting information in a form in which people typically convey it to each other. Thus we say that there is an impedance mismatch between computers and users, and an important problem for research is that of finding ways to compensate for this mismatch. The goal is to find ways for users and computers to exchange information at rates at least as great as those at which people communicate with each other, although it is not clear that it is impossible to do better than that.

Fairly simple things that have been done to facilitate typed input include the use of abbreviations and what is sometimes referred to as an autocompletion capability, whereby the user can type in the first few letters of a word and then strike a control key, whereupon the computer automatically completes the word (makes the entire word appear, for example, on a video display). The autocompletion feature is a common one and appears to be liked by users; there is some evidence, however, that it can increase error rates under certain conditions (Fields, Maisano, and Marshall 1978).

Speech and natural language are among the possibilities that first come to mind for increasing the input-output bandwidth of an interactive system. We will consider these individually in subsequent sections. There are other possibilities for increasing this bandwidth, which will be considered here.

A variety of structured techniques for inputing information to the computer have been developed. These include the use of menus, formatted (form-filling) displays, question-and-answer or prompting methods, constrained-syntax command languages, and special-purpose function keyboards. Input-output speed has not been the only consideration underlying the development of these techniques—ease of use has often been emphasized—but it has been one of them. There is a need for a surer knowledge of which option best suits a given condition; it is also desirable to mix several of these input methods in the same system, and many of the more versatile systems currently available do that.

Menu selection approaches are often favored for use with databases that are easily organized hierarchically (Thompson 1971; Uber, Williams, Hisey, and Siekert 1968). They have the advantage of not requiring they user to generate commands from memory, but simply to select the appropriate one from several that are simultaneously provided, and they manifestly work quite well. One drawback of menus is that they often are unnecessarily slow for people who can specify precisely what they want without being led through a series of explicit choices; the use of a light pen or mouse as a pointing device, however, can make the selection process quite fast. When the menu approach is appropriate, there is the question of the branching ratio that should be used. One can get to the same end point by a sequence of a few decisions with many choices at each decision point or by a sequence of more decisions with fewer choices at each point.

Some thought has been given to various unconventional ways of broadening the bandwidth of the communication channels between users and computer systems. Pressure sensing, as well as position sensing, of a stylus or finger on a display has been explored as a possible source of user input to indicate pushing, pulling, dispersing, or reorienting displayed objects with a touch (Negroponte, Herot, and Weinzapfel 1978). Negroponte, for example, has noted the possibility of using the fingers to manipulate objects on a display in much the same way as one uses them to manipulate real objects. A variety of other innovative graphical input techniques have also been investigated by the architecture group at MIT (Fields and Negroponte 1977; Lipman and Negroponte 1979).

The use of electrophysiological signals in order to bypass some of the time constraints arising from the slowness of the musculoskeletal system has been discussed as a long-term possibility. The measurement of eye fixation has also been proposed as an input method that might help increase input bandwidth by providing a channel that could operate in parallel with existing ones; it would be especially useful in situations where the user's hands are otherwise occupied. If the computer can determine where the user is looking at any given time, it can present an input menu and then let the user select items by, say, pressing a switch with a foot, or making a sound, while fixating the desired item. The method is of interest as a potential interface for people who have handicaps that prevent them from using more conventional input methods.

One device designed especially to serve in lieu of a keyboard for people with severe motor impairment is called the Eye Typer (*Computer-Disability News* 1984b). This device, which is manufactured by Sentient Systems Technology, Pittsburgh, Pennsylvania, has letters displayed on a panel in the familiar

QWERTY layout and the numerals in a row beneath the letters. Embedded in each letter or numeral is a tiny light. A camera in the center of the display catches the reflection from the user's eye of the light at which he is looking, and circuitry can then identify that character and have it appear on a liquid crystal display at the bottom of the panel. Thus, the user "types" by scanning the keyboard and fixing his gaze momentarily on each letter that he wishes to select.

There are many ways to measure eye fixations. Most of them, however, require that the user's head be kept in a fixed position; those that do not are quite expensive. A technique now being explored that might avoid both of these problems is based on the analysis of brain waves. When a person fixates on a blinking light, the brain waves produce a pattern that depends somewhat on the blinking pattern of the light. If a viewer were to gaze at an array of lights, all blinking with the same pattern but out of phase with each other, an analysis of his brain waves might reveal which light was being fixated at a particular time. Experiments with this approach are currently being done in the hope of developing an input method for people for whom more conventional methods are not possible (Becker 1983). Such techniques, if practicable, may give paralyzed nonvocal people a communication channel they have not had before. And if systems using these techniques are developed, it seems likely that people without disabilities will find them useful as well.

### How Important Is Typing Skill?

While much attention is being given to the search for effective alternative ways to get information into interactive computer systems, the most common input method, now and presumably for some time to come, remains typing. Moreover, the development of a variety of computer-based communication and information-management tools, including composition and text-editing aids, has made typing an important skill for researchers and managers as well as for secretaries and clerical staff. It is possible to use such tools through the intermediary of a skilled typist, of course, but their full usefulness as aids to composition is not likely to be realized unless the writer uses them directly. Lederberg (1978) has argued that "the author who does not interface directly with his own words with a text-display and editor is missing a powerful and precise organ of expression, which has no practical parallel in human communication today" (33).

Licklider (1965) once observed that there are very few well-developed skills that are both complex and widespread. Almost everyone can move about in three-dimensional space and speak and understand a natural language, but "relatively few people can do anything else that is even remotely comparable in informational complexity and degree of perfection" (99). The only remaining candidates Licklider saw for addition to this short list were writing, playing a musical instrument, and typing. "It is possible," he noted, "that, in future decades, typing will move up past music and that it will become almost as widespread as writing and more highly developed" (99). Presumably what would motivate the widespread acquisition of typing skills would be their utility in person-computer interaction. I do not know whether typing skills have become more widespread during the twenty years since Licklider's observations. We are only beginning to see personal computers become widely accessible, and it remains possible that they may serve both as a stimulant to the acquisition of such skills and as a means to facilitate that acquisition.

A long-range objective of people doing research on speech recognition by computer is the development of a speech-to-text system that will permit users to dictate directly to the computer and see the words appear on a visual display immediately as they say them. Until such a capability is developed, the question of whether people who use the computer through an intermediary rather than in a hands-on fashion suffer some loss of efficiency or power will remain an important one. Perhaps there is no definitive answer to the question; different types of use suit different work styles. It would be helpful if research methods could be developed that would reveal when hands-on use of a system is an important determinant of productivity and quality of output and when it is not.

### Individualized Interfaces

Ever since the Industrial Revolution, a basic principle governing the consumer prices of manufactured goods has been that unit costs are to be brought down by mass production. If the modern appliances that we take so much for granted had to be custom produced by individual craftsmen, very few people would be able to afford any of them. Some analysts believe that computer technology will, in time, free us of the need for many-of-a-kind production techniques. Customized design and the ability to produce goods economically in small quantities are expected to be the rule rather than the exception. David Evans, of Evans and Sutherland, puts it this way:

> All our industry has developed around the concept of producing large quantities of things. It was all based on the idea of interchangeable parts and standardized products. Now, as we have learned a great deal more about gathering and storing and communicating information, we may very well be entering quite a different era in manufacturing, one in which we can make relatively small quantities at low cost. That will really have a broad effect on how we live and on how we run our businesses. (Lewell 1983, 22)

One place where individualized design might be desirable is the person-computer interface. Birnbaum (1982) points out that the interfaces of today force people to adapt to the idiosyncrasies of the machine. A goal we should work toward, he suggests, is that of allowing users to design interfaces to their own preferences. There is also the longer-term possibility of providing a computer system with the ability to adapt to the styles and idiosyncrasies of its users (Negroponte 1975). The idea is that over time a machine would acquire a more and more accurate model of individual users (the knowledge they have, the specific ways in which they use the system, the kinds of mistakes they make) and would use that model to customize itself for those users, thus presumably increasing the effectiveness and efficiency of their work sessions. The development of such adaptive interfaces requires a better understanding of user styles and idiosyncrasies than we now have.

• • • • • • • • • • • • • • • • • • • • • • • • • • • • • • • • • • • • • • • • • • • • • • •

The physical interface between information systems and their users deserves a great deal of attention, and its improvement should be a high priority for developers of these systems. Improvement will come, as it usually does in technology, in two ways: through the refinement of existing devices and techniques and through the development of completely new ones. The mouse is an example of a new device that has emerged from efforts to enhance the usability of information systems.

Unlike the light pen, which bears some similarity to a conventional pen or pencil, the mouse has no obvious analog outside the domain of person-computer systems. It takes a little getting used to by beginners but is proving to be a very versatile device and one that has some distinct advantages over more conventional devices. Windowing is an example of a new display technique, the result of efforts to increase the versatility of computer displays. It, too, is proving to be a most useful development.

Interactive graphics holds great and largely unexplored possibilities for enhancing person-computer interaction. Now that the cost of displays—until recently a major obstacle—is getting low enough to encourage widespread use, the main constraint on the development of the potential of interactive graphics is the limitation of our imaginations regarding information representation and presentation. This is a point to which we will return in the following chapter. Suffice it to note here that there are two key questions associated with the presentation of information for any particular purpose: what information should be presented and how that information should be represented. These are the questions of content and format, of substance and structure. They relate as directly to the topic of the cognitive interface as to that of the physical interface. The characteristics of the available input-output devices set some constraints on what can be done by way of information representation; it is not clear, however, that the full potential of existing devices has been realized.

The development of input-output devices and techniques that will make possible a speedier and more natural exchange of information between person and machine is a high-priority goal of research and is likely to remain one for the foreseeable future. As versatile new devices broaden the range of input-output techniques, the possibility of individual adaptive interfaces becomes more and more real; there is need, however, for research both on the question of what kinds of customizing would be desirable and on that of how to make them possible.

# 7

## *The Cognitive Interface*

Fifteen years ago DeGreene (1970) noted that the main impediment to the development of symbiotic person-computer systems was our lack of understanding of psychological processes. The point is as timely now as it was then. We understand the capabilities and the limitations of machines far better than we do those of human beings, in spite of the fact that machines are evolving at a much faster rate. Moreover, as computer-based tools become increasingly pervasive, the cognitive aspects of the problem of matching systems to users will become increasingly important.

Sime (1981) has suggested that the development of the computer has created the need for a focus of human-factors work on cognitive aspects of person-machine interaction and has coined the term "cognitive ergonomics" to characterize this need. Nickerson, Myer, Miller, and Pew (1981) have used the term "cognitive interface" to suggest the types of interface issues with which the builders of interactive computer systems will more and more have to deal. It is clear that the physical interface and the cognitive interface are not independent of each other, so the distinction between them should not be drawn too sharply. The physical interface serves cognitive functions, and its design may have implications for the cognitive demands of the user's task. The idea of a cognitive interface is a useful one, however, not only because it calls attention to relatively invisible variables that are critically important to an effective user-system interaction, but also because it promotes the notion that one purpose of the physical interface is to ensure that the cognitive requirements of the user are met.

A person-computer interaction is an interaction between two information structures, one residing in the computer and the other residing in the user's head. The information in these structures is organized very differently, and in spite of arguments to the contrary there is no compelling evidence that it would be a good thing if both were organized in the same way. For the interaction to be meaningful, however, the two structures must connect, or articulate, in some way. For the interaction to be effective and satisfactory to the user, the nature of that connection must be consistent with the user's cognitive capabilities and limitations.

## Information Organization and Presentation

One effect of information technology is what Naisbitt (1984) has referred to as a collapsing of the information float, a reduction of the time that information must spend in a communication channel. It is easy to forget that until the invention of the telegraph less than two hundred years ago, the speed with which information could be transmitted was determined by the time required to transport it in some hardcopy form from place to place. It took days to get a message part way across the country, and weeks to transport one from one continent to another. Radio and television have made it possible to broadcast news of an event even as it is taking place. The telephone permits instantaneous communication between arbitrary points all around the earth, and electronic mail is beginning to make possible the point-to-point immediate transmission of lengthy written messages. There is probably not a lot to be gained by trying to reduce transmission delays much further. For practical purposes they are close to the absolute minimum now. The challenge for the future is more in the area of psychology than in communication technology. That challenge is to find more effective ways of organizing and presenting information for human comprehension, assimilation, and retention.

Some of the conventions that we follow in representing information derive from fundamental properties of language. Others, however, reflect the constraints and limitations of the paper media that have been used for the storage and transmission of information for several centuries. The book has served us extremely well for over four hundred years and is likely to continue to do so for a long time to come. But it is a means to an end—or to many ends. It serves a variety of purposes. The question is not whether it should be replaced, but whether any of the purposes that it serves might be better served by other means.

Independently of the computer, the amount of printed material (not necessarily to say information) that is being produced is overwhelming. It is now quite impossible for most scientists, educators, or professional practitioners to read carefully a significant fraction of the material generated in their fields. There is clearly a need for better ways of organizing and presenting information for human assimilation and use.

A specific goal of research should be the development of radically different ways of presenting information to people that would exploit the flexibility of the computer in this regard. One thinks immediately of computer graphics, and, in fact, computer-based graphical techniques are becoming widely used in various information presentation contexts in television, such as newscasting, weather reporting, and advertising (Madarasz 1983; Prince 1983b; Tolnay 1983). One thinks also of multimedia presentations (printed text, voice, graphics) and interactive media (systems that not only deliver printed material but also test the learner's comprehension, adapt subsequently presented information to the results of such testing, and permit one to ask questions).

Reflecting on such possibilities prompts a variety of questions about information representation more generally. Can one distinguish between a passage of text and the idea(s), information, or message(s) the text is intended to convey? Does it make sense to think of the text as a way of representing specific information, but as fundamentally different from what it represents? (People who write papers or books for instructional purposes often find themselves in lectures explaining what they "really meant" by a particular segment of text. If there were no difference between the text and the ideas it was intended to convey, this would be a meaningless exercise.) Is there *a best* way to represent a particular idea, principle, or relationship? While it may be impossible to identify the best way even if there is one, we ought to be able to determine whether some ways of representing an idea are better than others.

The claim is sometimes made that (spoken or written) natural language is a privileged medium because it bears a close relationship to thought. But this really begs the question. Is natural language a particularly useful medium because it bears a close relationship to thought, or does thought bear a close relationship to natural language primarily because natural language is the medium that we have typically used to convey our thoughts? If there were alternative media, might we not learn to use them effectively? Presumably the thought-life of congenitally deaf people whose primary language is sign language does not much resemble speech.

With the aim of facilitating the assimilation and retention of textual material, researchers in recent years have devised several techniques for representing the information contained in text in radically different ways. Some of these techniques make use of linked-node diagrams. One reason for the interest in this type of representation is the idea, stemming from the work of Quillian (1968, 1969), that information is represented in human memory as a semantic network, which in turn can be represented by linked-node diagrams. Various types of diagrams have been proposed, including networks and text maps (Anderson 1979; Armbruster and Anderson 1980; Dansereau 1978; Dansereau and Holley 1982; Long, Hein, Coggiola, and Pizzente 1978).

The objective of linked-node diagramming techniques is to identify the key concepts in a passage and to make the relationships among those concepts explicit. The technique developed by Dansereau and his colleagues, for example, involves clarifying the relationships among concepts by means of a network map that shows nodes (concepts) interconnected by lines (relationships). The network, if properly drawn, is supposed to reflect the conceptual structure of the textual material. Before attempting to produce such networks, a student is taught a set of specific relationships, each identified by a notational convention, to be used in connecting concepts.

The assumption behind the use of linked-node diagrams is that a reader who actively processes textual material will better understand and retain it. There is some evidence that students who employ this technique tend to comprehend and remember the material better than do students who simply read or study it in their own ways (Holley, Dansereau, McDonald, Garland, and Collins 1979). Whether the actual production of the diagrams has any beneficial effect beyond that which results from the fact that in order to produce such a diagram one must process the text deeply enough to understand it, is an open question, but perhaps one of more theoretical than practical interest.

If diagramming techniques are indeed effective aids to assimulation and retention, as the limited evidence seems to suggest, might the information be assimilated better if presented in the form of diagrams in the first place? One possible answer to this question is that diagram production facilitates learning because the act of producing a diagram forces learners to process the prose more deeply than they otherwise might. To produce the diagram one must identify explicitly key ideas and the relationships among them. If the ideas and the relationships were simply presented in diagram form to begin with, the learner would then be free to process them much more passively and the same benefits might not be obtained. This is a plausible possibility that should be investigated empirically. Another way to use linked-node diagrams would be to present them as supplements to the prose. A question that then would arise is whether the student should attend to the diagrammatic representation before or after reading the text.

Exploratory efforts to use linked-node (and other) diagrams as a means of presenting information on computer-based displays would undoubtedly evoke some new ideas about how to enhance the effectiveness of diagrams as a medium for conveying information traditionally conveyed by conventional prose. They would also promote some thinking about how to capitalize on the computer's ability to present diagrams in dynamic and interactive ways. One can easily imagine using an interactive display to build and modify diagrams and to present information via diagrams in ways that permit the learner to investigate the structure of a set of ideas at various levels of detail.

Two approaches that computer technology might greatly enhance are dynamic representation and interactive representation. Text could be combined, for example, with dynamic process simulations in such a way that instead of reading a description of a process, one would watch the process happen. Instead of the static diagrams we now have in books, there would be animated diagrams that could be activated at appropriate points in the text. Instead of being told to look at figure X, the reader might be told to press a certain control key or point at a certain position on a display and thereby trigger a dynamic simulation of a process that could be observed. The potential power of this approach for teaching about gear systems, stochastic processes, fluid dynamics, life cycle processes, plate tectonics, the behavior of the solar system, various chemical reactions, interactions among subatomic particles, and countless other things must be very great.

Imagine an electronic book whose purpose is to teach some fundamental concepts and principles about an area of physics, say mechanics. The book has stored within it a great deal of knowledge about the subject to be taught; it also has a model of how the various concepts relate to each other and how the understanding of a given concept depends on the understanding of specific other concepts. It has a plan that indicates the order in which the material should be taught. This plan is based on assumptions about what the student knows at specified points in the teaching schedule; the book has the ability, however, to test the student's knowledge of specific subsets of the material and to modify the teaching schedule accordingly.

The book gives the user a selection of type fonts, letter sizes, page formats, figure-ground contrast ratios, and so on. It can talk to the user (read itself aloud) when asked to do so, and it offers a choice of voices and reading speeds. It can present information in the form of conventional text and diagrams, in outline form, or in the form of linked-node diagrams at different levels of organization and detail. It is able to present clarifying or amplifying information on request; if, for example, the user does not know the meaning of a word, the book will give definitions, synonyms, and examples of the same word used in other contexts. It is able to restate or paraphrase sentences or segments of text that are not understood.

Our electronic book can simulate processes as well as describe them; it can enable the user to conduct simulated experiments. It is able to locate sections of text for the user on the basis of key words, quotes, or content descriptions. That is to say, it has an interactive index capability whereby it will indicate or "turn to" pages on which specified words, phrases, or topics may be found. It will, if one wants it to, test one's comprehension and/ or retention of what one has read (or heard). We are not close to the realization of this fantasy, but such a goal is sufficiently plausible to justify thinking about it and working toward it.

One of the fundamental differences between electronic information storage systems and other information storage systems with which we have had more experience is that the physical location of the information in the electronic system is largely irrelevant to its use or accessibility. In nonelectronic systems information is arranged in physical space. The books in a library are organized spatially more or less by subject area; the documents in a paper filing system are typically arranged in accordance with some categorical or topical structure; the books and papers in one's office may be assigned to physical locations, and even if there is no detectable mapping of location onto subject matter, people clearly do (sometimes) remember where things are located and use this information when they need to retrieve a document. As we have already noted, one way to think of a display terminal when it is used to provide access to an information store is as a window on that entire store. Given that the store contains much more information than can be seen at one time, a fundamental question is how to make it convenient for the user to move around metaphorically within the store.

Negroponte and his associates have devised a system in which the notion of "moving around" is less metaphorical than it might be. They have attempted to capitalize on people's ability to use location as a retrieval cue by providing an interface between a user and a database that permits the user to think of the data as being stored in physically different locations. The interface then gives the user a window into the spatially arranged data store and permits him to move around in that space (Lipman and Negroponte 1979).

How generally useful it is to impose a spatial organization on an electronic database remains to be seen. The fact that people sometimes remember where things are in a three-dimensional world is fortunate, because things indeed have locations in this space. Whether imputing a fictitious spatial organization to a database inside a computer will help users to avail themselves of the information in that database will have to be determined by research. The answer may depend both on the size of the database and on the nature of the other organizational schemes with which spatial organization is compared.

### Menus and Command Languages

Input techniques are constrained, of course, by available hardware; but for any given hardware configuration a variety of options exist. An important dimension along which these methods differ is the extent to which the computer leads the user through the interaction. The use of menus represents one extreme on this dimension and that of command languages the other.

In a menu mode of operation the computer presents to the user a set of options, often organized as a table, from which he can choose one by pointing with a light pen, moving a cursor with a mouse, or specifying the number of the option with a typed input. A consequence of having selected an option may be the appearance of another option menu. The interaction may proceed in this fashion through a sequence of menus. The menus typically reflect the hierarchical structure of an option tree, and by selecting from successive menus, the user works his way down the tree to some desired terminal action.

The great advantage of menus is that they minimize the user's memory load. At every step in the interaction all the options that are available at that point are explicitly identified. A disadvantage is that if the menus are based on a tree structure (which is not essential), an erroneous selection at any step assures that one cannot get to one's destination except by back-tracking through the tree. And apparently users often do make mistakes in selecting from tree-searching menus (Frankhuizen and Vrins 1980; McEwen 1981; Whalen and Latremouille 1981). Another disadvantage is that one must traverse every branch in the tree to get to a desired destination; even if one knows in advance exactly where one wants to end up, one must explicitly select all the intermediate options that will permit one to get there. In spite of these disadvantages, menus are very popular, and with current display technology they can be reasonably fast. A menu structure is considered to be especially advantageous for beginning or infrequent users of a system.

Command languages are special-purpose programming languages. With them the user can compose commands for the system. This requires that he be at least sufficiently familiar with the language's vocabulary and syntax to compose those commands that will permit him to accomplish what he wishes to accomplish. The preliminary learning required of the user is obviously greater in this case than in that of the menu approach. On the other hand, having learned the command language, the user may have considerable flexibility and a greater sense of control over the interaction. Command languages differ greatly in their versatility and complexity. More will be said about them in following sections.

There are a variety of other input modes that fall somewhere between menus and command languages with respect to who or what is in control of the interaction and how much preliminary knowledge the user must have to interact effectively with the system. Question-and-answer modes illustrate the point (Ehrenreich 1980, 1981). In this case the computer asks the user specific questions, which the user must answer in order for the interaction to proceed. Typically, a question requires a given *type* of answer, which may be more or less obvious from the statement of the question but may also require some knowledge by the user beyond what is explicit in the computer output. Form filling may be viewed as a special type of question answering. The computer displays a form, which may be partially filled in, and the user must fill in one or more indicated blanks. Again, it is incumbent on the user to know what constitutes an admissible entry and what does not.

## Programming Languages

Wirth (1984) points out that the use of the term "languages" to denote the formal notations used for programming is misleading, because programming is only superficially like writing. A better way to think of programming, in his view, is as the activity of designing a new machine. A program, in effect, turns a general-purpose computer into a special-purpose machine, the design of which is embodied in the program. I agree on this point but will use the word "language" here because this usage has been accepted essentially universally.

Computer programming may be divided into two rather different types of activity: (a) developing effective procedures for performing tasks and (b) expressing those procedures in a code suitable for machine execution. The second activity is considerably less interesting than the first and is necessary only because the ones-and-zeros symbology that is used to represent information within the machine is different from the language in which people customarily think. Thus, if a person is to communicate effectively with the computer, and vice versa, a translation from one symbology to another must take place. The history of the evolution of computer languages may be summarized by saying that the burden of this translation process, which originally rested entirely on the person, has been shifted more and more onto the machine.

No one has yet succeeded in developing a translator program (assembler, compiler, interpreter) that will accept as input a program written in "natural language," if that term is taken to mean the language of everyday speech. However, numerous languages have been invented that have more in common with natural language than do the binary codes in which information is represented within the machine. The history of programming languages includes a number of developments, each of which tended to reduce the disparity between the symbology in terms of which a person likes to think and that which the computer is able to accept as its input.

Perhaps the first step that was taken in this direction was that of representing programs with octal, instead of binary, numbers. Trivial as it may seem, this was a significant move in the direction of naturalizing the symbology with which the programmer must deal. Consider the problem of writing a program for a computer that has, say, an 18-bit word consisting of a 6-bit operation code and a 12-bit address. It is much easier to remember that "42" means "add" and "43" means "subtract" than to remember that these operations are coded as "100010" and "100011." Similarly, it is easier to write 3214 to refer to a particular memory register than to write 011010001100.

The next logical step was to do away with the numerical representation of instructions entirely and replace them with a symbology that was still easier for the programmer to use. If the octal number 42 could be used to signify "add," why could not the letters *a d d*? This meant developing a program that could accept *add* as an input, consult a table of such symbols for an interpretation of that particular one, and make the appropriate machine-code substitution. Thus, mnemonic codes made their appearance on the scene, and programmers started talking to the computer with words such as *lac* (load accumulator), *dac* (deposit accumulator), *add, sub, mul, and, not, jmp,* and so on. Clearly

lac 3214
add 3215
dac 3604

is a more convenient code to deal with than
203214
423215
243604.

If mnemonic codes could be used to represent instructions, then why not use them to represent addresses as well? Would it not be helpful to be able to say

lac A
add B
dac C?

This capability, although more difficult to implement than were mnemonic instruction codes, was developed in due course. The use of mnemonic operation codes and symbolic addressing were characteristics of the type of translator referred to as an assembler. Another defining feature of an assembler was that the transformation from symbolic to machine code was made on a one-for-one basis: each symbolic statement translated into a single machine-code instruction.

The next significant advance was the departure from the one-for-one translation principle, a step that opened the door to the development of translators that would be far more convenient and powerful than assemblers could be made to be. Now it became possible to forget about lacs and dacs and such things, and to write simply $C = A + B$, which, translated into English, means "add the value of (the variable called) $B$ to the value of (the variable called) $A$ and set the value of (the variable called) $C$ equal to the result." With this sort of symbology the translator program must be able to transform a single symbolic statement into several machine-code statements. Translators that had this capability became known as compilers. (In time *compiler* came to have a more precise meaning, because other types of many-for-one translators were developed, from which compilers had to be distinguished.)

Subsequent innovations have been too numerous and diverse to permit tracing them here; suffice it to say that hundreds of "higher-level" languages have been developed. Many of them were designed for specialized applications, but some were intended to be generally useful in a wide variety of contexts. In fact, a relatively small number of higher-level programming languages account for a very large percentage of all the programming that is done. These include COBOL, FORTRAN, Pascal, BASIC, and LISP.

Tesler (1984) notes that there have been many efforts to design a computer language "so complete and versatile that it could serve as the universal argot of programming," all of which have failed. He distinguishes between procedural or pre-scriptive languages such as COBOL, FORTRAN, Pascal, BA-SIC, and LISP, and descriptive languages such as VisiCalc and MultiPlan. Descriptive languages have been developed only re-cently. In contrast with a program written in a procedural lan-guage, which would specify how to get a particular result, one written in a descriptive language states the result desired but does not specify how to produce it; in this case the procedures have already been coded and are available to any program written in the descriptive language.

Languages, even those in a given class, differ from each other in many respects, and it serves little purpose to ask which of them is best in any general sense. Some are better for certain uses, while others are better for others. One approach that several investigators have taken in formulating principles of language design is that of manipulating specific features of languages and studying the effects of those manipulations (Gannon 1976; Ledgard 1971; Sime 1976; Sime, Green, and Guest 1973). Sime (1981), for example, had subjects write pro-grams using three specially designed "miniature languages" of the type invented by Ledgard. These languages differed in ways that were assumed to affect their utility for representing program structures. Two of them allowed the nesting of condi-tional statements, and the third provided for concatenating conditionals with a sequence of GOTO statements. One of the languages that permitted nested conditionals printed the state-ments in such a way that the depth of nesting was apparent from the program listing, whereas the other did not. Sime had both experienced and inexperienced programmers work with these languages and concluded from his results that the syntax and spatial structure of a program should make it easy for the programmer to trace the control structure, which is to say that the information required for such tracing should be overt and obvious in the program listing.

The intent of the microlanguage approach is to facilitate the isolation of effects caused by a feature of interest from the effects of other variables. One limitation is that in investigating the effects of features one at a time, one obtains no evidence regarding the role of interactions, and it seems likely a priori that the usefulness of a given feature will depend to some de-gree on the other features in terms of which a language is defined.

## Names

All programs make use of names, and all programming (and many command) languages make provision for the naming of various entities: variables, subroutines, objects, files. The conventions the user must follow in assigning names differ considerably, however, from system to system.

The two desiderata that are typically mentioned in connection with names are ease of learning and memorability (Barnard, Hammond, McLean, and Morton 1982; Barnard, Hammond, Morton, Long, and Clark 1981; Black and Moran 1982; Scapin 1981). The easier a name is to learn and to remember (with the correct association, of course), the better. To the extent that a name is mnemonic for, or suggestive of, the thing named, both the learning and the retention tasks are simplified, so efforts are made to find suggestive names (Rosenberg 1982). By definition names must be relatively short, however, so in selecting or inventing names that will serve this function effectively, one's task is often that of finding a way to pack a lot of information into a few letters.

Beyond the fact that mnemonic names are easier to use and to remember than are arbitrary names (Shneiderman and Mayer 1979), little is known about how best to assign names. What evidence there is indicates that there is not a high degree of correspondence between the names that different people select spontaneously for commands and files (Furnas, Landauer, Gomez, and Dumais 1983; Landauer, Dumais, Gomez, and Furnas 1982), although individuals do seem to have preconceived notions about what specific operations should be called (Black and Sebrechts 1981; Streeter, Ackroff, Taylor, and Galotti 1979).

Landauer, Galotti, and Hartwell (1983) illustrated this point convincingly in a study of people's preferences for the names of text-editing operations. They gave several typists a manuscript marked with proofreader's marks and asked them to type out brief instructions for someone else, describing the changes that were to be made. The required changes fell into five categories (insert, delete, replace, move, and transpose) and pertained to five types of objects (blanks, characters, words, lines, and paragraphs). The investigators did a frequency count of the verbs spontaneously used by their subjects to indicate the various operations on the several objects. One noteworthy finding was that in 24 of the 25 operation-by-object combinations the name most frequently used by the subjects was not the word used by the text editor against which they were compared (Ed of the Unix System).

A second interesting finding was a fairly low degree of consistency among the verbs used by different subjects in the same context, or even by the same subject on two different occurrences of the same operation-object combination. The probability that two users would use the same verb in the same situation, as estimated from their data, was .08. The probability that the same subject would use the same verb for two equivalent situations was .34. Cautioning that their frequency data must be interpreted carefully, given the low degree of consistency in the subjects' responses, Landauer and his colleagues note that the most frequently used verbs for the operations of insertion, deletion, replacement, moving, and transposition were, respectively, add, omit, change, put, and switch. They suggest that one reason why popular choices for command names might not be optimal is that popular terms tend to be imprecise.

In the same study Landauer, Galotti, and Hartwell had some subjects learn subsets of the command language of the Unix editor, Ed, while others learned names that had been more frequently given spontaneously to describe the operations. Use of the more "natural" names for commands did not facilitate learning in this case. In the investigators' words, "one important lesson from this is that intuitive guesses as to what is a 'common' or 'natural' name for a command are likely to be hazardous. One person's obvious name may not be another's" (502). The investigators also tested a version of the language in which the same command name was used whether the command referred to a single word or to an entire line, against another version in which different command names were used for operations on objects within a line and operations on lines. Learning was easier in the latter case.

Landauer, Galotti, and Hartwell concluded from their findings that the idea that the best words for command names are the words people use to refer to similar operations in noncomputer environments is "a somewhat naively differentiated view" and that rational selection of command names requires more than simple adherence to the slogan "make the language natural." While familiar words are easier to learn as associates, ease of associative learning is not the only requirement of command names. Another important requirement is precision of application. Common words frequently have more than one meaning and lack this precision, as was demonstrated in the results of the study: some words were suggested for all five operations.

Carroll has emphasized the importance of the careful selection of names for commands and files (Carroll 1982a, b, c; Mack, Lewis, and Carroll 1983). He notes too the importance of the *relationships* among names. What constitutes an appropriate name for an entity depends not only on the nature of the thing being named but on what other names are used in the system. *Naming paradigms* is the term he uses to convey the idea that there is a need for consistency or congruence among the various names in any given system. People have certain expectations, Carroll argues, both about how names relate to the things named and about how the various names that are used within a given context relate to each other. When these expectations are violated, people experience confusion and difficulty in associating names with their referents. Knowledge that typing the letter *L* would cause a cursor on a display to move to the left, for example, would lead one to expect that typing the letter *R* would cause it to move to the right. Carroll and his colleagues present other examples of such expectations and subsume them under the notion of congruence: "a command language is congruent if functional relations between command definitions are mirrored in structural relations between commands" (Carroll 1982c, 335). For example, in a congruent command language, commands that have contrary functions (oppositions) are given names that are semantic contraries in English.

The choice of names is also an issue in the design of menus for information retrieval systems. Dumais and Landauer (1983) point out that people's success at using such systems varies greatly from system to system. Success rates tend to be high with systems having relatively small domains (a few tens to a few hundreds of items) that are well known and readily partitioned into non-overlapping categories; they tend to be low, however, with systems that have large, less well organized domains. Interestingly, Dumais and Landauer present data that suggest that designating categories by means of a few (say three) examples of members may be an effective alternative to using names.

Closely related to the question of names is that of the use of abbreviations. Although there is wide agreement that being able to abbreviate typed input is a great convenience, there is some question as to how abbreviations should be constructed. Among the investigators who have studied the effectiveness of various ways of abbreviating words are Barrett and Grems (1960), Bourne and Ford (1961), Hodge and Pennington (1973), Moses and Potash (1979), and Moses and Ehrenreich (1981). Moses and Potash found that abbreviations produced by simple truncation were as good as or better than those produced by several other rules (including contraction by removal of vowels and the letters *h, w,* and *y*) and military-standard abbreviations that had been formed by consensus. Goodness in this case was judged by subjective ratings, ability of the subjects to produce the original word or term given the abbreviation, and ability to produce the abbreviation given the original term.

Streeter, Ackroff, and Taylor (1983) had some subjects learn to use rule-derived abbreviations for words by vowel deletion and truncation, while others learned to use abbreviations that had been given most frequently by other people. Performance was superior with the rule-produced abbreviations, even when these were identical with the ones most frequently produced. The implication is that knowing the rule by which abbreviations are produced is important independently of the nature of the resulting abbreviations. Moses and Ehrenreich (1981) summarize the findings from research on various abbreviation rules as follows:

Performance with abbreviations is enhanced when: (1) truncation rather than contraction is the abbreviation rule; (2) fixed-length abbreviations are used rather than ones depending on word length; (3) participants are taught the abbreviation rule rather than just learning abbreviations by rote; and (4) abbreviations that must deviate from the rule are specially marked when they occur. (3)

Some experimentation has been done to determine the importance of syntax, independently of the functionality of a command language. Ledgard, Whiteside, Singer, and Seymour (1980), for example, have reported an experiment in which they compared the performance of three groups of subjects using two text editors that were semantically identical but differed with respect to syntax. One of the editors was a slightly modified version of a commercially available editor supplied by Control Data Corporation with the operating system NOS. This editor has a syntax that does not resemble that of natural language. The second editor retained the functionality of the NOS editor, but its command structure was altered so that all commands were expressed by common English words organized as legitimate phrases. The three groups of subjects included one composed of users who had had less that 10 hours of experience with computer terrminals; one composed of users who had had between 11 and 100 hours; and one composed of users who had had over 100 hours of terminal experience.

The subjects were given as much time as they wanted to familiarize themselves with the command languages. They were then given structured editing tasks to perform. All subjects used both editors, but the order in which they encountered them was counterbalanced. Performance was measured in terms of percentage of task completed, percentage of erroneous commands issued, and editing efficiency. (The measure of editing efficiency was the ratio of the difference between commands resulting in improvement of the text and those resulting in degradation of the text to the total number of commands issued.) The performance of subjects at all levels of experience was better with the English-like editor with respect to all performance measures. Moreover, the differences were not only statistically significant but relatively large in magnitude. The error rate, for example, was twice as high with the notational editor as with the English-like editor. Subjects generally preferred the English-like editor, and more so after the experiment than before.

The investigators point out that while some of their subjects were experienced computer users, none of them had had much experience with the specific editors used in the experiment. Whether the performance differences they obtained would have been maintained following extensive practice with these editors is an open question. They argue, however, that even had the differences decreased, one could still make a strong case for using a language that simplifies learning. The case would be weakened, of course, if extended practice showed a reversal such that performance actually became better with the editor that had been the more difficult one to learn.

The other caveat that applies to this study has to do with the generality of the results beyond the specific editors with which they were obtained. While the results show quite clearly that one English-like editor was easier to use following limited exposure than one specific notational editor, it is risky to conclude from this that all English-like editors would be superior to all notational editors or that, in general, the more closely a command language resembles natural language, the more effective it will be. Much remains to be learned here.

It is well to end these comments on languages with a return to Wirth's observation that the use of the term "language" in this context involves a bit of poetic license. Computer languages are languages in the sense that they have vocabularies and grammars; used properly, they can be an effective means of communication. No computer language yet developed, however, comes close to human language in its complexity, versatility, or subtlety of expression. The proliferation of computer languages has been phenomenal, each new one testifying to somebody's dissatisfaction with those that already exist. To be sure, some new languages are developed for the purpose of facilitating the application of computer technology to specific problem areas. Many of them, however, are intended to be general-purpose languages.

We have not seen the end of language development. Nor is the development of natural-language understanding capability likely to make computer languages obsolete in the near future. People will continue to find existing languages constraining and will invent new ones; and new machine architectures—in particular, multiprocessor architectures—will call for innovations in language. We will return to this subject in the context of a discussion of standards in chapter 12.

## Natural Language

A prevalent assumption is that it would be a good thing if we could give "intelligent systems" the capability of behaving like human beings. People also tend to make the related assumption that the ideal way to interact with a computer is by means of natural language and speech (Giuliano 1974; Halpern 1967; Heidorn 1972). Before discussing these assumptions, we should note that natural language and speech are quite different, though not independent, capabilities. This distinction is not always made in the literature.

Interest in making it possible to interact with computers by means of natural language has been high for several years. If people could communicate with computers in this way, it is argued, the need to learn specialized languages would no longer exist; natural-language interfaces could make computer power available to people who are unable or unwilling to learn a formal computer language (Waltz 1983).

The term "natural language" is used in different ways by different writers. The Texas Instruments personal computer, for example, is said to provide a "natural-language" software package that permits the user to construct English-like commands by selecting words successively from a menu of options in each of several word classes (verbs, nouns, connectors). But the connotation the term has in this context is quite different from the one that people working in artificial intelligence usually give it. Woods (1977; see also Woods 1984) notes that as he employs the term, natural language means more than the use of the words and syntax of a language such as English: it assumes "understanding" on the part of a listener (or reader).

It [natural language] is characterized by the use of such devices as pronominal reference, ellipsis, relative clause modification, natural quantification, adjectival and adverbial modification of concepts, and attention-focusing transformations. It is a vehicle for conveying concepts such as change, location, time, causality, purpose, etc., in natural ways. It also assumes that the system has a certain awareness of discourse rules, enabling details to be omitted that can be easily inferred. (18)

Woods points out that natural language does not preclude the use of abbreviations and shorthand ways of representing concepts. Indeed, a language that did not accommodate abbreviations would have an important constraint that natural ones do not share. He rejects the idea that a system that uses English words in an artificial format should be considered a natural-language system, although the use of the English words may serve a mnemonic purpose and simplify the user's learning task somewhat. Woods does not foresee a time when the natural-language problem will be "totally solved" but notes that a total solution is not necessary to the implementation of systems with limited but useful natural-language capability. Winograd (1984), too, while predicting that the decreasing cost of computer hardware will pave the way for applications that are impractical today, states flatly that "software that mimics the full human understanding of language is simply not in prospect" (144). For an introductory treatment of natural-language processing, see Klinger (1973); for reviews of work on natural-language understanding by computer, see Addis (1977), Waltz 1982, 1983), and Woods (1977); for a reasoned discussion of the conditions under which use of natural language is advantageous and of what the prospective user of a natural-language system should look for, see Bates and Bobrow (1984).

Among the natural-language systems currently being offered for sale are Intellect, an interface for IBM data-management products (from Artificial Intelligence Corporation); Natural Link, a menu-driven front end to a database management system (from Texas Instruments); Savey, a front end for personal computers such as the Apple II and the IBM PC (from Excalibur Technologies Corporations); Straighttalk, an interface to a word-processing system (from Dictaphone Corporation); and Themis, a front end for the Vax Oracle (Frey Associates). At least one company (Cognitive Systems) is offering to develop customized natural-language interfaces to knowledge bases (Evanczuk and Manuel 1983). All of these systems are designed to work in limited domains of discourse, and none comes close to having the natural-language ability of a human being.

As long as natural-language systems have severe limitations with respect to what they can recognize, the argument that such systems minimize the learning requirements of users may be misleading. To be sure, users do not have to learn natural language; they do, however, have to learn the constraints of the particular natural-language system they wish to use. They have to be able to distinguish between those inputs it will recognize and those it will not, and the amount of learning required may be considerable. Failure to understand the constraints of the system with which one is dealing can lead to strange dialogs. Reitman (1984c) describes the following incident:

During a slow moment last week, my colleague Madeleine Bates stopped by my office. She and Rusty Bobrow, continuing their tracking of current language interfaces, had just returned from a trip to New York, to a firm with such a system. What follows is one interchange they had with it.

Bates:    Are all of the vice-presidents males?
System:   Yes.
Bates:    Are any of the vice-presidents female?
System:   Yes.
Bates:    Are any of the male vice-presidents female?
System:   Yes. (325)

Reitman notes that such a sequence of questions is a bit un-usual, but it clearly illustrates the limitations of current systems. The anecdotal archives contain many illustrations of the point. Reitman summarizes his view of the current situation in the following way:

1. You can now type English to a computer, and it can type English back; but unless you know your system well, you'd better not bet your job on what it tells you.
2. We may be watching the beginning of a bull market in artificial intelligence applications (I believe we are). But the early stages of a bull market can be treacherous: buyers beware. (326)

According to *Technology Trends Newsletter* (March 1984), there were three companies with estimated sales of natural-language software of $1 million or more in 1983: Artificial Intelligence Corporation ($6 million), Relational Technology ($3 million), and Cognitive Systems ($1 million). The same newsletter pre-dicted that the number of companies in this category would double in 1984 and noted that natural-language software sales are expected to grow at a compound rate of over 75 percent per year, making natural-language applications the area of fastest growth within the AI market.

Codd (1974) has suggested that an important component of a natural-language system must be its ability to query the user for the purpose of clarifying ambiguities in his inputs. The method proposed for doing this—presenting the user with multiple-choice questions relating to the ambiguities—presupposes sufficient language-processing ability to specify a set of alterna-tives that includes the user's meaning. How difficult this would prove to be in any given case would depend on the degree to which the user's inputs could be assumed to be constrained by the systems capabilities. The universe of discourse would be more constrained for a hotel reservation system, for example, than for a general question-answering system, and the resolu-tion of input ambiguities should be much easier in the former case than in the latter.

One might object, in principle, to the use of the term "natural language" to refer to communication between people and computers, because it involves the tacit assumption that what is natural for communication between people is natural for communication between people and machines. That assumption should be challenged, if for no other reason than to motivate the development of arguments that will support it. When one designs a machine to get people from one place to another, one does not necessarily do well to try to duplicate in the machine the human beings' "natural" means of locomotion. When one builds a machine to lift, pull, or push heavy objects, one does not necessarily do well to pattern the methods by which the machine exerts force after those used by human beings. When one designs an interface between a machine and its human operator, one should provide controls that the human being can operate effectively, but this does not necessarily mean building controls that look or act like the human hands or fingers that will manipulate them. Why should we assume that the machine side of an interface that has a large cognitive component should resemble the human side of that interface in specific ways? What is wanted is an interface that maximizes the usefulness of the machine to the user. Whether this means that the machine side of that interface should have human-like capabilities remains to be seen.

This is not meant to suggest that the common assumption that the machine should be given human-like capabilities is wrong. It may be quite right; the point is that it is an assumption, and one that has seldom been questioned in spite of the fact that it has very little empirical or theoretical support beyond the force of opinion.

A few investigators have questioned the assumption that natural language is the preferred way for people to communicate with computers (Hill 1972; Montgomery 1972; Small and Weldon 1977; Shneiderman 1980b). Hill has argued that something of value might be lost if computers became as imprecise in their use of language as are humans, and that instructing them would not necessarily be easier as a consequence. Indeed, he has gone so far as to make the somewhat heretical suggestion that communication among humans might be improved if we were taught to use some of the precision in communicating with each other that we are obliged to use when instructing a computer. Sammet (1966, 1969), on the other hand, has argued that its permitting ambiguities does not rule out natural language as an effective means of person-computer communication, and that the advantage of not having to learn a new notational system might outweigh any disadvantages in the long run. Winograd (1984) notes that a great advantage of natural-language front ends is that one can begin to use them without special training. He adds, however, that although the natural-language feature may promote early acceptance of a system, experienced users often move toward more stylized forms of language that they can use with the confidence that the machine will interpret their inputs correctly.

The assumption that person-computer interaction would be greatly facilitated if computers could understand natural language seems to involve the corollary assumption that people would find it easy to instruct computers in natural-language terms. Miller (1981) investigated the ability of people who had had no programming experience to write procedural instructions in natural language. The subjects were 14 undergraduate students, none of whom had had experience of any kind with computer systems. Six problems were described, each of which involved retrieving information from structured files and each of which was readily represented in a computer program. The various files contained information about the names, wages, employment dates, ages, marital status, and other "attributes" of employees. An example of a task would be, "Make a list of all those employees who are 64 or more years old who also have 20 or more years of experience. Lists should be organized by employee name." The subject's task was not to find the information from the files himself but to write a detailed set of instructions for someone else who was to do so.

The results of the study were fairly complex, and the reader is referred to the original report for details. On balance, the results did not lend much support to the idea that provision of a full natural-language interface would suffice to make people who lack training in programming effective users of computers. The instructions produced by the subjects tended to be incomplete, the more so the greater the complexity of the problem. Moreover, interpretation of the instructions required considerable contextual information and world knowledge. The results did provide evidence that limitation of the vocabulary to a few hundred words need not constitute a severe constraint on performance. Several comparisons between the natural-language procedures defined by the subjects and computer programs for accomplishing the same tasks yielded a list of important differences between them. In general, Miller concluded that the findings

would seem to remove from active consideration the notion of radically improving computer usability by a totally unrestricted natural language interface: the technolgy to accomplish this is simply not there, and probably will not be, even in approximate form, for a number of years. Aside from the technical difficulties, some other aspects of our study make us skeptical that merely (!) providing a natural language interface would permit anyone to become a programmer, capable of specifying procedures necessary to develop complex computer programs. We suspect that what would happen is that a lot of people would be able to generate easily lots of programs that did not do what they were supposed to—because the subtle conceptual complexities of the problem were not appropriately understood and dealt with. (211–212)

Miller is careful to point out that while his results support a negative conclusion regarding the effectiveness of unrestricted natural-language interfaces, they do not argue against some movement in this direction. In particular, he suggests two ways of making interfaces more natural than those that currently exist: "(1) implementing a natural language interface subject to several constraints and (2) modifying program languages to include more natural language features" (212). The types of constraints that might be imposed include the use of limited vocabularies (perhaps a few hundred words), limited syntax (restrictions on sentence types and lengths, prohibition of certain complex constructions), and assumed familiarity of the user with algorithmic procedures and their use in the applications area of interest. As for modifications that might be made to existing languages to make them more natural, Miller mentions the addition of contextual referencing capabilities (permitting the use of pronouns and other types of anaphoric reference) and providing the user with the capability of defining high-level procedures that could resemble natural language in some respects.

Giving procedural instructions is not the only occasion for interacting with a computer, and perhaps not the one for which the reasons for wanting a natural-language capability are most obvious. A major intended application of natural language is in the querying of information systems of various types. Presumably it should not be necessary to write procedural instructions in order to interact effectively with such systems, any more than it is necessary to give instructions to knowledgeable human beings when asking for some specific information. The effectiveness of natural language as a query language has been tested by Small (1983), who compared it with SEQUEL in a highly structured information retrieval problem. English and SEQUEL produced equally accurate queries, but subjects retrieved information faster with SEQUEL. Small was careful to point out that the advantage of the structured language in this case might have been peculiar to the task chosen.

Paradoxically, when we talk about "natural language" a problem arises from the ambiguity of the word "natural." It can connote something that has happened spontaneously, as distinct from something that is artificial in the sense of having been deliberately designed. It can also connote something that is easy to use as opposed to being "unnatural" or inconvenient to use. It has still other connotations as well. One might refer to the coded speech that a computer produces as natural in the sense of being indistinguishable from human speech with respect to its semantics and syntax, while describing it as unnatural with respect to the way it sounds.

Perlman (1982) views naturalness and artificiality as orthogonal features and urges the design of "natural artificial" languages in pursuit of the two objectives of facilitating concise and precise communication within limited domains and providing languages that are convenient to use. He notes that there is a tradeoff involved in the design of such languages, in that naturalness in a specific domain may be gained at the cost of a lack of generality to other domains.

What is the state of the art in natural-language interfaces? Waltz (1983) has summarized it this way: "Currently, we understand how to do a reasonably good job of literal interpretation of English sentences in static contexts and in limited, well-structured domains of application" (81). We may assume that research on natural-language understanding by computer will continue and probably accelerate. The present intense interest in the commercial potential of applied artificial intelligence assures a scramble to produce systems with natural-language front ends for various commercial markets. We may assume, too, without challenging Woods's and Winograd's assumption that a total solution to the natural-language problem is not in sight, that such systems will increase in power and scope. Waltz sees a number of applications of natural language in information utilities becoming possible within the next ten years, including "automatic directories of names, addresses, yellow pages, and so forth; electronic mail; on-line catalogs and ordering facilities; banking and tax services; routing directions; and access to books and periodicals" (82). Of great interest from a human-factors point of view will be the effect of limited natural-language capability on the quality and efficiency of the user's interaction with a system in any particular context. While we can probably assume the effect in most cases will be positive, it is not prudent to assume it will always be so.

A major practical objective of research in this area should be the design of languages that are natural in the sense of being convenient to use effectively for the purpose of interacting with computers. Whether such languages will turn out also to be natural in the sense that they are very similar to the languages people use in communicating with each other is an open question.

Efforts to provide computers with natural-language capability have had one clearly beneficial, albeit unplanned, effect. They have considerably enhanced our understanding of communication among people, affording us a keener appreciation of just how complex natural-language discourse is.

## Speech

Speech obviously is an effective communication medium for people. Moreover, it is apparently more effective than other media that people also use quite well. This was demonstrated in a series of experiments by Chapanis and his colleagues in which they investigated the effectiveness of various communication methods (typewriting, handwriting, speech, video, and combinations thereof) used by two people working as a team on structured problems (Chapanis 1973, 1975, 1981; Chapanis, Ochsman, Parrish, and Weeks 1972; Chapanis and Overbey 1974; Chapanis, Parrish, Ochsman, and Weeks 1977). This work has produced a large amount of data on the characteristics of interperson communication under controlled conditions; the most salient finding for present purposes, however, is the superiority, among the modes studied, of those that included voice over those that did not. Whether the superiority of speech over other possible communication modes will hold when the communication is between people and computers, and when the set of alternatives is broadened to include some less conventional possibilities, remains to be determined.

Work aimed at making speech a practical input-output medium is proceeding, and considerable progress is being made (Doddington and Schalk 1981; Flanagan 1976; Forgie 1975; Lea 1980; Rabiner and Levison 1981; Sherwood 1979). Speech products are being introduced into the marketplace very rapidly by a host of companies including Votan, Texas Instruments, Mitsubishi Electronics, Octel Communications, Speech Limited, Enter Electronics, Verbex, Voice Machine Communications, Bell Laboratories, Kurzweil Speech Systems, and numerous others. Product capabilities include word recognition of limited vocabularies, speaker recognition, speech compression, speech synthesis, text-to-speech conversion, voice message store and forwarding, and dialless telephone responding to voice commands. *VoiceNews*, a publication of Stoneridge Technical Services, reports monthly on recent developments in speech synthesis, speech recognition, and related voice technologies.

Most of the work in speech science that relates to the use of speech as a communication medium in person-computer systems is in one of three areas: speech coding, speech synthesis, and speech understanding. Speech coding involves the representation of human speech in digital form so as to facilitate its storage in computer systems and its transmission over digital networks; speech synthesis refers to the production of speech by computer; the objective of work on speech understanding is to give computers the ability to accept speech as input.

Either coded or synthesized speech may be used as speech output, but coding and synthesis are different processes and subject to different constraints. Coded speech is human speech that has been analyzed and represented digitally; synthesized speech is speech that has been generated by the computer in accordance with a set of rules for producing the elemental sounds and concatenating them into intelligible sequences. Coded speech can be made to sound as natural as desired; but the cost of naturalness is lots of storage space, and in a sense this defeats one of the purposes of coding, which is to limit storage requirements. Several coding schemes have been developed, and other possibilities are being explored; the goal always is to get the highest-quality speech for a given expenditure of computing or communication resources, or conversely to decrease what must be expended in order to get a given level of quality.

A disadvantage of using coded speech as computer output is that the only utterances that can be used are those that can be constructed from the words or sentences that have been stored; moreover, sentences constructed from individually uttered words tend not to sound very natural. The use of synthesized speech has the advantage that essentially any utterance that is desired can be provided about as easily as any other and without prerecording. The technology of speech synthesis has not yet progressed to the point where it can produce speech that sounds like that produced by humans. Whether this is considered a drawback, assuming the speech is intelligible, depends on one's point of view.

Smith and Goodwin (1970) report an early attempt to use coded speech as voice output in an interactive system. In this system the user entered inputs to the computer through a touch-tone phone, and the computer talked back to the user by formulating responses from stored words and phrases. The interaction had the character of what the investigators refer to as a "transactional sequence." The computer stopped talking after each brief output (composed of a few sentences or a short paragraph) and waited for the user to select one of several options. By pressing the appropriate key, the user could instruct the computer to continue or to repeat its last output. Sometimes the computer's output constituted a question that the user had to answer, perhaps with a number. Each output was terminated with a Morse-code-like signal that indicated whether the user was expected to make an input; if so, the computer provided a clue to the format of that input. Smith and Goodwin point out some limitations of systems that provide *only* auditory outputs from the computer, among them the absence of opportunities for browsing or skimming displayed material, the inability to present multiple alternatives simultaneously, and the lack of aids to memory.

Rabiner and Schafer (1976) describe several experimental applications of a coded-speech response system developed at the Bell Laboratories. The system concatenates words and phrases that have been coded and stored in the computer in adaptive differential pulse code modulation (ADPCM) form. The applications for which the voice response system has been tested are (1) computer-generated voice directions for wiring, (2) telephone directory assistance, (3) stock price quotation, (4) a data-set testing system, (5) flight information, and (6) a speaker verification system. These applications all require vocabularies of very limited size, and the range of messages they produce is fairly small. Rabiner and Schafer point out that waveform concatenation methods are quite useful for these types of applications but would not be practical for systems that require unlimited speaking potential.

A *VoiceNews* report claims that successful consumer products with speech output have invariably had three characteristics: quality (speech of cassette recorder quality at least), purpose (speech in a product that serves no real purpose will not be accepted by consumers), and pleasure (users msut be able to turn off speech output if they desire, and ideally they should be able to select from a variety of voices to fit the mood). What will really prove to be important to consumers in speech products remains to be seen, however. As *VoiceNews* puts it, to date "there has been exactly one unqualified speech synthesis success: the Speak 'N Spell product line, estimated to have sold over one million units" (*VoiceNews*, May 1984, 4). No other talking consumer product has yet attained more than moderate volume. The belief appears to be widespread, however, that the market for voice products is likely to expand rapidly, if not explosively, in the near future. The technology push certainly appears to be there. Perhaps when the right products come along, the market pull will become apparent as well.

In discussing the requirements for systems that produce or transmit speech, it is important to distinguish between speech that originates from a human being and speech that originates from a computer, because the requirements for acceptability may differ. In an ideal system, speech that originated from a human being would sound not only very much like human speech but in particular like that individual's speech. How much distortion and what types of distortions are acceptable to users probably depend on several variables. But what about computer-originated speech? The prevailing assumption is that it too should sound as though it was produced by a human being. That assumption may be correct, but there is some chance that it is not; certainly the assumption should not be accepted uncritically. Why should computer-generated speech sound human? And if it is to sound human, what human should it sound like? One wants it to be intelligible, of course; otherwise it would be useless. But intelligibility does not depend on sounding human.

A human-factors issue relating to speech synthesis that may seem humorous, but may also turn out to be a bona fide problem, is the question of how to decide what speech characteristics to give to talking machines. Whether or not speech becomes the primary means of person-computer communication, more and more machines are likely to have the ability to talk to us. Deciding what voice properties to give to a particular device is not just a matter of picking a voice that is intelligible and pleasing to the ear. One might want different devices to have different-sounding voices so that devices can be recognized by voice just as people are. One might also want a given device to have more than one type of speech for different occasions (Smith and Goodwin 1970). A distinctive voice could be used, for example, for purposes of warning or providing instructions in crisis situations. Developing useful guidelines for deciding what voices to give to which machines or functions is an interesting research problem that is likely to become a serious one in time.

Much of the work on speech processing today is motivated not only by the prospect of using speech as a means of person-computer communication but also by a desire to use computer and communication technology to enhance communication among people. There are several compelling reasons, for example, for digitizing and encoding speech for voice communication systems. Perhaps the most important is that by encoding speech one can decrease the bandwidth requirements and thereby reduce communication costs. Several ways have been developed for compressing speech (Andrews 1984; Flanagan 1972). All of them trade computation for transmission bandwidth. That is, the more one wishes to decrease bandwidth requirements, the greater the amount of computing one must be willing to do at each end of the communication link.

One consequence of large reductions in bandwidth may be some detrimental effect on the intelligibility or quality of the resulting speech. The magnitude of the effect generally scales with the magnitude of the bandwidth reduction. Digitized speech that has not been coded typically requires about 64,000 bits per second. The speech signal may be sampled 8,000 to 10,000 times per second, and each sample represented by an 8-bit code. This speech is essentially indistinguishable from the original analog waveform. Coding techniques have been developed that reduce the bandwidth requirements by a factor of about four with very little effect on intelligibility and quality. Speech that is encoded at less than 16,000 bits per second begins to show noticeable degradation. Although difficult to quantify, significant effects on quality occur before the level of intelligibility becomes unacceptable for most purposes.

The issues of intelligibility and quality are related but somewhat different issues. The first thing one wants of speech output is intelligibility. Assuming acceptable intelligibility, however, it does not follow that the speech produced by a computer system will be acceptable to users in all other respects (Beasley, Zemlin, and Silverman 1972; Zemlin, Daniloff, and Skinner 1968). A voice that sounds like that of a young child, for example, is likely to get a different reaction if it is believed to be coming from an adult male speaker than if it is believed to be coming from a child. Qualitative aspects of speech that are sometimes indicative of a speaker's condition or state (head cold, laryngitis, anger, fear) may evoke listener reactions independently of the intelligibility of the speech (Nickerson and Huggins 1977). There is much work to be done in the area of assessing speech intelligibility and quality and determining what qualitative changes are acceptable, or unacceptable, to various types of listeners in specific contexts.

For some applications it may be desirable to have a computer system speak at rates considerably faster than those typical of human speech. While measurements made on samples from radio announcements, lectures, and spontaneous speech have yielded speech rates varying from about 100 to about 250 words per minute, mean rates tend to be within the 125 to 175 words-per-minute range (Abrams, Goffard, Kryter, Miller, Sanford, and Sanford 1944; Gregory 1969; MacClay and Osgood 1959; Voekler 1938). While intelligibillity begins to fall off as speech rate is increased beyond about 150 words per minute (Abrams et al. 1944), the evidence indicates that it is possible for speech to remain highly intelligible at rates well over 200 and in some cases over 300 words per minute (Fairbanks, Guttman, and Miron 1957; Foulke 1971; Harwood 1955). A few efforts have been made to investigate the intelligibility of speeded and time-compressed speech and its dependence on the training, motivation, and listening strategies of listeners (deHaan 1977; deHaan and Schjelderup 1978; Gade and Gertman 1979; Lambert, Shields, Gade, and Dressel 1978).

The first attempts to speed recorded speech instrumentally involved simply recording the speech at one speed and playing it back at another (Fletcher 1929). Speeding speech up in this way changes its perceptual characteristics quite drastically; for one thing, it raises the fundamental frequency or pitch of the speech. Intelligibility declines greatly with speedup factors of less than two. More recently a variety of more sophisticated methods have been used. These involve snipping out alternating brief segments of speech and abutting the remaining segments (fixed-interval sampling), elimination of portions of between-word pauses, discarding alternate pitch periods of voiced sounds, discarding large portions of steady-state sounds, and other computer-based techniques.

The usefulness of speeded speech is likely to depend on the context and purpose for its use. It may be particularly serviceable for reviewing familiar material, for auditorily "scanning" recorded speech for segments of interest, or for refreshing one's memory regarding the dialog of a meeting. How much it will enhance person-computer interaction remains to be seen. To the extent that speech is used as an output medium, however, permitting the user to adjust the speed to his own preference would seem to be desirable. Of course, one would want to use speech compression methods that would yield high intelligibility at accelerated rates; and only time will tell whether speech scientists will succeed in developing compression techniques that will improve greatly upon existing methods.

Work on speech recognition by computer has been going on in parallel with work on speech coding and synthesis. Among the advantages that are sometimes cited for the use of speech input to a computer system are the following: speech is a skill that most people already have, so a new skill (typing, for instance) would not have to be learned; people can talk faster than skilled typists can type; speech leaves one's eyes and hands free to be used for other purposes; speech would be understood by other humans as well as by a speech-understanding machine and therefore would permit one to communicate simultaneously with other people and with machines; speech recognition capability would mean that telephones would instantly become useful input devices (Turn 1974).

Intensive research in this area was funded at several U.S. laboratories by the Defense Department's Advanced Research Projects Agency over a five-year period in the early 1970s. Considerable progress was made on the problem during that time (Klatt 1977; Lea and Shoup 1980), but it also became clear that the problem was more complex than had been realized. Research on speech understanding has continued since the termination of the ARPA project but on a much smaller scale and without the coordination provided by that vehicle.

Much of the work being done today is aimed at the development of commercial speech-understanding systems and is therefore proprietary to the companies involved. The results of these efforts are beginning to appear in the marketplace. In 1983 Texas Instruments introduced a speech input system under the trade name Speech Command. It accepts as many as fifty isolated words or phrases for each of nine vocabularies stored in the computer's memory. The user assigns a word or phrase to each of the commands that he wishes to be able to enter orally and provides the computer with a sample of each utterance. The sample is recorded for future comparison, and thereafter the user may enter these commands either by speech or by the keyboard. Numerous other speech recognition products are now commercially available. Most recognize only a small (less than 100-word) vocabulary of isolated words with high accuracy (Doddington and Schalk 1981; Moshier, Osborn, Baker, and Baker 1980), but systems with limited connected-word recognition are also beginning to appear.

Recently *Computer-Disability News* (1984c) reported the availability through a home construction company in Florida of a computerized house that "responds to vocalized requests made anywhere within its walls and also to commands made over a user pre-coded telephone line" (1). Details about the types of requests to which the system can respond are sketchy; the examples given include raising the temperature and dimming the lights. The report is interesting by way of its suggestiveness regarding possibilities for the future. Voice-activated control systems of the kind that would be useful in operating a household are of special interest to homemakers with certain types of mobility impairments.

The National Research Council's Committee on Computerized Speech Recognition Technologies recently characterized the state of the art in speech recognition as "not sufficiently advanced to achieve high performance on continuous spoken input with large vocabularies and/or arbitrary talkers," but "mature enough to support restricted applications in benign environments, with disciplined use under low-stress conditions" (Flanagan et al. 1984, 3).

Assuming that systems with moderate speech recognition capabilities become widely available during the next few years, there will probably be a need to train people to use them effectively. Can users readily learn to adapt their speech so as to make it recognizable by systems with limited word segmentation capability? Can they easily learn to use constrained vocabularies effectively and to produce speech with somewhat unnatural temporal properties?

The possibility of developing natural-language systems that operate with limited vocabulary and constrained syntax has had some attention (Plath 1972). And there is some evidence that people can learn to communicate effectively with each other using vocabularies limited to a few hundred words. Kelly (1975; Kelly and Chapanis 1977) had two-person teams collaborate on a problem-solving task while located in separate rooms. (In this case they communicated by teletype, but the results seem readily generalizable to speech.) Some of the teams were permitted to use their full vocabularies, others were restricted to vocabularies of either 500 or 300 words. In the latter cases the subjects were given prior training in the use of these limited vocabularies. The time required for solving the problems and the number of messages sent back and forth between the members of the team were found to be independent of the sizes of the vocabularies they were permitted to use.

It would be rash to conclude from such findings that restricting one's vocabulary does not materially impair one's ability to communicate. It is clear that that cannot be true in a general sense. Nevertheless, the demonstration that people can learn to use a relatively limited vocabulary effectively for the purpose of collaborating on certain types of problems is an interesting and non-obvious result. Seventy-five of the words in the limited vocabularies used by Kelly were task-related words. The remainder were function words. Presumably in any effort to use limited vocabularies it would be necessary to include some words specific to the particular task domain. Presumably too, however, many of the words one would have to have in the vocabulary would be independent of the task domain. It would be of some interest to attempt to identify a minimal adequate set of such domain-independent words.

In studies in which a listening typewriter (speech-to-text system) was simulated by having a human typist transcribe speech into text for the computer display, Gould, Conti, and Hovanyecz (1983) found that people were able to make effective use of (simulated) systems with vocabularies of 1,000 or 5,000 words and were able to accommodate to the requirement to enter the words with pauses between them. Interestingly, a system that recognized a 5,000-word vocabulary but required isolated word input was preferred to one that permitted continuous speech but recognized only 1,000 words. The subjects apparently considered the smaller vocabulary to be more constraining than the isolated-word requirement.

The use of speech input with a menu-driven system is a less demanding application of speech recognition technology than many others. In this case the speaker is constrained by the items of the menu, and the speech recognition system has the task, not of determining in an absolute sense what was said, but of deciding among a small number of known possibilities, which is much simpler.

Almost all of the work to date on speech as an input-output medium has been directed toward simply making it possible. The goal has been to develop the technology that would give computers the capability of generating high-quality speech on the one hand and of recognizing normal human speech on the other. Very little attention has been given to the types of questions that may arise if and when the technological objectives are achieved. Indeed, the prevailing assumption seems to be that there are no issues relating to the use of speech as an input-output medium except the technological problems associated with making it happen. I believe that this is not a reasonable assumption and that the existence of talking and speech-understanding machines will force attention to some fairly complex questions about how this technology should be used. Moreover, the technology is sufficiently advanced, and machines capable of using speech for both input and output in nontrivial ways are sufficiently close to reality, that it is not too early to begin trying to anticipate what those questions are likely to be. The unthinking application of this technology is likely to yield a great deal of silliness, and quite possibly some amount of unnecessary unhappiness. How to use it effectively will require some careful thought and probably some experimentation.

To date, speech has been a communication medium unique to human beings. We have shared it neither with other organisms nor with machines. We have very little idea what the psychological effects will be of having the world populated with machines that listen and talk. The experience gained with radios, television, and recording machines provides little insight into this question. One does not think of these devices as having the ability to talk. To be sure, they produce speech, but the speech they produce is perceived by the listener as coming from human beings, even if by indirect routes. Systems that have the ability to generate language and synthesize speech are in a different category altogether. If, as many anticipate, machines acquire the ability to converse with human beings (or with each other), we must wonder what the implications will be for people's perception of themselves. We have yet to see how people will adapt to the reality of interacting with machines that listen and talk.

In addition to serving as a means of conveying semantic information between a computer system and a user, speech is of interest to system builders for the purposes of user verification and identification. Verification is the simpler of the two functions, because in this case the system has only to decide whether or not the speaker is who he claims to be; it has only to determine whether the characteristics of the speaker's speech are a close enough match to a stored description of that speaker's speech to be accepted as coming from the claimed source. In the case of speaker identification, the system must decide which (if any) of a number of speakers is speaking. If the number of possibilities is very large, this can be a difficult problem, and of course there is always the possibility that the speaker may be someone other than one of the possibilities being considered. Both verification and identification capabilities are of considerable interest, however, because of their applicability to problems of security and system-access control.

## On *"Conversational" Interaction*

The term *conversational* is often used to characterize the nature of the interaction that occurs between an on-line system and its user. What is meant by this term is not always clear, inasmuch as person-to-person conversations have many distinctive features not all of which are seen in conversations that occur between people and computers. A question of some practical as well as theoretical interest is whether human-like conversational capability is an appropriate goal toward which developers of interactive systems should strive. I have suggested elsewhere (Nickerson 1976) that the answer to this question may be no, and that whether an interaction resembles the conversation that might take place between two persons is irrelevant to the more important question of whether it is satisfying to users and effective in helping them realize their goals.

Interactions that are both satisfying and effective may resemble conversations in some respects, but they may differ from them in others. A two-directional, mixed-initiative capability is clearly desirable for many purposes. Rules of transfer of control are essential. The development of new input-output techniques that will increase the communication bandwidth and especially the bandwidth of the user-to-computer channel seems indisputably desirable. All of these features are in keeping with the notion of a conversation. On the other hand, there may be ways in which one would prefer that the user-computer interaction differed substantially from typical interperson conversations. For example, one might want the person-computer interaction to be less symmetrical in various ways than most interperson conversations. One might want the computer to respond, insofar as possible, instantaneously to requests from the user while being infinitely patient regarding delays originating with the user. One might want the computer to be very precise in its use of language but tolerant of a certain amount of imprecision on the user's part. One surely wants the computer to be able to do things that no human participant in a conversation could be expected to do, such as produce complex graphical or pictorial displays on demand, search rapidly through very large databases and output the results in organized ways, purge its memory of incorrect, obsolete, or useless information, and so on.

It can be argued that the model that likens human-computer interaction to interperson conversation is inappropriate and misleading, not to say fundamentally undesirable. A preferred model, at least to some tastes, is that of a person making use of a sophisticated tool. Sometimes the human's role in the interaction is described as that of a *partner* and sometimes as that of a *user*. The second term strikes me not only as more descriptive of the relationships that real (as opposed to conceptual) systems permit but also as preferable because it implies an asymmetry with respect to goals and objectives that *partner* does not. *User* is not a term that one would normally apply to a participant in a person-to-person conversation, but it seems like precisely the right term for the case in which one party to an interaction is a machine.

Whether or not one considers *conversation* to be the appropriate designation for the type of interaction between people and computers that would be desirable, no existing system permits anything like a substansive interperson conversation today. Also, regardless of one's attitude toward the appropriateness of human-like conversation as an objective, everyone will undoubtedly agree that improvement on the types of interactions that are possible today—in terms of bandwidth and fluency—is certainly a desirable goal. Several investigators have focused on the problem of designing computer-to-user messages (such as error messages, control requests for input, answers to queries, results of program execution) as one that needs more attention than it has received (Dwyer 1981a, b; Golden 1980; Shneiderman 1982).

## Friendliness and Usability

What makes for "friendliness" on the part of a computer-based system? Presumably anything that would improve the quality of the interaction, reduce the probability of catastrophic mistakes, and make it easier for users to get the help from a system that they need. What follows is a list of features and capabilities, found at present in one or more systems, that represent the variety of ways in which system designers and implementers have addressed the problem of making their systems friendlier than they otherwise might be.

- Command confirmation. There are certain commands that one particularly wants to avoid executing at the wrong time. Examples include commands that would trigger extensive operations, such as printing out a very large file, or that have irrevocable effects that could be unfortunate (such as deleting or expunging a file, or distributing a message to the wrong recipients via an electronic mail system). Some systems deal with this problem by asking the user who has just given such a command to confirm it explicitly before it is executed.

- "Undo" commands. Another way of dealing with potentially disastrous commands is to delay their execution until the very end of a work session (that is, when the user logs off the system) and to permit the user to negate a command by issuing a different command such as "undo" (Myer and Barnaby 1971; Teitelman 1972).

- Corrections of misspelled commands. Some systems have the capability of recognizing a misspelled command, correcting it, and executing the intended command (Mooers 1982; Teitelman 1972). It is important, of course, when a system accepts a misspelled command, that it provide feedback to the user regarding the way it is interpreting the command that it assumes has been misspelled.

- "Your turn" signal. It is important for the user to be able to distinguish between when the system is waiting for an input and when it is otherwise occupied. Customarily this is accomplished by the computer displaying or printing out a symbol that is interpreted as "ready" or "your turn." The JOSS system, one of the earliest interactive systems for general-purpose use, provided two colored pilot lights on a control box to keep the user informed about who (the computer or the user) was in control ("talking") at any given moment (Bryan 1967). The user then was never in doubt as to whether the computer was waiting for an input.

- "Forget it" command. Often one realizes partway through the issuance of an instruction to a system that one does not really want that instruction carried out. It is a convenience in such cases to be able to signal the system to ignore the partially completed command. Again, some systems provide this capability through a special function key.

- "Enough" command. It should be possible for the user to interrupt a computer output and take control. Sometimes an action by a user will trigger a lengthy output (such as an error message or the typing of a file), and the user will realize partway through the output that it is not necessary to look at all of it. At such times the user should be able to signal the computer "enough" and have the output stop immediately.

- Default keys. When a system requires input from a user and that input is highly predictable (for example, some form-filling activities), it may be useful to provide a "default key," which when struck will produce the single most likely character or sequence of characters at that particular stage in the interaction. Thus, when several forms are likely to have many entries in common, such a default key can simplify the inputting process considerably (Pew and Rollins 1975).

- "Help" facility. Many systems permit the user to request help at any point in the interaction. The request for help usually produces some on-line documentation relating to the specifics of the interaction at the moment the help is requested. Some systems can respond not only to a general help request but also to requests to describe or explain specific system features or capabilities.

There are numerous ways that system developers have found to make their systems easier to use and to increase the fluency of the interactive session. (Making a catalog of these features in existing systems would be a worthwhile undertaking.) For the most part, such features owe their existence to the recognition by individual programmers and system developers that their incorporation would indeed make the systems more convenient to use. They have not been prescribed by a theory to which system designers all subscribe. Nor do they reveal a strong consensus among designers regarding how convenience, fluency, and friendliness are best obtained. As a group they do, however, represent the collective wisdom of developers on the question.

Shackel (1981) has discussed the concept of usability, particularly as it applies to computer-based systems, at some length. He defines usability as "the capability in human functional terms to be used easily and effectively by the specified range of users, given specified training and user support, to fulfill the specified range of tasks, within the specified range of environmental scenarios" (5). Whether a system is usable, or how usable it is, will depend, Shackel suggests, on the dynamic interplay of four system components: user, task, tool, and environment. He points out, however, that how usability relates to or depends upon these four system components is not yet well understood. Usability will be an especially important determinant of whether or not a system is used by people whose jobs do not require computer use.

Developers of computer-based systems are probably precisely the wrong people to evaluate their usability. It is too easy for sophisticated users to forget that the wealth of minutiae they have acquired in the course of developing and using a system is not known to the novice (Stewart 1976). The evidence is fairly compelling that the system designers on the whole underestimate the difficulty that novice users are likely to have with their systems (Gould and Lewis 1983; Gould and Boies 1983). As a bit of practical advice to designers, Branscomb and Thomas (1983) suggest that clean separation of the user interface from the rest of a system should be an architectural objective. The purpose of such a separation is to facilitate the making of improvements to the interface without requiring major revision of the underlying system.

• • • • • • • • • • • • • • • • • • • • • • • • • • • • • • • • • • • • • • • • • •

The designing of the interface between a computer system and its users presents new challenges, in large part because such interfaces typically have very significant cognitive aspects. What is being exchanged across the interface is information in (at least from the human's point of view) nontrivial amounts. The question of how to design interfaces that are well suited to users' cognitive capabilities and limitations will take on increasing importance as computer systems acquire more cognitive capabilities themselves. While it may be a very long time before computers have the full range of cognitive capabilities that human beings have—indeed it is not clear that they will ever have it—there can be little doubt that systems will be developed that have increasingly powerful capabilities and display more and more aspects of intelligence and expertise. What the limits of machine intelligence are, we really do not know. We can be sure, however, of continuing efforts to push technology closer to those limits. As the capabilities and versatility of computer systems are extended, the cognitive interface will surely change. Exactly how is difficult to anticipate. The trick will be to ensure that those changes are in the direction of making the demands on the users of such systems increasingly natural, which is to say consistent with their capabilities and limitations. Whether "friendliness" is the right concept is perhaps a matter of taste. "Usability" strikes me as the more appropriate and completely adequate concept; in imputing the quality of friendliness to a machine, one is diluting the meaning of one of the most pleasant of words.

# 8

## Software Tools

Manufacturers of computers know that computer sales are determined at least as much by the software available as by the characteristics of the machines themselves. Consequently every major manufacturer offers an extensive library of programs of various types. Independently of the computer manufacturers, other companies are making a business of marketing special-purpose software that will run on one or more small machines. As a result, the number of software packages available to the general consumer is growing very rapidly. In the words of Green, Payne, and Van der Veer (1983, 4), "Commercially produced programs are falling on us like leaves in autumn. Word processors, data bases, spread sheet calculators, financial packages, symbolic algebra programs, program generators, graphics packages—the air is full of them." As the authors point out, although some of these packages are much easier to use than others, we have yet to identify the factors that determine ease of use and to understand how those factors relate to each other.

### Statistical and Data-Management Packages

Several publishing companies are offering general statistics and data-management software packages. These programs offer similar though not identical functionality. Most will do simple descriptive statistics as well as various types of analyses and tests, including time series, correlation and regression analysis, analysis of variance, and a variety of multivariate analyses. Most also have the capacity to represent the results of analyses graphically—with histograms, pie charts, function graphs, and the like.

Illustrative of what is commercially available by way of statistical packages is one produced by BMDP. The promotional literature for this package emphasizes ease of use and documentation, as well as functionality. The user operates the programs with "English-based instructions," although what this seems to amount to is using "is" and "are" instead of ":" or " = ." The BMDP package was designed to run on mainframe machines. Recently, however, BMDP (1982; undated) announced a desktop computer, the Statcat, which it classifies as a supermicro. Forty programs in the BMDP package are available for the Statcat and are advertised as identical to those that run on mainframes. This system also comes with a UNIX operating system, several languages, including FORTRAN and Pascal, some other software (including a screen editor), and an Ethernet connection. The role of human engineering in integrating the statistical package with the computer is stressed. BMDP publishes a newsletter, *BMDP Statistical Software Communications*, several times a year to keep users posted regarding software developments, scheduled short courses, and other items of general interest to the user community.

RS/1 (Research System/1) provides data management, statistical analysis, and graphics. The system runs on the VAX (VMS and UNIX), the PDP-11 (RSX-11M), and the DEC Professional 350 (P/OS), and it provides a built-in programming language, RPL (for Research Programming Language), for users who wish to add to its functionality by writing new programs of their own. The basic data structure in RS/1 is the two-dimensional table. Users can create as many tables as they wish, of any desired size; they can refer to them by name and have a variety of operations performed on their contents. The command language permits the manipulation of entire tables, specified rows or columns of tables, or portions of tables identified by data values. The admissible operations are those that one would typically wish to perform on tabulated data: make a table; display a table or portion thereof; enter or edit data; add, insert, or delete columns or rows; sort, merge, compare, or transpose tables. The graphics capabilities provide for the specification of x/y plots, bar graphs, and pie charts. Curve fitting, statistical analysis, and analytic modeling capabilities are also provided (Finman, Fram, Kush, and Russell 1983). RS/1 is marketed by DEC and is used widely. Courses are offered to new users, and a newsletter is published and distributed to the user group.

Numerous project-management software packages are now available on a variety of PCs, and more are being introduced almost daily. Those that are currently available include the LisaProject Manager from Apple, PM/580 from Pinell Engineering, Project Management from Peachtree Software, Project Planning and Analysis (PAL) from Micromatics, and Harvard Project Manager from Harvard Software (Strehlo 1984). Each of these packages includes some variant of one of two much-used approaches to project planning and control, the Critical Path Method (CPM) and the Project Evaluation and Review Technique (PERT), but they differ considerably in functionality and interface characteristics. There are, as yet, no very helpful guidelines to assist potential users of these packages to select among them.

Among the more widely used programs for data management, both for businesses and for individuals, are the "spreadsheet" programs. The better known of these programs include Lotus 1-2-3, VisiCalc, SuperCalc, MicroPlan, and SuperPlan. These programs differ in functionality and ease of use, but they are similar in concept and share many features. They are all designed to deal with data that can be organized in two-dimensional arrays, or tables. The entry in any cell of the array can be a number or a formula that expresses the value of that cell as a function of entries in other cells. The marginal cells, for example, might be formulas for the sums (averages, standard deviations) of row or column entries. Preprogrammed functions (such as future value, net present value, investment rate of return) also are typically provided by these programs.

One of the great advantages of spreadsheet programs is that the implications of changes in the values or formulas in individual cells can be seen instantly in all other cells whose values are affected. This makes it easy to investigate the implications of various sets of assumptions, objectives, or decision alternatives on financial projections and other variables of business or personal interest.

While spreadsheet programs are simple in concept, becoming an effective user requires a significant investment in learning. Mastering the details of specific programs is only part of the challenge; the other part is discovering what analyses and numbers are really useful to one's purposes. Spreadsheet programs are quite capable of generating more numbers, by orders of magnitude, than anyone could ever use in a rational

way. My own intuition is that beginning users of these programs often inundate themselves with useless numbers simply
because they are so easy to produce. The users who get the
greatest benefit from them not only have learned how to tap
the capabilities of the software but have also acquired a good
understanding of their own information needs.

### Authoring and Editing Tools

Nobody guessed, when electronic computers began to be a reality in the early fifties, that within a few years a major application
of these machines would be to help people do things with
words. It is difficult to imagine anything farther from the applications that the inventors of computers had in mind than the
writing, editing, and formatting of documents. Today computer-based word processing is so commonplace that it is taken
pretty much for granted, although much remains to be learned
about exploiting the technology effectively in this domain
(Embley and Nagy 1981; van Dam and Rice 1971).

When computer-based systems were first used to facilitate
document preparation, most user terminals were paper
oriented. Consequently the text-editing tools that were developed were designed to accommodate the constraints of such
terminals. As graphics terminals have become more and more
prevalent, text editors have been evolving so as to take advantage of the greater flexibility that such terminals provide. One
finds examples, however, of editors used with graphics terminals that were really designed for use with paper output devices. Such editors fail to exploit the versatility of the electronic
display.

Graphics terminals offer many obvious advantages over paper-oriented devices. They permit one to bring up considerable
amounts of text quickly and to identify portions of the text that
are to be modified by pointing or by moving a cursor, as opposed to specifying line numbers; and they permit one to display figures and graphs. One of the most important advantages
of such terminals is that the effects of edits are apparent immediately. Deletions really disappear, and insertions appear
where they are supposed to; control characters need not clutter
the display.

Roberts and Moran (1983) have described a procedure for
conducting "standardized evaluations" of text editors. A stan-

dardized evaluation is distinguished from a "specific evaluation": whereas the latter is tailored to a specific purpose or situation, "a procedurally standardized evaluation attempts to address the most fundamental issues and is thus applicable to a variety of editors" (266). The method involves the use of benchmark editing tasks, the measurement of user performance in terms of three dimensions (time required by experts to perform basic editing tasks, the cost of errors for experts, the learning of basic editing tasks by novices), and an analysis of the editor's functionality. To exercise the procedure, Roberts and Moran have identified 212 editing tasks that potentially can be performed by a text editor. Within this large set of tasks they have also identified a small set that they consider to be "pure editing tasks" in that these can be performed by any editor and are the most common tasks typically performed in actual text editing.

The investigators conceive of text editing in terms of a set of core editing tasks. They identify five types of text entities (characters, words, sentences, lines, and paragraphs) and eight kinds of editing functions (add, remove, change, transpose, move, copy, split, and join). Any editing function can be applied to any text entity, excepting certain nonsensical combinations. Discounting the exceptions, this conceptualization recognizes 37 different editing tasks that one might wish to perform.

Roberts and Moran applied their evaluation method to nine text editors: TECO, WYLBUR, EMACS, NLS, BRAVOX, BRAVO, a Wang word processor, STAR, and GYPSY. The nine editors were all used on the same set of editing tasks and evaluated in terms of the same dimensions. The number of mental "chunks" required to represent a specific editing operation or series of operations was found to be a reasonably good indicator of how difficult it was to learn to use an editor. The number of chunks was taken to be an indication of "procedural complexity." The editor requiring the largest number of chunks to represent a given procedure was considered to be the most complex with respect to that procedure.

A major difference that emerged from the study, not surprisingly, was that display-based editing systems were, on the average, about twice as fast as nondisplay systems. No single editor proved to be superior on all dimensions, however—a result that led the investigators to conclude that a choice of editor must involve tradeoffs among the desirable features.

While some editors are undoubtedly more suitable than others for certain uses and users, any one of a fairly large number of them can be a very useful writing tool.

Card, Moran, and Newell (1983) also report the results of comparisons among several text editors (some paper oriented and some display oriented) on several benchmark tasks. The subjects were experienced and regular users of the editors (POET, SOS, TECO, BRAVO, and RCG) that were compared. The tasks included typing and editing letters, assembling text by merging files, and typing tables. Task completion times varied more with different editors than with different subjects; the variation could be largely, but not completely, accounted for by differences in the number of key strokes required to accomplish the tasks (those used in this study were relatively short, about six minutes in the longest case). Which editor was best, in terms of task completion time, depended somewhat on the task; but, as in Roberts and Moran's study, display-oriented editors generally were much better than editors designed for paper-oriented terminals.

Not surprisingly, given the potential market, word-processing programs for PCs are being produced in abundance. No two of them are quite alike, of course, but each represents the developer's guess as to what will be most attractive to users, given the constraints of the hardware systems for which the program was designed. Most of these systems are alleged by their developers to provide all the needed document preparation and editing functions, and to be easy to use. The many ways in which they differ attest to the fact that there are differences of opinion about what is needed and how best to provide it. There is no widely accepted set of criteria in terms of which word-processing systems might be evaluated or compared; however, a look at some reviews of word-processing software products reveals many of the features and factors that users consider to be important (see Grupp 1984; Martin 1984; Perrone 1984; Seybold 1983). Questions that are often raised include the following:

- How much does the user have to learn before the system is usable? (Menu-driven systems generally require less initial learning than do command-driven systems.)
- Are the words that are used to identify actions descriptive of the actions they identify?

- Can commands be abbreviated?
- Does the system permit one to operate on (delete, insert, copy, modify, search, replace, relocate) linguistic units of various sizes: words, sentences, paragraphs, files, arbitrarily delimited blocks of text?
- Is it easy to move the cursor, especially to positions to which one is likely to want to move frequently (such as the top of the page, the last character of the last line typed, the beginning of a new paragraph)?
- Does the system provide for a variety of type fonts (especially boldface and italic) and the mixing of them in the same document?
- Does it provide an adequate set of text formatting capabilities (pagination, idention, right margin justification, centering)?
- Does it give the user easy control over its formatting options? and line lengths?
- Does tabulation treatment permit the alignment of tabular columns on the decimal point?
- Is it easy to set up standard formats for frequent use (as in the generation of form letters)?
- Is it easy to change the format of a document already on file?
- Can pages be broken at logical places?
- Can running heads and feet be generated easily?
- Are footnotes easily accommodated?
- Is hyphenation permitted so as to achieve good line justification?
- Does the text, as it appears on the display screen, look exactly as it will when printed on paper? (Does it contain control characters that will not appear in the final output?)
- Is it possible (with split screen or windowing) to display simultaneously two versions of a block of text?
- Is the user documentation adequate?
- Does the manual have a good (complete, extensively cross-referenced) index? (Very important if the manual is to be used for reference as well as training.)
- Does the system permit scrolling across pages?
- Can it generate a table of contents automatically?

- Does it have a good spelling checker?
- Does it permit the printing of specified segments of a text (as opposed to the entire text)?
- Can text be filed and easily retrieved for future use?
- Is the file system conceptually simple?
- Is the file space adequate?
- Does the system maintain an accessible file directory?
- Can files be merged easily?
- Is it easy to retrieve material (text, data, figures, tables) from other documents for the purpose of inserting them in a text in preparation?
- Does the system help prevent the inadvertent destruction of a working document (by providing an *undo* or *undelete* command, by reminding the user to save a file before printing it, by requesting confirmation before performing an irrevocable action such as overwriting an old file)?
- Does it give useful error messages?
- Does the system have any objectionable constraints (such as page-limited operations, excessive function-initiation time, inability to bypass menus, too severe limitation on file name length)?

Spelling correctors were among the first widely used composition aids. Such a program typically contains an expandable dictionary of words and checks for spelling errors by searching the dictionary for each word in the text. If it finds the word in the dictionary, it passes it as correctly spelled; if it does not find the word, it assumes it was misspelled and changes its spelling to that of the word in the dictionary that most nearly resembles it. Obviously the utility of such a program is limited by the size of its dictionary, inasmuch as real words that are not in the dictionary will be treated as misspellings. Also this technique cannot detect incorrectly spelled words that are themselves words, such as *principal* used instead of *principle, further* for *farther, cite* for *site.*

Digital Equipment Corporation's spelling-correction system (DECspell) for its VAX line has a 70,000-word dictionary and is claimed to be the first such product that can correct multiple spelling errors in the same word. Rather than simply searching the dictionary for the word that is closest in spelling to the nonword, the system constructs a phonetic rendition of the nonword and checks for a real word that sounds like that. Thus *ofishel* should be recognized as *official,* and *chewkago* as *Chicago.*

A variety of text composition, editing, and formatting programs have been developed (Callahan and Grace 1967; Engelbart and English 1968; Magnuson 1966; Reid and Walker 1979), differing considerably in functionality and in the details of their operations. Most text-processing systems today require that text be prepared in two stages: first the user inputs text and edits it with the help of a text editor, and then the file created by this process is run through what is usually called a formatter—a program that formats the file for final copy. The file that is produced during the first stage includes not only the text itself but also commands to the formatter. Thus if one wants to have an equation appear in the text, the equation will, in most cases, be expressed as a linear sequence of symbols and words. Most of the words are commands to the formatter and will not appear in the finished copy; they might include, for example, "equals," "sup" (superscript), "sub" (subscript), "over" (indicating a division line), and names of Greek letters. This is a concession to the constraints of many existing terminals, especially hardcopy terminals. Unfortunately the effects of these constraints are also found in software that operates on video terminals. Ideally one would like the text one is working on to appear on the display exactly as it will appear on the finished product. Systems are beginning to be available for which this is more or less true, and one can expect it to become the norm in time.

Publishers of books and journals all have their own style conventions. Programs exist that can take a reference list (or a whole manuscript for that matter) prepared according to one set of conventions and automatically change it to make it consistent with another set. Bahil (1983) has described one such system in use at Carnegie Mellon University. The system contains a database of references. When a paper is cited in a text in preparation, a formatter program enters the citation and the reference in the format that is appropriate for the journal for which the paper is being prepared. The system accommodates 20 different formats.

Most electronic mail systems provide not only the capacity to send and receive messages but tools to help compose, edit, and annotate them as well. Again, these systems differ both in functionality and in the ways in which that functionality is made available to the user. One recently announced approach to providing assistance in letter composition involves supplying a large number of categorized examples of letters that one can use as models while composing one's own letter. This is the Einstein Letter Series, designed to be used with the Einstein Writer word-processing package (Perrone 1984). The sample letters are organized in several major categories, which are broken down into a number of subcategories, each containing several examples. A "congratulations" category includes letters appropriate for a birthday, an anniversary, an engagement, a marriage, and the birth of a child. The Einstein Writer features a split-screen presentation, so users can select, via menu, examples of the types of letter they wish to compose, and have one displayed on one half the screen while working on a composition on the other half.

One of the most extensive packages of writing aids is provided by the Writer's Workbench, a program developed at the Bell Laboratories (Frase 1983; MacDonald 1983; MacDonald, Frase, Gingrich, and Keenan 1982). It is capable of flagging some types of grammatical errors, as well as redundant words or phrases, cliches, overused expressions, and excessive use of the passive voice. It checks spelling and punctuation and has some ability to critique written material in English. It can, for example, provide editorial comments on text abstractness and estimations of overall readability in terms of reading grade levels.

The Writer's Workbench makes use of programs included within the UNIX operating system, also developed at Bell Laboratories. Of course, the program reflects its developers' views about what constitutes good writing style. Some of the thinking and attitudes that lie behind its design have been spelled out by MacDonald (1983). The Writer's Workbench software does not automatically make stylistic changes in a text, but provides the writer with suggestions that he may or may not choose to follow. Some effort has been made to assess the effectiveness of the Writer's Workbench software as perceived by users (Gingrich 1983). The results indicate that many writers do find it helpful. There is little objective information, however, regarding what effects the use of such tools has on the quality of the writing that is done with their help.

Researchers at IBM's Thomas J. Watson Research Center are

developing a software system called EPISTLE (for Evaluation, Preparation and Interpretation System for Text and Language Entities) (Heidorn, Jensen, Miller, Byrd, and Chodorow 1982; Miller 1982; Miller, Heidorn, and Jensen 1981). The long-term goal in this case is to provide office workers, and especially middle-management personnel, a variety of tools to assist in their interaction with natural-language texts. A short-range objective is to facilitate the processing and preparation of business correspondence. Operations the system is intended to be able to perform on incoming letters include "synopsizing letter contents, highlighting portions known to be of individual interest, and automatically generating document index terms based on conceptual or thematic characteristics rather than 'keywords'" (Miller, Heidorn, and Jensen 1981, 649).

One function of the EPISTLE system is to give suggestions for the improvement of outgoing letters. In the initial versions the suggestions focus on grammatical errors and violations of "good style." As of 1982, EPISTLE was able to diagnose five classes of grammatical errors: subject-verb disagreements, wrong pronoun cases, noun-modifier disagreements, nonstandard verb forms, and nonparallel structures (Heidorn, Jensen, Miller, Byrd, and Chodorow 1982). To detect such errors, it first attempts to parse a sentence using its full set of syntax rules. If the parse fails, a second attempt is made with one or more of the grammatical restrictions relaxed. If the sentence parses this time, the program may have enough information to specify and locate the error.

Most of the stylistic problems that EPISTLE is programmed to detect can be found by syntactic analysis and do not depend on semantics. An exception is the detection of certain "overworked, outdated, stilted, unnecessarily lengthy, excessively formal or obscure" phrases. Miller, Heidorn, and Jensen estimate that there are perhaps 600 to 1,200 such phrases, possibly even fewer if highly similar patterns (such as phrases that only differ in that the noun is singular in one and plural in the other) are combined. They believe that EPISTLE is capable of recognizing most of the major phrases of this kind. Other types of stylistic problems that EPISTLE can detect include the use of "business-ese," jargonistic phrases, too many qualifiers of a noun, overly lengthy sentences, double negatives, and overuse of the passive voice.

An interesting feature of EPISTLE's style critic, which was

developed by reviewing a number of tutorial books on writing, and especially letter writing, is that its sensitivity to various stylistic problems can be adjusted by moving thresholds for certain characteristics up or down. One could, for example, make the system more or less sensitive to the number of words intervening between the subject and its verb by adjusting the threshold for the detector of this particular variable. Presumably, therefore, EPISTLE could be fine-tuned to be especially sensitive to the stylistic characteristics that a specific writer wished to avoid. Longer-range objectives of the group developing EPISTLE include the ability to analyze passages semantically as well as syntactically, to represent the meanings of sentences as series of related propositions, and to evaluate sequences of propositions for consistency and continuity.

The EPISTLE system is built upon a general-purpose natural-language processing system called NLP (Heidorn 1972, 1975), which includes a dictionary, an input decoder (which translates text into internal representations consisting of lists of attribute-value pairs), and an output encoder (which translates internal representations into natural-language output) (Miller, Heidorn, and Jensen 1981). The decoding rules by which input is parsed are known collectively as an *augmented phrase structure grammar* (APSG) and, in distinction to some other approaches to grammatical analysis, are essentially context free (Heidorn 1975).

The EPISTLE-user interface is described in Heidorn, Jensen, Miller, Byrd, and Chodorow (1982). It is similar in many respects to a variety of graphics-oriented text editors. It uses windowing and provides the user with action options via a menu with which he interacts through a light pen. A suspected grammatical error is flagged on the display by having the questionable segment of text boxed with a red background. A brief description of the error and a suggested correction are shown in a "fix" window immediately below the error. This window has a background of a different color than the color that boxes the error itself. When a suspected error has been identified and a fix suggested, the user can take any of four actions: (1) retain the original wording; (2) request additional information about the error (in this case the system provides the additional information in a "help" window that is superimposed on some other portion of the text); (3) accept the system's suggested fix; or (4) replace the error with a correction of his own.

A question that arises with respect to automated critics of this kind is whether they are more likely to foster originality and creativity in writing or to inhibit it. To the extent that they force letters to correspond to some standard or "ideal" pattern, one might guess that they would inhibit originality. On the other hand, to the extent that they identify clichés and stilted or overworked phrases, one might expect them to facilitate it. Like the Writer's Workbench, EPISTLE makes no changes automatically, but simply identifies problems and suggests possible changes, so the writer has the final say on how a document is worded. Any advice giver or suggestion maker that works on this principle need not be foolproof, inasmuch as the user is free to accept or reject the advice it gives.

Miller (1978) has argued that the development of better editing facilities provides the greatest opportunity for enhancing computer use. Editing is what people do more than anything else when using a computer: in one study involving 3,712 terminal hours of use of a time-sharing system by 375 users at the IBM Watson Research Center, over three-fourths of all the commands issued to the system were text-editing commands (Boies 1974). Other areas that Miller sees as having great potential for facilitating computer use are file manipulation and information-partitioning displays. What is needed in the former case, he suggests, is software that will allow the user to perform such file operations as the following in a natural way: naming, describing, storing, copying, distributing, annotating, and retrieving. With respect to information-partitioning displays, what is needed is the development of techniques that will make it easier to display different kinds of information simultaneously and will permit remotely located users to share common work spaces.

One wants to work toward an eventual document preparation and processing system that has a broad range of capabilities including the following:

- speech to displayed text;
- a multimode control interface, including keyboard, speech, and an analog positioning or pointing device such as a mouse;
- optical character recognition capability, allowing input of printed material without retyping;
- the ability to input, store, and transmit nontextual graphics;
- a set of document preparation and editing tools;

- multimedia mail;
- customizable information storage, retrieval, and management tools;
- "what you see is what you get" displays;
- high-speed printer capability;
- the ability to make color prints or slides of displays.

Word-processing systems and document preparation aids are changing the way papers get written. The ability to edit documents by deleting or inserting arbitrary segments of texts, by rearranging paragraphs, by merging fragments from several sources, is making it possible to evolve documents in a way that is not practical without the use of computer-based tools. It remains to be seen what effect the use of these tools will have on the quality of the writing that is done with them. It is worth at least considering the possibility that in some cases it may be negative. One surface-level effect is the appearance of certain types of errors that are unlikely to occur with conventional typing. These include the repetition of words or word sequences; the garbling of a segment of text as a result of the insertion of a word sequence in an inappropriate place; the dislocation of entire paragraphs; the appearance in the text of a reminder to oneself to fix or add something.

The more serious question is whether the quality of the writing will be affected at a deeper level. Novelist/columnist Ernest Herbert estimates that by using a PC he has quadrupled his writing speed but notes that the quality of his first drafts has declined. Because it is so easy to produce text with a PC, he anticipates that the number of people writing bad novels is likely to increase (Latermore 1984). Another professional writer puts it this way: "I have the impression we are heading toward a future filled with the emperor's new words, where word processing cranks fast-food prose, becoming to writing what xerography has become to the office memo: a generator of millions of copies of contentless words assembled for appearance's sake—rarely read, much less reflected upon" (Sandberg-Diment 1984).

When people first started to write on papyrus instead of on wet clay, one might have wondered whether it had suddenly become so easy to put words down that scribes would become careless about their art. One might have had a similar worry when typewriters came along and made it that much easier to put ideas in visible form. Word processors and other computer-based tools for facilitating the preparation and editing of documents simplify further the writing task, and again one wonders what their overall effect on the quality of writing will be. The concern that they will facilitate the production of writing of less than superb quality is not baseless, in my view. On the other hand, they should facilitate high-quality writing as well. Moreover the tools that are being developed have the potential, I believe, to help writers who wish to do so to improve the quality of their writing in substantive ways.

### Computer-Aided X

One reads of computer-aided instruction, computer-aided decision making, computer-aided design, computer-aided manufacturing, and computer-aided activities of various other sorts. Sometimes the event has not matched original expectations. This has certainly been true, for example, of computer-aided instruction. Computers have not yet had the effect on education and training that many of the projections made fifteen or twenty years ago led us to expect. It is possible, of course, that the original projections were overly optimistic only with respect to timing. We have already noted the rapidly increasing presence of computers in U.S. public schools; we should note too the growing involvement of computers in industry training programs. We will come back to the topic of computers and education in a later chapter.

The idea that computer-based systems might be developed for the express purpose of assisting the process of decision making in various contexts has been promoted by numerous writers (Briggs and Schum 1965; Edwards 1965; Licklider 1960; Shuford 1965; Yntema and Klem 1965; Yntema and Torgerson 1961). Examples of experimental systems that are intended to provide some degree of decision aiding include AESOP (Bennett, Haines, and Summers 1965; Summers and Bennett 1967; Doughty 1967), RITA (Anderson and Gillogy 1976; Waterman and Jenkins 1977), and ROSIE (Fain, Hayes-Roth, Sowizral, and Waterman 1982; Waterman, Anderson, Hayes-Roth, Klahr, Martin, and Rosenschein 1979). Much of the work on expert systems (about which more later) involves computer-aided decision making in a sense.

Some of the decision-aiding programs that have been developed were designed for use in essentially any decision-making situation. That is to say, they are intended to be useful independently of the specific decision context. Newsted and Wynne (1976), for example, implemented a system to help decision makers retrieve from their own memories the information that is relevant to any decision they must make and to organize it in a useful way. Most experimental decision-aiding systems, however, have been focused on specific application areas, such as the military (Blackledge 1974; Bowen, Feehrer, Nickerson, and Triggs 1975; Freedy, Davis, Steeb, Samet, and Gardiner 1976; Freedy and Weltman 1974; Hanes and Gebhard 1966; Kibler, Watson, Kelly, and Phelps 1978; Levit, Heaton, and Alden 1975), air traffic control (Whitfield 1976; Whitfield and Stammers 1976), business (McCosh and Scott-Morton 1978; Rockart and Treacy 1982), or medicine and health (Kush 1981; Meindl 1982; see also chapter 15).

Decision aiding is an area that is rich with research possibilities. Moreover, although it is perhaps natural to think first of problem areas such as the military, medicine, and business for applying decision-aiding tools, all of us make decisions, and presumably we might make better ones if given some appropriate help. One possible general objective for interactive decision-aiding systems would be to counterbalance some of the more common reasoning errors that people make. The tendency to seek evidence that is consistent with a favored hypothesis rather than to look for information that would refute it is fairly well documented, for example, as are a number of other similar reasoning limitations and deficiencies (Nickerson, Perkins, and Smith 1985; Nisbett and Ross 1980; Tversky and Kahneman 1974). Decision-aiding systems designed to compensate for such weaknesses should be useful.

While interest in computer-aided instruction and computer-aided decision making remains as keen as ever, relatively more attention has been given in recent years to computer-aided design and computer-aided manufacturing. Computer-aided design (CAD) involves the use of software (typically graphics oriented) to facilitate the design process. CAD systems are now in widespread use in industries with products ranging from electronic circuits to automobiles and airplanes. Computer-aided manufacturing (CAM) carries the involvement of computers beyond the design process into the actual building of prototypes and products. Given the existence of computer-aided design tools and computer-aided manufacturing tools, it would be surprising indeed if an effort were not made to link them in such a way that a designer, having developed a design that satisfied criteria the industry had imposed before attempting prototype production, could thereupon pass the design off to a CAM system and have the prototype produced automatically (Hudson 1982). It is easy to oversimplify this problem. Tooling to produce a complicated product may be a major undertaking. Some movement in the direction of integrated CAD-CAM systems, however, is certainly possible, given the state of the art, and is in fact occurring. Noting that sales of CAD-CAM systems increased from $99 million in 1977 to $1.2 billion in 1983—a 52 percent per year increase—Predicasts (1984) has predicted that the CAD-CAM market will reach $12 billion by 1995.

The problem of assuring that what has been learned by human-factors research will be applied to the design of CAD-CAM systems is one that has not yet been solved. Begg (1984), for example, concludes,

It seems the lessons of ten years of research into dialogue design for interactive computer use have not always been learned by those who designed CAD systems. In the author's experience the most frequent complaints made by CAD users have been about menu systems, on-line documentation, and error messages, where the application of a few simple principles of software design were all that was needed to render the system far more user-friendly. (34)

Mention of computer-aided design prompts the observation that there is an interesting positive feedback loop in computer technology. Today's computers are essential to the design and development of tomorrow's computers. Whitted (1982) refers to this as a case of technological "tail chasing" and notes that it is illustrated nicely by the design and use of VLSI circuits. The inexpensive processing power provided by VLSI is essential to further improvement in computer graphics, and, conversely, advances in VLSI design are facilitated by the use of graphical design aids.

The process of designing an integrated circuit can be broken roughly into four components: (1) logic design (specification of arithmetic and logic operations, data registers, control, instruction decoders, timing), (2) circuit design (translation of the logic into circuits with the appropriate electronic components), (3) layout design (positioning of the components so as to optimize intercomponent communication paths), and (4) engineering drawing (production of precise scale drawings that can serve as fabrication blueprints). Given the very large number of components on modern chips, this is an extremely complex process with the opportunity for creativity at the first three stages and for error at all four. Moreover, the complexity of the process increases, probably nonlinearly, with the packing density of the circuit components. The latter has gone from less than ten active-element groups (logic gates or memory cells) per chip in 1960 to a few thousand in 1970, to several tens of thousands at the beginning of the present decade. It is expected to go to over a million before 1990.

The problem of managing this kind of complexity represents a significant challenge to the microelectronics industry. One approach to the problem has been the use of predesigned structures, such as programmable logic arrays, that simplify the designer's task at the expense of making suboptimal use of chip space. Another has been the use of generic parameterized circuit subunits that are hardware analogs of computer subroutines. This approach also buys simplification of the design process at the cost of some efficiency in the final circuit layout. Still another approach, and perhaps the most promising one for the long term, has been to try to develop computer-based design aids (Mead and Conway 1980; Stallman and Sussman 1976; Sussman and Stallman 1975). The first such aids automatically checked layouts for design rule violations, such as insufficient spacing between certain types of components. More complex aids are now being developed to facilitate other aspects of the design process. Much remains to be done along these lines, and the ever increasing circuit complexity that decreasing component sizes are making possible will continue to provide motivation for the development of increasingly powerful aids.

We may expect that computer aiding will be applied to an ever increasing range of activities in the future. It seems likely also, however, that the tendency to refer to any activity X that is aided by computer involvement as computer-aided X will diminish; as the use of computers becomes more pervasive, their involvement in any particular activity will cease to be remarkable.

# 9

## Communication and Information Services

Information technology should increase our ability to acquire, use, and transmit information. That is its proper role. There are many ways in which it is likely to play this role; we consider here a few of them. The new communication and information services considered illustrate the point made in chapter 2 that the distinction between communication and computation is becoming increasingly blurred. All of these services are communication services in that they involve the transmission of information over communication channels of one sort or another, but they also depend very definitely on computing resources of various types.

### Computer-Based Message Technology

In any discussion of computer-based information systems, it is important to distinguish between person-computer communication and communication between people via computer-based systems. A primary purpose of many of the systems that have been developed is to facilitate interperson communication. Facilitation in this context includes both quantitative and qualitative dimensions. That is, one can increase access of people to other people with whom they wish to communicate and one can also facilitate the communication process itself by making it more effective and less error prone.

The methods that have been devised over the millennia to get information from one place to another are many, ingenious, and remarkably diverse. Cyrus the Great of Persia, we are told, established in the sixth century B.C. a system of signal towers radiating in several directions from his capital. Men stationed on these towers relayed messages by shouting from one tower to the next (Flanagan 1972). Other time-honored, and perhaps less taxing, methods of relaying information over relatively long distances have included the use of drums, smoke signals, whistling, and various means of transporting the media on which messages are stored. It is easy to forget that such methods sufficed until the telegraph, telephone, and radio were developed within the last 150 years.

The application of computer technology to the storage, processing, and transmission of messages, which will be referred to here as computer-mediated message technology, could have profound and far-ranging effects on our lives—in particular, on the ways in which we communicate and handle information. The basic innovative idea in this technology is that of the interposition of a computer system between the sender and the receiver of a message, and the provision to both sender and receiver of a set of computer-based tools for composing, editing, modifying, forwarding, storing, retrieving, and otherwise manipulating messages.

Computer-based message systems are not simply electronic versions of the postal system that we have had for many years. They are qualitatively different from anything we have known in the past. Sending mail by rail and later by air did not represent qualitative changes from the pony express. These advances reduced the time a letter spent in transit but did not otherwise change the method of correspondence. In all of these cases a message was written on paper, placed in a parcel or envelope, and carried from one point to another. The method by which it was conveyed from point to point had no implications for the way in which it was prepared by its sender or for what was done with it by its recipient.

The basic functions of an electronic message system are to send and receive mail, but these are not the only ones. Most systems today provide text composition and editing aids, selective broadcast capabilities (automatic distribution of the same message to all the addressees on a list), the ability to annotate and forward messages, and a variety of message-management capabilities (including the ability to search a message file on the basis of such properties of messages as sender, receiver, topic, and date). In short, computer-based message technology represents a radically new way of not only sending and receiving messages but also preparing, storing, retrieving, and using them. It will have implications both for our communication with other people and for our communication with ourselves Electronic mail services are expected to add a variety of capabilities that will make them increasingly attractive to potential users. These capabilities include access to news wire services, stock quotations, electronic funds transfer, and electronic sales transactions.

Computer-based message technology illustrates quite nicely that not all significant technological developments are planned. It came into existence when a programmer (I believe it was Ray Tomlinson of Bolt Beranek and Newman) wrote two programs, called SNDMSG and READMAIL, that would permit the sending and receiving of messages on the ARPANET. These programs quickly became widely used, and the potential that computer networks had to assist interperson communication became apparent. Soon efforts were under way to develop more formal computer-based message systems with considerable functionality and flexibility. One of the first to be developed was the Hermes system, which was developed over several years at Bolt Beranek and Newman under ARPA sponsorship (Dodds 1983; Mooers 1982; Myer 1980; Myer and Mooers 1976). It is a versatile system with considerable functionality, but it is also complex and few of its users are aware of its full capability. Mooers (1983) has described the kinds of changes that were made to Hermes over the years in the interest of making it more user-friendly. She notes the importance of designing programs so they will be easy to change, given that even when great care is exercised in the initial design process, it may not be possible to anticipate what users will like or dislike about a system. Largely as a consequence of the development of computer-based message systems, computer networks, which were intended to make possible the sharing of computer resources and thereby to facilitate computing, have come to play an important role in interperson communication.

Lederberg (1978) has coined the term *eugraphy* to denote the emerging form of communication that we have referred to here as computer-mediated message technology. His description of the advantages of a *eugram,* a message or text that is composed, stored, and transmitted electronically, is worth quoting at length.

The EUGRAM . . . has all the advantages of digital storage and accessibility to archiving, sorting, and searching mechanisms that are far easier to implement, and require far less bandwidth than do voice messages. The EUGRAM itself can be composed quickly with a text editor on the user display, where it is readily rehearsed, corrected and reedited before being transmitted. The same EUGRAM can be fanned out simultaneously to a large number of recipients, or it can be revised and perfected through several versions with similar broadcast, or with selective distribution. From the receiver's perspective, he has the advantage of a literate spatially oriented medium. In contrast to the time-fluent telephone, radio, or TV, he has the option of perusing his mail at his own pace, or interruption, backtracking, and crosschecking the text, even of marking it for reexamination and further rumination. He retains mastery of the use of his own time, and can coordinate attention to a coherently chosen set of tasks. He is liberated from the tyranny of synchronizing his own mental processes to those of the external actor. This freedom of course reduces the impact of that actor, just in proportion to the responsible autonomy it returns to the reader. In framing responses, entire messages or selected extracts together with added comments can be forwarded to others, or returned to the sender—lending focus to a "discussion" and providing unambiguous texts for the development of a consensus. EUGRAMs can be filed and retrieved efficiently, or transcribed into hard copy as required. Text editors may be embellished with elaborate formatting aids, spelling correctors, even an on-line thesaurus to aid in composition. When quantitative calculations are in question, numbers can be mechanically copied directly from program outputs, avoiding pestiferous typographical errors. The same computer is likely to be the user's research tool and give access to shared databases: the EUGRAMs can then refer to common files by names that are themselves machinable. The user will also have access to other conveniences, such as desk-calculator-like programs for the checking of figures. He can even track the growing size of a EUGRAM-script (like this one) to be sure it fits into the assigned space. These word-processing capabilities can of course be consummated with hard copy sent through mails, but with some additional effort, and the degradation of the machinability of the product at the other end. (1315)

Drawing upon several studies of computer-based message systems (among them Carlisle 1976; Turoff and Hiltz 1977; Uhlig 1977), Bair (1978) lists the following as some of the benefits of this technology:

1. Permanent, searchable, stored record. A permanent, automatically stored copy of all communications is available for retrieval by keyword, sender, date, etc.

2. No simultaneous activity necessary. The sender and receiver(s) do not need to interact simultaneously. Messages may be originated, sent, received, and read at different times (asynchronously).

3. No meeting schedule necessary. Topics can be introduced and responded to by all communicators at their convenience and over an extended period of time.

4. Optimum time for composing, reading, and responding may be selected. This may be done when the communicators are best able to respond, having the needed resources and disposition.

5. Physical collocation not necessary. This permits interactions in situations in which telephoning is too costly and the mail service is too slow and/or unreliable.

6. No interruptions of meetings and conversations.

7. One action for general information distribution. Simple communications, such as an announcement, require only one action.

8. Fast delivery at low costs. Mail delivery time for written communications is reduced from days to seconds, and the need for phone contact is minimized.

9. Automatic distribution. Distribution requires only specifications of recipients' names for action or information, or the name of a recipient list.

10. Automatic headers. Data about messages (e.g., date, length, and sender) are included in automatic formatting. (735)

Uhlig (1977) describes how computer-based message technology was introduced to managers and their secretaries at the U.S. Army Materiel Development and Readiness Command beginning in 1974. The system was the Office-1 facility developed by the Stanford Research Institute. The initial user group of seven data-processing-oriented managers was expanded over a period of about two years to approximately 200 users. Uhlig notes that the initial reaction from top management to the new capability was positive. He notes too that one of the effects of introduction of the computer-based message system was increased communication among the users. Among the other benefits he reports are these: improved communication between people in widely separated time zones; reduced turnaround time on decisions; ability of the receiver of messages to control when he receives them; ability to send or receive messages independently of location; being "connected" to the office no matter where one is; ability to send multiple copies of a message in lieu of making several phone calls; ability to coordinate actions among people independently of location. Uhlig sees the "connectivity" of users as a particularly important accomplishment of the computer-based message system.

If I were to try to capture the advantages of message technology in a single phrase, the phrase would be "increased accessibility." What it accomplishes, or promises to accomplish, is to make it much easier to get information—messages—from place to place and at relatively low cost. A message sender's access to an addressee is independent of the addressee's location at any given time. To send a message to a specific individual, one always uses the same fixed address within the system. Conversely, the intended recipients of messages have access to them more or less instantaneously and independently of where they are—assuming, of course, that they have terminals and the means to connect to a computer network. The point is perhaps best made by illustration.

*Example 1:* You are the director of a project team of ten individuals. Some information has just reached you that has some serious implications for the future of the project, and it is imperative that you pass this information on to the members of the project team as quickly as possible. Three of the team members are within shouting distance; three others are reachable by internal phone; two are on a field trip and also presumably reachable by phone, albeit not conveniently; one is in transit on a business trip, and his precise location will be uncertain for a couple of days; and one is working at home. You type your message into the computer-based mail system, address it to "Project Team," and rest assured that each of the group will have read the message at latest by some time on the following day, assuming that all team members habitually read their computer mail at least once a day.

*Example 2:* Your office is in Cambridge, Massachusetts. You are traveling in Los Angeles, California, and an electronic message that you received from a business colleague some six months ago turns out unexpectedly to be relevant to the purpose of your trip. You cannot remember some important details of the message, so you get your computer terminal out of your briefcase, dial up the local port to a network, retrieve the message that is stored in your computer-based mail file in Cambridge, and have it typed out on the terminal in Los Angeles. Assuming there is a telephone readily accessible, the time required for this process is comparable to what would be required to retrieve the letter manually, if you were located in your office and had a conventional filing system.

*Example 3:* You are a research scientist at a university in the Midwest, collaborating on the writing of a paper with two colleagues, one of whom is a professor at a university in the North-east and the other of whom is with a government laboratory on the West Coast. Each of you has taken the responsibility of developing one section of the paper and of playing critic with respect to the other sections. You are sharing the responsibility of integrating the pieces and assuring that they are mutually consistent and collectively coherent. As each of you adds some material to your own section, or wishes to make a suggestion with respect to a colleague's section or the overall organization, you enter the addition or suggestion into a file that is im-mediately available to your colleagues.

*Example 4:* You are the manager of a sales force and wish to have daily reports from your sales staff, which is scattered over a large geographical area. Each of your staff carries a portable terminal that can be used to transmit data over standard phone lines. At the end of each day, or at appropriate times during the day, each member of the staff enters the desired informa-tion concerning the day's activities into his terminal and trans-mits it to your message file; every morning upon arriving at work, you have an up-to-date report of what has occurred the preceding day. If you wish to get messages to your staff, you can transmit them via the same system and be assured that any messages you send will be read before the day is over.

*Example 5:* You are deaf, and consequently the telephone is of no use to you. When you want to communicate with a friend, you type a message into your terminal. If your friend happens to be home at the time, you can carry on a real-time conversa-tion through the system. If he is not at home, your message will be in his "mailbox" to be read the next time he uses his terminal.

Of the various ways in which people communicate, the two that are most likely to be affected by the emergence of com-puter-mediated message systems are the telephone and the postal service. The impact on the latter will be very great in the long run. More and more of the information that is now trans-ported will be transmitted digitally via computer networks. The impact on telephone systems will probably not be so noticeable, at least in the near future, although for some applications elec-tronic mail may even now be more economical than the use of a phone. There undoubtedly will always be a demand for real-time voice connections to permit person-to-person conversations. Even this form of communication will probably make use of digital transmission, however, and the ability to use voice as both input and output for a computer-mediated message system is already a reality.

It should be noted, however, that voice mail systems, which permit the sending and storing of digitized voice messages, were introduced in the early 1980s but to date have not been well received. Seaman (1983), who lists 26 vendors of voice mail systems, says bluntly that "the technology has been a flop" (188). There are reports of successful experiences with voice mail systems in a few companies, but in general these systems have not lived up to the expectations of the companies that developed and marketed them. There is little but speculation by way of an explanation. Among the possible factors:

- A general reluctance that people have to talk to a device or machine.

- The lack of convenient editing facilities. It is not possible, for example, to prepare a letter by dictating a rough copy and then cleaning it up in an editing pass.

- Lack of a visual output. People may find it difficult to respond effectively to lengthy messages that they are unable to see.

The second and third of these factors will become nonfactors when speech recognition technology has advanced to the point of making it possible for a system to accept speech input and then show that input in written form on a display. The voice mail systems that have been developed to date cannot do this; they may store a digital representation of the speech waveform, but they do not have any recognition capability.

Electronic mail systems or services are now being offered commercially by at least 65 vendors (Kull 1983), and it seems inevitable that their use will grow. Among the major vendors are the Computer Corporation of America (Comet), Tymeshare (OnTyme), and GTE Telenet (Telemail). Initially many companies limit the use of electronic mail to internal communication via company-owned networks. We can expect that in time its use for intercompany communication will grow. The Predicasts Research Group has estimated the total number of electronic-mail messages sent in 1980 at 930 million and has projected the annual figure by 1995 to be approximately 11 billion (Pomerantz 1982). The Yankee Group has predicted a roughly tenfold increase in the number of electronic mailboxes for users of computer-based message systems (from about 130,000 to over 1.5 million) over the period 1981–1986 (Burstyn 1983). International Resource Development gives estimates that are reasonably close to those of the Yankee Group over the same period; it extrapolates farther into the future with an estimate of 21 million users of electronic mail (about 50 percent of all white-collar workers) by 1991 (Pomerantz 1982).

The use of this technology by the general public is subject to a "critical-mass" effect that will work against it initially but is likely to stimulate its expansion eventually. It does one little good to be a user of an electronic mail system if few of the people with whom one would like to communicate are users. Conversely, the expansion of a community of users has a positive-feedback effect: as more people become users of the technology, the incentive for the rest to become users increases by virtue of the growing number of people with whom one is able to communicate by this means. We should note too that the situation in which a message system is used by *everyone* in a community of communicators (office, business organization) is qualitatively different from one in which the system is used by only some of the members of that community. As long as there are some people who are not accessible via the computer-based system, one is forced to use conventional methods in at least some cases, and many of the advantages or potential advantages of electronic systems are lost.

In thinking about the future of this technology, two facts should be borne in mind. First, the term "electronic mail" does not really capture the essence of it. The technology offers much more than simply the ability to send a letter electronically rather than by more conventional means. As is clear from Lederberg's description, it provides the user with a set of electronic tools that help in the management of messages and information more generally. Second, while most of the computer-mediated message systems currently in use provide only the ability to transmit text, systems are being developed that will permit the mixing of text with video (facsimile) and speech:

Imagine if paper documents had sound tracks like films. Whenever someone started reading a document, a voice would speak up adding further explanation. In addition, instead of scribbling comments in the margins, the reader could touch a spot on the paper and speak his comments, which would then "stick" to the paper and be heard by anyone who touched that spot in the future. (Pollack 1984)

Thus begins a recent *New York Times* article on multimedia message systems. The article goes on to describe the concept of a multimedia document, which in this context is a document that permits the use of voice and images, such as graphs or pictures, as well as text (Forsdick and Thomas 1982; Thomas, Forsdick, Crowley, Robertson, Schaaf, Tomlinson, and Travers 1985). The addition of these capabilities will persumably make the technology more versatile and of greater interest to prospective users.

Are people likely to say things to other people via a computer-based message system that they would be unlikely to say either face to face, over the phone, or in a letter? Are they likely to express themselves in a different way? One need not accept without reservation McLuhan's dictum that the medium is the message in order to believe that the medium may condition the message to some degree. Clearly postal correspondence differs from phone conversations in a variety of ways. At the syntactic level, written communication differs considerably from oral communication in general; and one suspects there may be semantic differences as well. People may find some things easier to say by letter than over the phone or face to face. Conversely, it is probably easier to be impulsive in a conversation that is taking place in real time than in a message that one is composing at one's leisure. Perhaps messages that are exchanged via computer-based message systems will take on their own unique characteristics. Personal experience suggests that some users of these systems will develop a much less formal style—reflected in the ways in which they do or do not use capital letters, punctuation, and other grammatical conventions—than they use when writing letters. These messages, then, would seem to fall somewhere between oral conversations and postal correspondence with respect to formality. Certainly many exchanges have a distinctly conversational flavor. Reder and Conklin (1984) suggest that as the telephone extended the possibilities for conversations over space, electronic mail extends the possibilities for conversations over both space and time. They note too that electronic mail represents a new genre of communication, and one whose social functions overlap partially but not completely with those of other channels.

Uhlig (1977) has noted that computer-based message systems do not preclude misunderstandings and may, in some instances, foster them. Because the communication is in writing, the nuances of speech and the nonverbal cues of face-to-face communication are lost. Because a user can send a message immediately upon composing it, one also loses the safeguard that the usual delay in posting a letter provides against hasty responses to incoming messages that might have been misinterpreted on a quick reading. On the other hand, it has been suggested that electronic mail systems tend to break down the barriers to communicate and that their immediacy reduces the chances for misunderstanding (Spinrad 1982). Some investigators believe that computer-mediated communication systems tend to promote equality among users, so that status and control hierarchies tend to be less stable than when the communication occurs face to face (Hiltz, Johnson, Aronovitch, and Turoff 1980; Vallee, Johansen, Randolph, and Hastings 1974). Uhlig asserts the need for an etiquette of computer-based message systems, which may be somewhat different from that appropriate to voice conversations or letters.

Most of the speculation about computer-based message technology has focused on what are expected to be the beneficial effects of its use. This is perhaps not surprising, inasmuch as most of the writing about the technology has been done by enthusiasts. And indeed the effects may well be largely beneficial. It is probably a good idea, however, to consider the possibility of undesirable effects as well. Moreover, in some cases, whether a particular effect is beneficial may take some time to tell. One effect computer-based message technology has had on some users is a great increase in the quantity of mail they receive. Accessibility can be a mixed blessing: while it is convenient to have access to many other people, it is less clear that an individual's productivity or quality of life is enhanced by virtue of a great many people having easy access to him.

### Computer-Mediated Conferencing

The possibility of using computers to mediate conferences among people who are in different locations has been of interest for several years and has stimulated some research (Bavelas, Belden, Glenn, Orlansky, Schwartz, and Sinaiko 1963; Carlisle 1975; Forgie, Feehrer, and Weene 1979; Hiltz and Turoff 1978; Johansen, Vallee, and Spangler 1979; Kerr and Hiltz 1982; Short, Williams, and Christie 1976). Teleconferencing systems are not yet widely utilized, and they do not seem to threaten the demise of the conventional face-to-face confer- ence in the very near future. There is some indication, how- ever, that hotels are beginning to give serious thought to teleconferencing facilities and to worry not only about the elec- tronic equipment involved but about the special requirements of teleconference rooms with respect to acoustics, lighting, and the positioning of cameras. The problems of acoustics and lighting are greater than in the case of conventional meeting rooms, inasmuch as designers must be concerned with the ef- fects of these variables both on the people in the room and on the quality of the audio and visual signals received by the re- motely located conference participants (Kristal 1983).

The idea of teleconferencing centers in hotels may seem in- congruous, considering that one of the main reasons for inter- est in teleconferencing has been the assumption that it would reduce the need for travel. But one of the many forms that teleconferencing can take is that of collecting the conferees in several well-equipped centers that can be linked electronically, rather than having them all travel to the same place. This ap- proach does not do away with the need for travel altogether, but it could substantially decrease the average distance traveled per conference. Teleconferencing has other attractions as well: scheduling, for example, would presumably be facilitated, in- asmuch as busy people would find it easier to commit to a con- ference that did not require distant travel than to commit to one that did.

What is probably needed to make teleconferencing a pre- ferred mode of holding meetings, however, is the development of tools that would assist group problem solving even if the conferees were in the same room. Such tools might include shared, dynamic, "smart" blackboards; information storage and retrieval procedures; group decision aids; polling soft- ware; and modeling and simulation programs. Systems with such facilities could, in theory at least, support not only short- term conferences but also long-term collaborative efforts by remotely coupled project groups.

In spite of the research that has been done on the subject, it is still not known how to do teleconferencing well. Many of the original questions remain unanswered, such as how important video is, how to control the dialog, and whether participants will find this method of interacting as acceptable as conven- tional face-to-face meetings. Most of the research and discussion about teleconferencing has concentrated on making it cost-effective and on devising systems that will permit conferees to communicate as effectively as they do in a face-to-face meeting; although some investigators believe that teleconferencing improves decision making by minimizing social influences and increasing the likelihood that participants will focus on the task and not on irrelevancies (Hiltz and Turoff 1978; Johansen, Vallee, and Collins 1978). Whether decision making is improved is likely to depend on the specifics of the system's design (Murrell 1983).

The assumption that people travel to conferences only because there is no effective alternative, and would really prefer to stay at home, is open to debate. People attend conferences for many reasons, not the least of which is to get away temporarily from their usual places of work. Moreover, as Falk (1975) has pointed out, much of the important communication and information exchange that occurs at conferences may occur outside the formal sessions:

Some observers believe that formal conference sessions and presentations serve mainly as a mechanism to bring together people with common interests and concerns. The truly valuable and meaningful conference exchanges, according to this view, take place across the hotel breakfast or lunch table, in brief personal contacts outside conference rooms, and en route to an evening's entertainment. Mutual confidence developed through these personal contacts makes the most useful flow of information possible, both at the conference and afterward. The key question may not be whether clear voices and images can flow between conferees, but whether trust, and just plain liking each other can be built by remote control. (42)

In short, it seems fair to say that teleconferencing has not yet fulfilled the expectations that some of the earlier investigators had for it. There are several conjectures as to why that is the case, but we do not really understand it completely. What teleconferencing technology must offer before it is widely adopted as a preferred way of having "meetings" remains an unanswered question.

A particularly promising direction for the future involves the merging of teleconferencing with decision-aiding and group problem-solving techniques. There is a need for software that would help explicate consensual views, converge on corporate decisions, and generally facilitate collaboration on problem-solving tasks. In face-to-face meetings decisions are often swung by a member of the group who happens to be particularly articulate and persuasive, more or less independently of the objective merits of his position. How might group decision making be done so that the views of the various members of the group, including those who may be less articulate but highly knowledgeable, are all weighed in some equitable and rational fashion? Perhaps further exploration of teleconferencing methods can help to answer this question.

### Consumer Information Services

It was suggested almost fifteen years ago that the time was right to begin thinking about implementing a "national information bureau, whose sole purpose would be to provide information to any citizen on any issue" (Yourdon 1971, 24). I believe that such an information resource should be a national goal. One may make a convincing case that the greatest causes both of technological progress and of the development of democratic forms of government have been those events, discoveries, and inventions that have made information more accessible to the average person. The kind of general and comprehensive information resource that Yourdon suggests is indisputably a very large order. As an objective toward which we should work, however, it seems just right.

In the meantime we can expect that many types of information services will be accessible from the home in the near future. These will provide educational courses, consumer advice, travel and health information, investment services, and information about sports, gardening, hobbies, and numerous other topics of potential interest to the general consumer. Consider, for example, the possibility of a system that would provide information to help a homemaker in planning meals and shopping for groceries. Among the objectives one might try to realize simultaneously are the following: (1) providing a diet that is nutritionally well balanced, (2) obtaining maximum value for the money spent, (3) satisfying the individual tastes and dietary restrictions of those for whom one buys, and (4) minimizing waste.

It is easy to imagine an information system that could be of considerable help here. Such a system would contain (1) a relatively permanent database of basic dietary information, (2) current price lists (by brand name when appropriate) for participating local markets, (3) a set of information-manipulating algorithms, and (4) a user file containing data regarding the preferences and dietary restrictions of the family or group with which the planner is concerned. Using such a system, one might compare what a list of groceries would cost at each of several markets. Or one might obtain from the computer a list of meats available for less than some specified amount per pound. Or one might get a list of foods that have a high content of a particular vitamin, that are low in some ingredient that must be avoided, or that simultaneously satisfy some set of such constraints.

One might plan in interaction with the computer a family menu for, say, a week. The interaction could take a variety of forms. The user might propose a set of possible menus, asking the computer in each case to determine costs, specific quantities of necessary ingredients, and so forth. Alternatively, the user might specify a set of constraints (a cost limit, certain items that are to be worked into the menu, items that are to be avoided, any special vitamin or nutrient requirements) within which the computer could be given the task of suggesting possible menus. One can imagine that the final production of a menu might involve several cycles of interaction between computer and user: the computer proposes a menu; the user reacts by suggesting changes (forgot to mention that the family does not like lamb); the computer responds either by modifying the proposed menu or by specifying the implications (in terms of the budgetary and dietary constraints originally imposed by the user) of substitutions that the user suggests.

As another example of how an information system could be of help to the average consumer, consider the convenience of a system that could answer a question such as, What is available for entertainment within a ten-mile radius of Mytown on the evening of August 12 that is appropriate for a family of five (ages 6, 9, 13, 40, and 42) and for which the total cost would be less than a specified amount? Or, Are there any antique auctions within a twenty-mile radius of Mytown on either Thursday or Friday night of next week?

Or consider a "buy, sell, or swap" system. It would be interesting to know the monetary value of useful goods that are in the possession of people who no longer want or need them, but retain them simply because the process of unloading them is too bothersome. A prospective seller may assume, in most cases, that there are people out there somewhere who could use that stove or refrigerator or rug, but the problem is finding them. And the prospective buyer's problem, of course, is that of finding the individual who has the stove or refrigerator or rug that he needs.

The conventional methods for bringing such potential buyers and sellers together work only marginally well. Prospective buyers must invest time and effort in seeking out potential sellers, and often they find themselves pursuing leads that were dead before they discovered them. Prospective sellers typically announce publicly (via a newspaper advertisement, for example) that they have something to sell. Because there is no effective way to let the world know when a sale has been transacted, a seller may find himself apologizing to disappointed inquirers for days after he has parted with a particular article. An interactive classified-advertisement system could maintain a file of items for sale, prices, and sellers' names (or codes). A prospective buyer could interrogate the system with such questions as, "Refrigerator, not more than 5 years old, between 16 and 20 cubic-foot capacity, white, not more than $300." In reply, the system would present the relevant information about any listed refrigerators meeting these constraints. Part of the relevant information in each case would be the number of prospective buyers already in an inquiry queue for that particular item.

Such a system would have some obvious advantages over presently available mechanisms. The task of searching for the information that the prospective buyer wants would be given to the computer. The buyer would define a particular interest and would see only information that was directly relevant to it. An item could be removed from the list immediately upon the conclusion of a transaction, thus assuring that only live items were listed.

Other possible information resources that could be useful in the home include programs to

- help optimize the layout of a family garden to make the best possible use of a specified area, given the climatic and soil conditions, shade/sunlight situation, and the desired yield (types of vegetables in what relative quantities);
- provide instructions on home repair;

- give information about first aid and home treatment of injuries or illness;
- help with budgeting and financial record keeping;
- maintain mailing lists and records of correspondence;
- design a house interactively, with instant feedback regarding the cost implications of specified aspects of, or changes in, the design.

Such examples could easily be multiplied. Systems of these types are well within current capabilities technically, the organization of the required databases presenting no very difficult problems. This is not to say, however, that they would be cost-effective at the present time. A major problem associated with some of the types of systems imagined is that of keeping the database current and sufficiently complete, inasmuch as their utility is limited by the accuracy and completeness of the information they contain. Moreover, the information with which some of them deal tends to become obsolete very rapidly. The development of mechanisms for updating the database on a more or less continuing basis could prove to be the most difficult of the practical problems involved in making such systems work. Another problem, of course, is that of the interface. If systems of the type imagined are to be used by the average consumer, the interaction protocols must be designed with that fact in mind.

Many information sources and reference books are currently accessible through computer networks. These include the New York Times Information Bank, the *Encyclopaedia Britannica*, *Newsweek*, the *Federal Register, Books in Print,* and several professional journals (including sixteen of the journals of the American Chemical Society) (Lancaster 1982). Some writers believe that it is only a matter of time, and a relatively short time at that, until print on paper will give way to electronics as the primary medium for storing and distributing the information that traditionally has been contained in books, magazines, and newspapers (Kubitz 1980; Lancaster 1982).

The 1984 edition of Telenet's *Directory of Computer-Based Services* lists nearly 140 service providers and over 300 databases that are accessible on the network (GTE Telenet 1984). In addition to conventional time-sharing and remote batch resources, electronic mail, data management, word processing, statistical analyses, and a variety of programming languages and applications programs, the available services include computerized ticket purchasing services, weather forecasts, movie reviews, credit checking, publication and bibliographic searches, adoption services, investment information (such as quotes on stocks, bonds, and commodity futures, historical and summary data, insider trader information, foreign currency exchange rates, information regarding limited partnerships), on-line encyclopedias, accounting services, order entry via computer terminal, and a host of others, providing information on everything from aquaculture to zinc production. Some of the services are intended for the general public. Many are designed for specific groups: librarians, investors, insurance agents, business managers, policemen, firemen, lawyers, publishers, real estate agents. The number of resources available through the network is actually far larger than the 140 providers and 300 databases listed, inasmuch as some of the services that are available provide access to numerous resources. Source Telecomputing alone, for example, provides access to some 800 databases and services. Other sources of information on accessible databases are Cuadra Associates (1984) and Williams, Lannom, and Robins (in press).

Several pioneering attempts to provide general information services via telephone and television equipment are also being made. The British teletext systems Ceefax ("see facts") and Oracle (Optical Recognition of Coded Line Electronics) provide for the transmission of text and graphics in an unused portion of the television signal carrier. A special adaptor is required to make these transmissions visible on one's TV. (The principle is the same as that used to transmit and display captions for deaf viewers.)

The British Post Office's Viewdata makes it possible to transmit over telephone lines information that can be passed to a television display through a special decoding device and then viewed on the television screen. Telidon, the Canadian version of Viewdata, has improved the graphics by adding processing power to the decoder. Such systems are known as Videotex systems in the United States and Continential Europe. Because they use the telephone line as the communication channel, Videotex systems are two-way systems: the user can query them as well as simply view preprogrammed information (Bown, O'Brien, Lum, Sawchuck, and Storey 1979; Bown, O'Brien, Sawchuck, and Storey 1978). Videotex systems are currently being developed in the Federal Republic of Germany

(Bildschirmtext), the Netherlands (Viditel), Sweden (Datavision), Finland (Telset), France (Teletel and Antiope), Canada (Telidon and others), Japan (Captain), Switzerland (Videotex), and the United States (various private projects) (Hawkridge 1983).

Among the major suppliers and vendors of financial and business information via computerized databases are Data Resources, Business International, and Automatic Data Processing (Houghton and Wisdom 1980). These companies provide a variety of types of information regarding finance, economics, marketing, and products. Other information services being offered commercially include The Source, Compuserve, Delphi, The Dow Jones Information Service, and The Knowledge Index. These systems vary considerably with respect to the types of information and the range of services they provide, but as a group they offer electronic mail and bulletin boards, conferencing, document preparation and editing tools, selected news, information searches and question answering in limited domains, shop-at-home services, and games. Most of these services expect to expand as the technology permits and the market demands.

Such services are offered to anyone who wishes to have them. Typically one pays an initial membership fee (usually under $100) and an hourly use charge. For equipment one needs a keyboard terminal (which can be that of a home computer) and a modem with which to connect the termial to the phone. A standard TV console can be used as a video output device. (This also has to be connected to the phone with a modem.) There is considerable variation among services as to ease of use, thoroughness and clarity of documentation, and distribution of update information. Some systems use menus, for example, while others depend on user-initiated commands. Most provide on-line prompts and help, but not equally effectively. All provide some printed documentation for the user. Some issue newsletters or magazines to announce new features or capabilities and to provide hints regarding their use.

Today's information services will surely be considered primitive in a few years. But they are pioneering efforts to use computer technology to make information more accessible to people who want it. How these services will evolve is anybody's guess; *that* they will evolve is certain. At the end of 1984 the total number of publicly accessible electronic databases was estimated to be 2,800 and growing (Williams 1985). The potential impact of powerful special- and general-purpose information services on our lives is clearly great.

### Information Seeking

The primary purpose of all information systems, no matter how they may otherwise differ, is to make some type of information available to their users. Perhaps the most common way of doing this is by automatic distribution of formatted reports to preselected recipients. This is the standard operating procedure for the vast majority of information systems now in existence. A disadvantage of this approach is that people frequently receive a great deal of information that is of dubious value to them. As more and more people acquire on-line access to information systems through terminals in their offices or homes, there arises the possibility of their obtaining information by requesting it rather than by having it sent to them automatically. There is some evidence that this method is gaining in popularity over the more conventional automatic distribution approach (Hedberg 1970; Morton 1967). A risk associated with this method is the possibility that users may fail to see information they really should see, either because they did not think to ask for it or because they intentionally avoid certain types of information (Ivergard 1976). The challenge will be to design systems that neither inundate users with irrelevant information nor let them fail to receive the information they really need.

This is the classic problem of information retrieval. In signal-detection terms, the problem is to design search procedures that will yield a high signal-to-noise ratio—a large number of relevant returns relative to the number of irrelevant returns (Swets 1963). One might expect that interactive systems would have some advantages over search systems for which all the search parameters must be specified in advance, and there is some evidence that that is the case (Atherton 1971). They should be less likely to swamp the user with masses of irrelevant material, because can one monitor the output and modify the search parameters on the basis of what is being found.

Ideally one would like an information retrieval aid to be able to find and retrieve information that one does not already have on a specified topic. This would require, among other things, that the system be able to determine what the user already does or does not know. Whether this would be worth doing would depend on whether the cost or inconvenience to the user of having his knowledge assessed would be greater or less than that of having to filter out from the information retrieved that which he already knows. Moreover, making such an approach practical would require the development of better techniques than now exist for assessing what one knows. A good reason for trying to develop such techniques, apart from their usefulness in information retrieval systems, is the potential for their use in computer-assisted-instruction systems.

Most of the human-factors research on information retrieval systems or information-seeking aids has focused on the question of how to make an interactive retrieval system easy to use. Design guideliness for on-line information retrieval systems have been proposed by several writers, including Lowe (1966); Clarke (1970); Martin and Parker (1971); McAllister and Bell (1971); and Martin, Carlisle, and Treu (1973).

Numerous command languages have been developed to facilitate a user's interaction with computer databases. These languages, which are usually referred to as query languages, include SEQUEL, Query-by-Example, EASYTRIEVE, CUPID, SQUARE, QUEL, and numerous others. Such languages provide users with the means of specifying subsets of information that they wish to retrieve from a structured database. Typically the user identifies the information that he wishes to retrieve by composing a formatted query that contains names of data sets, qualifiers, and various Boolean operators. The languages differ in many particulars, and much of the research on the use of query systems has focused on comparing the relative effectiveness of different languages (Greenblatt and Waxman 1978; Reisner 1977; Reisner, Boyce, and Chamberlain 1975; Welty 1979) or on trying to determine why specific languages work as well or as poorly as they do (Gould and Ascher 1975; Thomas and Gould 1975).

Information seeking is a pervasive human activity, and we need to understand it better in a general sense. Parenthetically, one could make a compelling argument that a course called "Information Seeking" should be offered at the college or pre-college level, the objective of which would be to sensitize students to the central importance of information seeking to most other activities, and to acquaint them with the different types of information repositories that exist and with various ways of accessing these repositories. In their chairmen's report from the Computers in Education Conference, sponsored by the Office of Educational Research and Improvement of the U.S. Department of Education, Lesgold and Reif (1983) point out that one of the implications of recent technological changes for education is a devaluation of memorization skills and a growing emphasis on the importance of information finding and use.

> The well-educated future citizen will be adept at selecting information, reasoning abstractly, solving problems, and learning independently. To teach such skills effectively is a major educational challenge. Excellence in education can no longer be measured by counting the number of facts a student has memorized. Rather, the criterion must be the ability to sort through bodies of information, find what is needed, and use it to solve novel problems. (12)

In looking for information in a conventional repository such as a library, one uses a variety of aids: the topical organization of the libary stacks; the card catalog; book lists; the Science Citation Index; abstracts; tables of contents, indexes, and reference lists in books. What kinds of aids might be developed to help people search for and find information in very large electronic databases and information banks?

Such a question leads to even more fundamental ones. How should we conceptualize the information that exists in, say, a scientific article? Does it make sense to think in terms of the article being composed of a set of information elements? What is an information element? An idea? A concept? A fact? A hypothesis? An assumption? A conclusion? Inasmuch as what constitutes information depends on one's prior knowledge, or ignorance, the information elements of a piece of text will differ from reader to reader. This does not invalidate the idea of information elements, however; it just recognizes that the same text might decompose into different elements for different readers. A possible way around this problem is to think in terms of what the information elements are for the reader whom the writer presumably had in mind. In any case, the larger question is, How *should* we think about text if our interest is in developing methods for extracting from it the information it presumably contains?

Getting large amounts of information (for example, the information in the Library of Congress) into a single random-access memory, and giving people access to that repository via computer networks and remote wireless terminals, represent great steps in increasing information accessibility. It is not clear that one could do much with such a system, however, in the absence of a rather sophisticated software interface between the user and the database. What should that interface be like? What functionality should it have? How should it appear to the user?

The effective use of a very large database requires performing each of the following tasks:

- locating specific items of information;
- making decisions about relevance, importance, and credibility;
- organizing, aggregating, and integrating information;
- making inferences from data;
- redirecting the search more of less continuously on the basis of results to date;
- keeping track of leads to further information that one may not be able to pursue at the moment, so as not to forget to do so when one has the opportunity.

How many of these tasks could, and should, be given to the computer to do, in whole or in part?

Sometimes information-seeking activity is focused, and sometimes it is diffuse. That is, sometimes one is seeking quite specific information or answers to specific questions, whereas on other occasions one may be browsing for information that is "relevant" to some problem or simply "interesting," and may not know exactly what one is looking for until one has found it. It is helpful, then, to distinguish two types of information retrieval aids: question-answering systems and browsing aids. A question-answering facility would produce answers to specific questions (assuming the database contained the relevant information): What was the GNP of Japan in 1972? How long ago was the world population one-half what it is now? What are the amino acids from which proteins are made? Who was the head of state in Bulgaria in 1939?

The function of a browsing aid would be that of a friendly librarian: to help one find one's way around a database, locating items of interest and responding effectively to changes in the focus of interest as a consequence of the browsing. An alleged advantage of books and libraries is that they facilitate browsing. There is no reason why one could not browse through an electronic book or an electronic library as effectively as one browses through a conventional book or library. It is quite possible that browsing aids will be developed that will make electronic browsing even more effective and satisfying than is browsing as we now know it.

The development of radio and television increased greatly our ability to convey information from place to place and to keep the general population informed about major events, in many cases even as they are happening. The telephone has greatly increased the ability of individuals to stay in touch with each other, even when separated by great distances. These technologies have their limitations, however, some of which may be removed or reduced by developments such as those discussed in this chapter. While the telephone puts us in personal touch with eather other, its use depends on both parties to a conversation being at the phone at the same time, and on one of them (the recipient of the call) being at the particular phone whose number was dialed by the other. Electronic mail has neither of these limitations. Radio and television are superb media for broadcasting information, but not for delivering it selectively in accordance with the interests and schedules of individual recipients. Computer-based information services have the potential for increasing very substantially the range of options that people have with respect to information acquisition. They should also make it much easier to acquire specific information in accordance with one's own interests and time constraints. The personalization of news and information services is a worthy goal and one that seems attainable, at least to some degree.

### Data Banks and Privacy

The close relationship between information and control has always been appreciated by individuals in positions of power; with the prospect of systems that can acquire, organize, store, manipulate, and distribute huge amounts of information, it is now becoming part of the public awareness. Computer technology has made possible the management and control of information on such a scale that questions concerning who should have access to what information are likely to be hotly debated for some time (Westin and Baker 1972).

On the one hand, there can be little doubt that large databases containing incidence statistics on diseases, information on the effectiveness of drugs and other forms of treatment on specific illnesses, and the relative frequencies of occurrence of specific symptoms conditional upon the presence of specific disease processes, as well as the medical histories of individual people, could have enormous benefit; and indeed such databases are beginning to be developed (Rosati, McNeer, and Stead 1975; Weyls, Fries, Weiderhood, and Germano 1975). On the other hand, the immense potential for abuse of personal information raises Orwellian visions, and it is well not to be insensitive to the possibility.

In 1981 the Office of Technology Assessment of the United States Congress reported the findings of an overview study on the use of computer technology in national information systems and related public policy issues. "National information systems," as defined by the authors of this report, means "systems that are: (1) substantially national and geographic in scope (i.e., multi-state); (2) organized by government or private organizations or groups to collect, store, manipulate, and disseminate information about persons and/or institutions; and (3) based in some significant manner on computers and related information and communication technology" (Gibbons 1981, 10). The study focused on "large interconnected national systems where a substantial national interest is involved." As examples of such systems the report mentions the computer-based National Crime Information Center, operated by the Federal Bureau of Investigation; the FEDWIRE electronic funds-transfer network, operated by the Federal Reserve System; and the nationwide electronic mail services operated by several private firms and soon to involve the U.S. Postal Service.

Such systems raise the public-policy issues of privacy, freedom of information, First-Amendment rights, computer crime, proprietary rights, and several others. The report points out that policy issues arise from certain tensions that have roots in the distinction between private, public, and commercial information. Conflict may exist, for example, between freedom-of-information laws and the right to privacy. Tensions may also result from conflict between commercial and public interests or between commercial interests and individual privacy. The national information-system issues judged by the Office of Technology Assessment to be among the most important over the next few years include innovation, productivity, and employment; privacy; security; government management of data processing; society's dependence on information systems; constitutional rights; and regulatory boundaries.

Particularly difficult problems associated with data banks arise when public and private interests are, or appear to be, in conflict. Consider the case of computerized law-enforcement identification and intelligence systems. Advocates of such systems argue that they can improve the administration of criminal justice while enhancing, rather than eroding, civil liberties. Immediate access to centralized fingerprint files and criminal histories could mean not only more rapid identification of perpetrators of criminal acts but also the speedier establishment of the innocence of possible suspects. Critics are concerned over the possibility that the "criminal" histories on file may contain records of arrests that resulted in dropped charges or acquittals or that were politically motivated. They are also apprehensive about the gluttonous appetites of such systems for more and more information and about the eagerness of their developers and users to feed them.

It is perhaps this tendency for small information systems to grow, and for specialized systems to broaden their scope, subject only to technological limitations, that constitutes the most threatening aspect of data banks. The possible merging, generalizing, and centralizing of data files seem to be the main bases for alarm. Miller (1971) put it this way:

> ... numerous reporting services currently performed by a large number of independent companies and investigators someday may be provided by a few unregulated conglomerates having unbridled power to vend their vast information store without regard to the purposes to which it will be put. Thus, there is a risk that enormous quantities of financial and surveillance data garnered from a variety of sources will be made available to anyone who wishes to reconstruct an individual's associations, movements, habits, and life style. We are only beginning to perceive the intrusive uses to which data bases of this type may be put. (79)

Miller did not end his comments on this dour note, but went on to offer numerous concrete suggestions for managing the information managers. The issues are too involved to permit a thorough discussion of them here. Suffice it to say that the question of information privacy will be debated for some time to come, and we cannot afford to be sanguine about the outcome.

The development of increasingly powerful techniques for gathering, processing, and using information for control purposes is bound to produce a variety of difficult moral dilemmas. The following fantasy—which is less fantastic now, technically, than it was a very few years ago—illustrates the point.

Imagine that everyone were forced to wear (or had implanted) a device that continuously reported his location. Now imagine an enormous computer system, capable of tracking the whereabouts of every individual in some geographical area, say the continental United States. Suppose further that this system had sufficient storage capacity to store a "track" for every individual in that area, a track being a record of an individual's location within, say, an $n$-kilometer radius that is updated every $m$ minutes. Whenever a serious crime was committed, the system could immediately partition any group of suspects into two subgroups, one of which could be ruled out from further consideration on the basis of not having been in the vicinity of the crime at the time the crime was committed.

The question of whether such a system will ever be feasible is a technical one. A more difficult question to answer, however, relates to the social implications of the implementation of such a system. Clearly, given sufficient resolving power, such a system could be very useful in protecting some innocent persons, if not in identifying persons guilty of specific crimes, and might well serve as a deterrent to some types of crime. (It would also be useful in locating missing persons, kidnap victims, and fugitives.) The idea of such a system has a distinctly Orwellian aspect, however, and to permit some system, agency, or person to know of one's whereabouts at all times certainly seems like an invasion of privacy and a threat to individual freedom of the most profound sort.

While this particular system might be considered too fanciful to be taken seriously, it illustrates the type of problem that is bound to become more and more of an issue as information technology spawns increasingly sophisticated capabilities that can be used for law enforcement and crime detection. Such capabilities will frequently involve making more information readily available to law enforcement authorities; and making information available may appear to be, perhaps really will be, an invasion of the privacy of some of the people to whom the information pertains.

All of us get frustrated from time to time by the inefficiencies that we see in the way our society is organized and governed. One is led to wonder, however, whether perhaps this has been our salvation. The human species may not be sufficiently mature, morally, to cope with a truly efficient information and control system. In Norbert Wiener's words,

As engineering technique becomes more and more able to achieve human purposes, it must become more and more accustomed to formulate human purposes. In the past, a partial and inadequate view of human purpose has been relatively innocuous only because it has been accompanied by technical limitations that made it difficult for us to perform operations involving a careful evaluation of human purpose. This is only one of the many places where human impotence has hitherto shielded us from the full destructive impact of human folly. (1964, 64)

The issue is not *whether* data banks and sophisticated information-gathering techniques should be permitted; the complexity of our society requires them. The issue is how to ensure that their emergence and growth do not erode our constitutional and common-law rights and freedoms, and destroy our status as "private" individuals. The resolution of the issue will involve coming to grips with numerous specific questions. Who may collect what kinds of information? What responsibility do the collectors have for assuring the reliability of the information they collect? Who may have access to what information? Who, or what, should determine the uses to which information can be put? What rights does the individual have to determine who may collect or have access to information regarding himself and his personal affairs?

Numerous suggestions have been made concerning methods and measures for safeguarding against the possible misuse of data banks. How many of these suggestions are reasonable and workable remains to be seen. There is little doubt that some new principles, guidelines, and regulations will have to be developed to deal with this aspect of the new information technology. Specific proposals must be thoroughly aired in public debate, however, before being implemented, if we are to avoid the pitfalls of extremism of either sort.

Most of the debate about information systems and data banks

has focused on the individual-privacy issue. There has been remarkably little discussion of the potential that information technology holds for increasing the public accountability of both public and private institutions. What would be the implications for the quality of life of the existence of an information system that would provide for any individual, immediately on request, the voting records of federal, state, or local legislators on specified issues; customer-service ratings (the consumer's counterpart to the merchant's credit ratings) for suppliers of a specified type of merchandise in a particular area; an objective comparison of specified models of automobiles with respect to specified features; "return on investment" comparisons for specified "small investor" opportunities; itemized records of expenditures of public monies; interpretations, in language meaningful to the layman, of industrial product warranties; fault and failure statistics on specified consumer products, such as automobiles and household appliances; data regarding industrial sources of specified environmental pollutants?

Consider the possible effects of a system that had a complete record of all the public speeches and statements of political figures, government officials, and other individuals in positions of public trust, and could retrieve quickly what an individual had said with respect to specified issues. What effect might such a system have on the political process in this or any other country? Would people be more careful about what they say if they were aware that any public pronouncement would be stored in a public memory that could be accessed easily by anyone who wanted to access it?

# 10

## Information Technology and Jobs

Information technology will affect jobs in many ways: it will make some existing jobs obsolete; it will create some new jobs; it will change the way many tasks are performed. It may fundamentally alter our attitudes about the character and function of work. For many people it will afford the freedom to choose when and where they will do their work. Changes in work patterns will have secondary effects on the ways in which people relate to each other and on how they spend their leisure time. It is, of course, expected that computer technology will increase productivity; that is the reason for its introduction into any workplace. Whether it will increase the satisfaction that people get from their jobs is another question altogether, and one that deserves a great deal of attention.

### Automation: Threat or Promise?

Few aspects of computer technology have caused greater apprehension than its potential for displacing labor through automation. What is often overlooked in discussions of this topic is that automation, and science and technology more generally, have already profoundly affected the way the labor force is employed. As we have already noted, seven out of ten people born in the United States 150 years ago spent their lives working on farms. With the mechanization of farming and the development of effective fertilizers and pesticides that raised the yield per farmed unit of land, the number of people who make their living by farming has fallen off steadily until today it represents only about 3 percent of the labor force.

The shift away from farming was initially accompanied by a corresponding increase in the proportion of the labor force that was engaged in industry and manufacturing. By the end of World War II nearly 70 percent of the labor force occupied this sector of the economy. During the past 35 or 40 years a second major shift, no less dramatic than the first, has occurred, in which a steady decline in the percentage of people engaged in production and manufacturing has been accompanied by a compensating increase in the percentage of those involved in service and information-handling occupations. Included in this category are medicine, law, education, retail businesses, entertainment, tourism, financial services, transportation, publishing, law enforcement, health and welfare services, sports, and the arts. Naisbitt (1984) notes that in 1956, white-collar workers outnumbered blue-collar workers for the first time in American history. Porat (1977) has estimated that "information jobs" produced nearly half (46 percent) of the gross national product and over half (53%) of the income earned in the United States as of 1967. The general trend toward a greater and greater proportion of the labor force being involved in service or information jobs is seen not only in the United States but throughout the industrialized world. The likelihood is that the demand and opportunities for employment in the service area will continue to increase at least over the near future, if not indefinitely, while the opportunities for employment in manufacturing will continue to diminish until it, like farming, accounts for a very small fraction of the work force.

One might suppose the natural consequence of automation to be a decrease in the number of jobs available for people: as more and more jobs become automated, the need for human labor becomes less and less. It is not that simple. To be sure, automation has eliminated or very nearly eliminated many types of jobs. Some industries have become so highly automated that they use today only a fraction of the human labor that they did just a few years ago. But the technology that has given us automation has also created many jobs. Numerous new industries have emerged during this century as a consequence of technological developments. The automotive industry is a case in point, as is the television industry. The most spectacular example of a rapidly growing technology-based industry that did not exist fifty years ago is the computer industry itself.

Many writers have pointed out that before the domestication of animals and the harnessing of elemental sources of energy, such as wind and water, humans were a major source of the energy with which they accomplished physical work. The decline in our role as power generators accelerated with the advent of internal combustion engines and electric motors; today an individual's worth as a power source has shrunk to insignificance. Estimates of how much muscle-power output an individual can deliver in a year vary somewhat; even the most liberal estimate, however—about one megawatt hour—comes to less than 100 dollars' worth, given the going rate (including cost of fuel adjustment) of about 8.5 cents per kilowatt hour. It is not that individuals are less capable of generating power today than in times past, but rather that we have invented more effective machines than our own bodies for doing so. (We still retain some advantages over machines, however, even as power sources, by virtue of our cordless mobility and the flexibility we have in being able to transform energy into useful work in a great variety of ways.)

As a consequence of the Industrial Revolution and the attendant proliferation of machines that could do many of the things that only human or animal muscle could do before, the human being's primary role shifted toward that of user and controller of power machines. With the advent of servomechanisms and automatic control systems, that role in turn has given way, in many instances, to that of dial watcher and monitor. Attentiveness and a high tolerance for boredom have replaced manual dexterity as prime requisites for many jobs, just as manual dexterity replaced physical strength. With increasingly sophisticated and reliable automated systems, even the monitoring functions themselves will be performed more and more by machines. More generally, we are now seeing an increasing amount of intellectual work of many types being performed by machines. One must ask, if machines get to be very good at doing most of the intellectual work now done by people, what will be left for people to do?

I believe there are two answers to this question. The first challenges the assumption that seems to follow from the observation that machines have replaced people and animals as the primary sources of energy for physical work—namely, that there is consequently less opportunity for people to engage in physical work. It is not at all clear that that is a valid assumption. What is clear is that as a consequence of the existence of power machines, much more physical work gets done today than did before. There are still many opportunities, however, for people who wish to do so to engage in physical work. Quite conceivably there are more opportunities—a greater variety of possibilities—than there were in the past, both by way of making a living and for avocational purposes. To be sure, power machines have made muscle worthless as a means of propelling ocean-going ships, plowing acres of soil, and moving mountains of earth, but few people will rue that fact. On the other hand, just because of the existence of power machines and the ingenuity of people in finding ways to use them, there are many jobs today requiring skilled human labor that did not even exist before the machines came along. If machines become very good at doing much of the intellectual work now done by humans, it does not follow that there will be fewer opportunities for humans to do intellectual work. It could very well be that those opportunities will increase in number and in type. The information machines may help to create some possibilities for intellectual work that did not exist before.

My second answer to the question is that there is one function the machines do not threaten: that of setting goals, of deciding what we want and what we and our machines should attempt to achieve. The machines, especially the information machines, can play very significant roles in helping us to understand what our options are and to make rational choices. The choices, however, will remain ours. Machines can help us see the tradeoffs, but it is up to us to decide on the kinds of trades we wish to make. They can help us understand the tradeoff between energy consumption and resource conservation, or between the convenience of private transportation and its costs, or between automation and employment. This is not to say that machines will simplify the decisions; in fact, they may make them more difficult. Consider, for example, the dilemma of highway safety and speed limits. Suppose we knew, without doubt, that by lowering speed limits by $n$ miles per hour and strictly enforcing the lowered limit, we would save $x$ lives per year. More generally, suppose we were able to plot the tradeoff function relating the number of highway deaths per year to speed limits. We would then be in a position to decide, quite explicitly, how many lives we would be willing to trade for a few miles per hour on the highway. We make such decisions now, but only by default; and one wonders if, given the choice, we would not prefer to remain ignorant of such matters.

Ernst (1982) points out that one of the main challenges of the mechanization of work is to manage the required restructuring of the labor force in a more humane way than was characteristic of the Industrial Revolution. He notes too that when the transition to a mechanized system takes place slowly or in a growing industry, the displacement of labor is unlikely to be severe and employment might even increase. This is illustrated by what has happened in the air transportation industry. In spite of (perhaps in part because of) increased automation, employment in this industry increased by about 15 percent between 1970 and 1980. The rapid growth in this industry (the number of passengers on U.S. airlines rose by a factor of 5 between 1960 and 1980) is attributed in part to the introduction of electronic reservation systems in the early 1960s. On the other hand, when a mature industry mechanizes rapidly, displaced workers, especially the least skilled, are likely to suffer disproportionately.

It is impossible to tell what the effects of automating information-handling jobs will be and, in particular, whether it will, on balance, increase or decrease job opportunities. It seems clear, however, that the job mix will continue to change, as it has been doing for a long time. The percentage of professional and technical jobs is expected to rise significantly over the next few decades, especially in the area of computer science, whereas the percentage of clerical jobs is expected to decline (Science and the Citizen 1984). We can also be reasonably certain that the *types* of things that people will do within a job category will change significantly. In many offices the tasks of a secretary, for instance, are very different today from what they were even five years ago. We know also that the cost of human labor is going up while the cost of computing resources is continuing to fall. If both these trends continue, this cannot but have implications for the ways in which both human labor and computing resources are employed.

Whether, on balance, the demand for human skills will grow or diminish during the next few decades must still be considered an open question. Several trends do seem clear, however: (1) as already noted, a decreasing fraction of the population will be involved in subsistence occupations—the production of food and other necessities; (2) expanding educational opportunities, combined with trend (1), will give individuals more freedom in choosing how to spend their lives; and (3) leisure will increase.

Each of these trends has been evident for some time. We may expect to see an acceleration in all of them in the next few decades as automated techniques are introduced into more and more segments of both the blue- and white-collar sectors of the economy.

One of the implications of these trends is the need to reassess some of our long-standing attitudes concerning the function and significance of work. The necessity for manual labor was once viewed not only as God's way of keeping us out of mischief, but as a prerequisite to our general well-being. Writing for the *World* in 1755, before the Industrial Revolution was well under way, Edward Moore said,

Providence has therefore wisely provided for the generality of mankind, by compelling them to use that labor, which not only produces them the necessaries of life, but peace and health, to enjoy them with delight. Nay further, we find how especially necessary it is that the greatest part of mankind should be obliged to earn their bread by labor, from the ill use that is almost universally made of those riches that exempt men from it. . . . It was a merciful sentence which the creator passed on man for his disobedience, by the sweat of thy face shalt thou eat thy bread; for to the punishment itself he stands indebted for health, strength and all the enjoyments of life. (200)

We might smile at this apotheosis of *manual* labor today; however, our general attitudes regarding the importance of productive work probably have not changed that much. Industriousness is still perceived as a cardinal virtue, and productivity is still one of the commonest yardsticks by which society judges a person's worth.

It is interesting to contrast the opinion expressed by Moore in 1755 with one stated somewhat more recently by Norbert Wiener (1950):

It is a degradation to a human being to chain him to an oar and use him as a source of power; but it is an almost equal degradation to assign him a purely repetitive task in a factory, which demands less than a millionth of his brain capacity. It is simpler to organize a factory or a galley which uses human beings for a trivial fraction of their worth than it is to provide a world in which they can grow to their full stature. (16)

Surely we must agree with Wiener on this point; but what are the characteristics of a world in which all individuals can grow to their full stature? And what part does work play in a world in which the labor of a small fraction of the population is sufficient to produce the essentials of life, and many of its comforts, for the whole?

Not only has industriousness been perceived as a virtue by our society, but work has been considered a duty. Western culture has operated on the principle that every able individual has the moral obligation to earn his keep. The Apostle Paul endorsed this principle rather pointedly when he admonished the early Christians that if a man would not work he was not to eat. But what does it mean to earn one's keep in a society in which most of the essential work is performed by machines?

Maddock (1983) has pointed out that the association in recent Western thinking between work and purchasing power has been so strong that the provision of employment has been viewed as more important than increasing the total availability of goods and services. "It is regarded as better to continue inefficient and uncompetitive enterprises because they provide purchasing power, albeit a small one, rather than to improve efficiency and the ability to compete in home and foreign markets" (167).

Work has played still another role in Western culture. People have found meaning and dignity in their labor. One's need to be needed has been filled in an obvious way. If he saw no other purpose to his life, a person had one at least that was clear and concrete—namely, to provide for his own material needs and for those of his family. In a truly automated society, it may be more difficult for individuals to perceive their existence as absolutely essential to their family's, or anyone else's, physical well-being. As the "provider" role diminishes in importance—if it does—what will take its place?

As a society, we are beginning—but just beginning—to feel the strain of maintaining the work ethic on the one hand while attempting to preserve the economic value of labor on the other. The latter we do by trying to keep the supply of labor artificially small, so as not to exceed the demand. The unemployment rate is perceived as second only to the gross national product as an indicator of the health of the national economy, and there are two ways to reduce it: increase the demand for labor or decrease the supply by removing potential labor from the labor market. Among the means for accomplishing the latter goal are increased opportunities and pressures for advanced education (which delay the entrance of youth into the labor market); forced or induced early retirements (which remove people from the labor market, often during a still-productive period of their lives); shorter work weeks; and government-imposed ceilings on the wage earnings of Social Security recipients. If the demand for labor decreases drastically, as it may, such means for keeping down the supply and thereby maintaining a low unemployment index will not suffice. There will be a need for new attitudes toward the unemployable and for the development of procedures for assuring an equitable share of the society's wealth to those who find themselves excluded from the labor market.

These are difficult problems, and they may take a long time to be resolved. One may find reason to be either as optimistic or as pessimistic as is consistent with one's general confidence in our ability, if not to manage our affairs, at least to adapt to our circumstances.

I personally see grounds for optimism. If a greater and greater fraction of the essential work is done by machines in the future, it should be possible for many more people to spend much more time in creative pursuits. The rebirth of craftsmanship is an attractive possibility. There should be time to master skills in the performing arts—music, theater, dance. If most of the physical necessities of life can be provided by machine, psychological ones cannot. There are many needs in the latter category that are going unmet today for want of people to meet them, and a major reason that people are not available to meet them is that they cannot make a living doing so. It should be easier, with increasing automation, to find the human resources necessary to provide adequate services for ill people, elderly people, and other people with special needs. It should be possible to have much smaller student-teacher ratios throughout the educational system.

On the other hand, the sanguine view that the short-term disruptions caused by increasing automation promise to be temporary and that the long-range benefits to society in general will make them worth bearing, overlooks the probability that certain individuals and groups will have to shoulder a disproportionately large share of the cost of transition, and that not by choice. The argument that for every job that is abolished several are created brings little comfort to the individual who has lost one of the abolished jobs if he has no prospects of getting one of those that were created. Moreover, the people who tend to get hit first and hardest are those who are least equipped to absorb the blow: the inexperienced, the unskilled, the illiterate, the disadvantaged.

Until very recently technological progress has been almost universally perceived within Western culture as beneficial to humankind in general. It is not my purpose to suggest that technological progress is not a good thing. Although we sometimes look back nostalgically to the good old days when life was simpler, the air was cleaner, and wars were more manageable. I suspect that few twentieth-century people would trade today's problems for those of the days before electricity, penicillin, the automobile, and indoor plumbing. Nonetheless, we are becoming increasingly aware that technological developments, in spite of their apparent beneficial effects, have all too often had detrimental second- and third-order effects that were not foreseen and, in some cases, not discovered until considerable damage had been done. Automation holds great promise for humankind. It also presents the prospect of many profoundly difficult problems, the solutions to which will not simply appear of their own accord. They will require the best efforts of people of ability and good will who are not intimidated by the thought of radical change and who are equal to the challenge of thinking through the long-range implications of change, in order to minimize the unanticipated casualties of technological progress and assure the preservation of human values and dignity.

### Telecommuting

Much has been written about transportation-communication tradeoffs and about the possibility of bringing jobs to people instead of vice versa. Some writers have suggested that in the future a significant proportion of the white-collar work force will perform their tasks at home with the aid of telecommunications and information-manipulation tools. There has been some speculation about the desirable and undesirable effects that home-based work might have. Little, if anything, is really known, however, about what the effects would actually be.

One of the bases for interest in bringing jobs to people electronically is the prospect of thereby conserving a significant fraction of the energy that is now required to transport people to jobs. It is estimated that approximately one-quarter of the gross national energy use is for transportation (Hirst 1973; Stanford Research Institute 1972) and that slightly more than one-third of this is associated with urban automobile use (Nilles, Carlson, Gray, and Hanneman 1976). Ninety-seven percent of urban passenger traffic uses the private automobile (Hirst 1973), and more than 40 percent of all trips to and from the city are made for the purpose of commuting to and from work (Automobile Manufacturers Association 1971). Thus, according to these estimates, commuting accounts for about 15 percent of U.S. transportation energy consumption and about 3.9 percent of total U.S. energy consumption (Nilles, Carlson, Gray, and Hanneman 1976).

Nilles and colleagues have used the term "telecommuting" to connote the use of telecommunications technology to bring jobs to people. Regarding the implications of this technology for energy conservation, they assert:

The replacement of urban commuting by telecommuting (or at least telecommuting which is carried out at telephone bandwidths) would result in a net reduction of U.S. energy consumption of at least 8.2 billion kilowatt hours annually for each percentage of the urban commuting work force which engages in the substitution. As another measure, a 1 percent replacement of urban commuting by telecommuting would result in a net reduction of U.S. gasoline consumption of 5.36 million barrels annually. (1976, 83)

There are other potential benefits in addition to conservation of the energy now used in transportation: reduction in the need for expensive office space in urban areas, savings in time normally devoted to commuting, reduction of air pollution from automobile exhaust (assuming that the saved commute is not offset by driving for other purposes).

The potential drawbacks of telecommuting are perhaps less apparent. It is not safe to assume, however, that there would be none, and prudence dictates that some effort be made to identify them before telecommuting systems are implemented rather than have them identify themselves after the fact. It is not at all clear, for example, that all people who believe they would enjoy working at home would do so in fact. Going to work is a deeply ingrained habit for most of us; and while going to *work* is ostensibly what is really important to us, we should not overlook the possibility that *going* to work is more important than we think. Widespread working at home would change very significantly the social aspects of one's work space.

There is one group of people for whom the possibility of working at home must surely be seen as positive, namely those people for whom there are no alternatives because they are, for one reason or another, confined to their homes. There can be little doubt of the desirability of making jobs available to such people through information technology. A note of caution here, however: For many people with disabilities, the opportunity to participate as fully as possible in society's mainstream is of the utmost importance. Any developments or policies that would tend to diminish access to facilities and opportunities that are available to people without handicaps, anything that would tend to segregate to set apart people with disabilities from those without them, would be in direct opposition to the widely accepted goal of removing existing barriers to integration rather than erecting additional ones. There is some fear that making it easy for handicapped people to work at home might be perceived as relieving society of its responsibility to make conventional workplaces accessible and to address the problem of transportation.

Sensitivity to the needs of all people, handicapped or not, for social contact ought to guide us here. Clearly, the possibility of working at home is a good thing for a person for whom there is no feasible alternative in prospect. When there are feasible alternatives, they should be pursued and work at home should be perceived as an option among others, as it is likely to be for many able-bodied persons. When technology is used to make it possible for a disabled person to work at home, care should be taken to exploit the same technology for the purpose of expanding his opportunities for communication with other people. In general, exploitation of technology in the workplace should be done with a view to humanizing job situations.

### Job Quality

The authors of the 1981 Office of Technology Assessment report on computer-based national information systems believe that the applications of information technology are likely not only to increase productivity but also to bring about "better product quality, improved work environment and job satisfaction, and longer-term social benefits such as improved job

safety and greater opportunities for on-the-job learning and
career development" (16). The prospect of information technology making jobs more interesting and rewarding to those
who perform them is a welcome one. Perhaps it will be realized
spontaneously in the natural course of events. This is not a safe
assumption, however, and the question of how to assure that
the effect of the technology is indeed that of enriching and
humanizing job situations is worth considerable attention.

That the way in which new technology is introduced in the
workplace may significantly influence the degree to which it is
accepted and the effectiveness with which it is used is not only
easy to believe on intuitive grounds but also easy to document.
In 1967 McGraw-Hill established a word-processing center in
its New York headquarters, installing 22 word processors; the
intention was to reduce the number of secretaries in the headquarters office. In 1970 the center was discontinued because of
lack of use ("The Office of the Future" 1975). One conclusion
that was drawn from this experience was that word processing
is "a giant step backward for job enrichment" (70). An alternative way to view such abortive attempts to introduce technology
into the workplace would be to focus not so much on the technology itself as on the particular ways in which it has been
introduced. Perhaps what people objected to in the McGraw-
Hill experience was not word-processing capability but the idea
of an impersonal word-processing "center" that threatened to
replace at least some individual secretarial positions with a
monolithic automated resource.

What is it besides income that people look for in work? A
sense of accomplishment, a sense of doing something that
needs doing, an opportunity to interact with other people,
structure, a routine, a challenge, an opportunity to learn, recognition, status, a reason for getting up in the morning, a sense
of personal worth? Jobs play complex roles in our lives. The
question of what makes people feel good about their jobs is not
likely to have a simple, or simply discovered, answer. Moreover,
the answer undoubtedly differs for different people. In spite of
the considerable research that has been done on job quality and
worker satisfaction, I believe that our understanding of these
topics is still primitive and that our ability to predict how
changes in work situations will be perceived by the people in
those situations is very limited. Information technology is going
to change many jobs, some quite drastically. We should attempt

to anticipate the consequences of these changes for job quality and worker satisfaction, and to do what is possible to assure they will be positive. It would be risky, however, to assume that the effects will always match our expectations. The ability to monitor and to adjust will be very important, if the effects of unpleasant surprises are to be minimized and the potential of information technology for improving work situations is to be realized.

• • • • • • • • • • • • • • • • • • • • • • • • • • • • • • • • • • • • • • • • • • • • • • •

The effect that information technology will have on the number of job opportunities in the future is not clear. It is clear that some jobs will be abolished, some created, and many modified. Demands on workers will change. Society will be less dependent on human labor to produce food, shelter, and consumer goods. One result could be a greatly increased opportunity for people to engage in creative pursuits and in person-oriented vocations such as the arts, teaching, health care, and service to people with special needs.

The possibility that computer networks may make it feasible to substitute, to some degree, the transmission of information for the transportation of people and goods is an interesting one. The prospect of bringing jobs to people instead of people to jobs, as captured in the notion of telecommuting, is attractive in some respects but also has its questionable aspects. It remains to be seen how pervasive the practice will become and how satisfying it will be. More generally, the fact that the demands of so many jobs are changing as a result of the effects of information technology in the workplace compels greater attention to the question of what determines the attitudes that people bring to their jobs and the satisfaction they derive from them.

# 11

## Information Systems in the Office

Much has been written about the "office of the future." Typically "the future" is not very precisely defined. No one can say with any certainty what the office will be like even a few decades from now. We can assume, however, that it will be very different and that the information manipulation tools that office workers will use will be an order of magnitude more powerful than any we have today. Existing systems already provide word processing (document preparation, editing, and formatting tools), message processing (electronic mail), data management, calendar and scheduling, spread sheet, and administrative support, although most of them provide only a subset of these functions. Eventually, in order to remain competitive, a system will no doubt have to provide these functions and others as well—such things as speech-to-text systems that permit a user to see immediately what has been dictated; effective ways to access information from very large databases; teleconferencing techniques that permit effective broadband communication among remotely located participants; shareable electronic work spaces that permit geographically separated collaborators to draw on the same surface; message systems that accommodate voice and facsimile as well as text; and a variety of other capabilities that we cannot now foresee.

However else the office of the future might be characterized, it is appropriate to think of it as an information system—a collection of data structures and activities that are performed on them. The data structures include letters, memos, reports, ideas in people's heads; the activities include generation, modification, storage, transmission, and destruction. This conceptualization prompts one to think in terms of the tools and procedures that exist or might be developed to facilitate performing the desired activities on the various types of data structures. Some of these tools are relatively hard (typewriters, dictaphones, file cabinets, telephones), some are relatively soft (operating procedures, computer programs); and the soft ones are increasing in number and importance.

### The Virtual Office

One of the many common words to which computer technology has given a new meaning is "virtual." Often put in quotes (presumably to warn the reader that the word is being used to mean something different from what it typically does, or possibly to indicate that what it is being used to mean is not entirely clear), "virtual" in computer jargon seems usually to mean *appears to be but really is not*. Thus when a user is said to have access to a "virtual" memory of a certain size, this means that he may behave *as though* that much of the computer's primary memory were assigned to him, even though in reality it is not. The computer works a trick to take care of the fact that one does not really have what one thought one had and provides something that is equivalent for practical purposes.

Giuliano (1982) speaks of the "virtual" office that is made possible by those aspects of information technology that permit an individual to communicate with computing resources by means of a portable terminal and perhaps over telephone lines from essentially anywhere. If one has access to one's information files and to the tools normally found in a traditional office, then one's virtual office is wherever one happens to be. How people will accommodate to virtual offices remains to be seen. Working in a virtual office may have implications not only for how one accesses information resources but for how one relates to people as well.

### Computer-Based Office Tools

Historically, office workers have not been thought of as tool users. The per capita investment in equipment for office workers has been considerably lower than the per capita investment in equipment for manufacturing or agriculture. Schoichet (1981) gives $2,500 for the former figure and $15,000 to $20,000 for the latter. The situation is rapidly changing, however, with the introduction of word processors, printers, electronic mail systems, and other aspects of communication technology in the office workplace. It is not yet clear whether office work will become as capital intensive as manufacturing or farming—it may well not because of the steadily declining costs of computing power—but there can be no doubt that it will involve the use of increasingly powerful tools. It also seems clear that whether one works in a virtual office in the future or in a traditional one, the nature of the work one does will be changed. The introduction of computer-based tools has already had significant effects on the nature of office work. As more powerful software becomes available, and more and more managers and executives become hands-on users of information manipulation tools, there will be greater changes still.

Word-processing systems are now commonplace in offices, and word-processing software is available for all of the most popular personal computers. PCs are beginning to appear in offices in lieu of special-purpose word-processing systems, because they offer much greater functionality, including spreadsheet software and graphics.

Computer graphics are finding many applications in the business world. One reads of *executive information stations* that permit executives to interact with information on their companies' operations in ways that they prefer, which typically means the use of bar charts and graphs; *presentation support stations* for corporate graphic art departments; and *conference support stations* that make graphics accessible to groups for display on conference screens (Klein 1983). A well-equipped graphics-oriented business system might contain—in addition to numerous terminals or interconnected personal computers—printers, copiers, slide makers, video projectors, phototypesetters, and plotters.

The printing methods used in offices are expected to change considerably over the next few years. Printers with daisy-wheel or ball-shaped type elements are adequate for conventional typing but not for graphics or unconventional symbology. Dot-matrix impact printers are fast and have some graphics capability, but they are also noisy and do not offer high resolution. Nonimpact printers (lasers, ion-implementation, ink-jet) provide multiple-font capability, facilitate the mixing of text and graphics, have high resolution, and are fast and quiet.

Their main drawback is their price, but this is expected to fall over the next few years, and in the meantime they are shareable among several workstations. Anderson (1981) anticipates two office printer markets, one for the less expensive impact devices for internal and personal use, and another for the more expensive nonimpact devices that will be used primarily to produce documents intended for external distribution.

It is also possible to equip an office with what Anderson calls an "advanced image/text processor": a system that integrates "imaging technologies with optical character recognition and image scanning, on-line communications capabilities, and bulk electronic storage." Such systems "combine copier and printer functions with telecommunications to transmit and receive information and produce hardcopy on demand, locally or remotely" (Anderson 1981, 121). Xerography devices that will copy graphics from a CRT display, while still expensive, are beginning to be used as well.

The difference between an office automation system and a collection of computer-based office tools hinges on the notion of integration and access to tools through a common, consistent interface. What exist today, for the most part, are collections of tools; truly integrated multifunction systems remain to be developed. An element of such a system would be what Spinrad (1982) describes as the *electronic desk*, which would include, say, a display screen, a keyboard, a pointing device (such as a mouse), and perhaps a printer. When such a facility is connected to an office network or to a wide-area network, it becomes not merely a desk but a terminal on a communication system.

While there is some concern that the increasing application of computer-based tools in the office environment may dehumanize office work, there is also some expectation that these tools will enrich office jobs and provide new opportunities for growth and advancement (Murphy 1983). It is clear that there are and will be new skills to learn. Perhaps there will be less human involvement in the mechanical aspects of office work. Electronic message systems will give people greater access to other people independently of where they sit in an organizational hierarchy. Organizations may in fact become less hierarchically structured as a consequence of the continuing evolution of information technology and its use in the workplace.

## Productivity

Substantial improvements in productivity have been reported as a result of the introduction of word-processing technology in the office (O'Neal 1976). But how best to measure office productivity remains an open question. The number of documents or words produced per unit time seems an inadequate measure in the absence of some credible indicator of the quality of what is produced. Borgatta (1983) claims that organizations that are acquiring word-processing systems expect from this technology "lower costs in processing documents, less time in preparing and distributing them, and improved text quality" (42).

Cost savings in the secretary/typist area will have relatively little effect on the overall operating expenses of business, according to Bair (1978), inasmuch as secretary/typist costs amount to only about 6 percent of the total business labor costs in the United States. A similar observation has been made by Branscomb (1982), who claims that the real leverage is with systems that are intended to help managers and professional information workers. Such systems, he suggests, provide at least twenty times more potential for improving office productivity than do today's systems designed to help secretaries.

It is very difficult to determine exactly how office workers, perhaps especially managers, spend their time, because the only techniques available for getting such data are likely to affect the behavior of the people on whom the data are being collected. Self-observation techniques have the same problem. An individual who is keeping a detailed record of how he spends his day may well spend it differently from when he is not keeping such a record. Questionnaire data lack credibility because there is no compelling evidence that people are accurate judges in retrospect of how they have spent their time, unless they have highly structured jobs.

In spite of these difficulties it seems clear that no matter how one measures office work, one will discover that a large fraction of the time is given to communication in one way or another. Mintzberg (1973) has estimated that managers spend about 95 percent of their time in activities that can be classified as communication; the bulk of this time (about 75 percent) is spent in meetings and on the phone, and the remainder reading and writing. Nonmanagerial professionals spend somewhat less time in communication activities, but still a very significant amount (63 percent). Teger (1983) reports 80 to 96 percent as the amount of time managers and professionals spend communicating or managing information. According to Poppel (1982), "knowledge workers" spend about 46 percent of their work time on the phone or in face-to-face meetings and another 21 percent reading or writing. Bair (1978) estimates that managers lose roughly half an hour a day to shadow functions (making phone calls that fail to connect, getting meetings under way, and so on). He estimates further that effective use of computer mail systems could save as much as two hours of the labor of nonclerical office workers per day and has translated this savings into $62.25 billion per year nationwide.

Cost reduction and increased productivity are not the only indicators of success of an office innovation. How are we to tell whether the introduction of new technology or new procedures into an office has positive effects on balance? Giuliano (1982) offers some criteria:

If new information technology is properly employed, it can enable organizations to attain the following objectives: a reduction of information "float," that is, a decrease in the delay and uncertainty occasioned by the inaccessibility of information that is being typed, is in the mail, has been misfiled, or is simply in an office that is closed for the weekend; the elimination of redundant work and unnecessary tasks such as retyping and laborious manual filing and retrieval; better utilization of human resources for tasks that require judgment, initiative, and rapid communications; faster, better decision making that takes into account multiple, complex factors; and full exploitation of the virtual office through expansion of the workplace in space and time. (164)

The "if" in Giuliano's observation is very important, of course, and the challenge for research is to flesh out the notion of what it means for this technology to be "properly" employed.

How will the new computer-based office tools change the training and skill requirements for office personnel? How will they affect the need for people in the office? Will they create more jobs than they abolish, or the reverse? What will be their effects on job satisfaction and the quality of the workplace? Will jobs be more enjoyable, dignified, and fulfilling, or will they become more tedious, dehumanizing, and boring? How will the new tools change the nature of office operations? What things will be doable that could not be done before? How will the quality of the work performed in offices be affected? How will the ways in which office personnel communicate with and relate to each other be affected? These are only some of the questions that remain to be answered.

### Organizational and Operational Effects

Fox (1977) has pointed out that one of the reasons the potential for medical computing has not been more widely realized is that the introduction of a computer into an organization often causes some organizational disruption. The same observation could be made with respect to organizations other than medical ones. There seems to be general agreement that the introduction of computer-based systems can affect the way organizations structure themselves; however, it is not yet clear what the nature of that impact has been or is likely to be. We have already noted the speculation that the widespread use of electronic mail among the members of an organization could increase one's access to people at different levels in an organizational hierarchy and might, as a consequence, tend to promote a greater sense of equality within the organization. Whether it will, in fact, work that way remains to be seen. It does seem likely, however, that any developments that facilitate direct communication between individuals within an organization, independently of organizational position or status, has the potential to effect structural change.

The introduction of information-management systems and computer-based decision aids can affect not only the structure of an organization but also its modus operandi. It can change traditional procedures for problem solving and create new functions to be performed. Adjusting to changes in organizational structure and procedures may be made more difficult for an organization's staff when the introduction of computer-based tools involves the recruitment of new staff to operate those tools. Problems of role changing and displacement may result, and may lead in turn to distrust and resentment among existing staff toward both the new personnel and the new operating procedures.

Naisbitt (1984) has argued that as a consequence of computer technology the pyramidal or hierarchical structure that has characterized management systems during the industrial era will give way to "the network model of organization and communication, which has its roots in the natural, egalitarian, and spontaneous formation of groups among like-minded people. Networks restructure the power and communication flow within an organization from vertical to horizontal" (281). The claim is not that companies will abandon all formal controls on communication but that a network *style* of management is evolving that will be characterized by informality and equality, and lots of lateral, diagonal, and bottom-up communication. One effect of the use of a computer network, Naisbitt suggests, it that users tend to treat one another as peers. Perhaps this is because of the direct access they have to each other independently of their status within a company.

• • • • • • • • • • • • • • • • • • • • • • • • • • • • • • • • • • • • • • • • • • •

Office work will be different in many respects from what it has been. Office workers will have different tools, some of them quite sophisticated by current standards. These tools will be portable, and many of the resources one will need to get one's work done will be accessible from a distance. Consequently, the office itself need not be in a fixed place—indeed it is quite possible that *office* may cease to be the appropriate concept. What these changes will mean with respect to productivity, and to the ways in which people relate to each other, remains to be seen.

# 12

## Designing Interactive Systems

The term *design* can mean two rather different things. Sometimes it connotes a process, or the product of a process, that precedes an attempt to build something. The purpose of a design, in this sense, is to specify in some detail the characteristics of the thing to be built. At other times the term connotes simply the characteristics of something that exists. Whether the design in this case preceded the thing to which the design refers is incidental. To say an artifact has a poor design in this sense is simply to say that it has undesirable characteristics.

It is a popular misconception that products of engineering come into existence by means of a sequence of clearly demarcated phases, beginning with design in the first sense of the word. This misconception is nowhere farther from reality than in the domain of software systems. To be sure, some system developers begin with an effort to specify an intended system's functionality, and perhaps its gross structure, before making any effort in the direction of implementation, but my belief is that this is the exception to the rule. More typically, the design process goes on more or less in parallel with implementation. Ideas about specific system features usually lead the embodiment of those features in the emerging system, but not by far. Seldom are those ideas explicitly written down or expressed in a sufficiently tangible form that they can be subjected to criticism and careful review.

### Designing for Users

It is often argued that the appropriate starting place for building an information system is the determination of the information needs of the prospective user (Bentley 1976; Epstein 1981,

1982; Mjosund 1975; Peace and Easterby 1973). The idea has some logical force, and one would be hard pressed to defend the notion that the user's needs should be ignored. Some efforts have been made to develop techniques to facilitate the identification of user requirements (Peace and Easterby 1973), but these requirements have proved surprisingly elusive. Moreover, needs may change as a consequence of the introduction of new tools. While some writers have emphasized the fact that the needs of users of information systems are always subject to change (Cavanagh 1967; Sheil 1982), too often the tacit assumption seems to be that they are fixed. But what one "needs," by way of information or any other commodity, is determined in part by what it is possible for one to get. The introduction of sophisticated information-manipulating devices and procedures into the workplace will make available to users information they could not have before and therefore did not "need." Once they have it, they may quickly come to depend on it; but until they have it, they should not be expected to be able to volunteer that they need it.

Wright and Bason (1982) have argued that attempting to design a system so that it "fits the user" rather than forcing the user to learn new procedures and new ways of conceptualizing the task is not always the best approach. They describe an experiment in which two software packages were designed to facilitate data analysis by a group of intermittent users. One of the packages was designed in accordance with users' descriptions of how they wished to think about the data. The other was designed to take advantage of certain characteristics of the programming language for organizing data. In the latter case users were told how to think about the data and were required to learn new procedures for analyzing them. The second package proved the more usable of the two. It does not follow, of course, that it is never a good idea to attempt to match the system to users' expectations or preferences. The study does show, however, that attempting to achieve such a fit is not a guarantee of optimal system design from the point of view of usability.

In order to understand better the process of design, several investigators have observed real designers working on real design problems (Carroll, Thomas, and Malhotra 1979; Malhotra, Thomas, Carroll, and Miller 1980). Thomas and Carroll (1981) point out that the design process is seldom a linear-sequential activity in which the large problem is decomposed and the sub-problems are solved independently of each other. Rather, the process tends to be cyclical and evolutionary. Solutions of some aspects of the problem are likely to result in recasting or redefining the overall problem and redirecting the subsequent design effort. Thomas and Carroll note too the importance of communication and collaboration between designers and intended system users in the design process. And they make the thought-provoking point that the statements of goals by the intended users of a system are typically focused on symptoms of current difficulties; aspects of the current approach that are working well are likely to be overlooked and may inadvertently be omitted from the new system design.

Hammond, Jorgensen, MacLean, Barnard, and Long (1983) interviewed several experienced system designers in an attempt to determine what types of decisions they made during the design process and on what these decisions where based. Their findings are difficult to summarize, but the quoted comments of the designers provide useful glimpses of the design process and of the attitudes and assumptions regarding users and human-factors issues that some designers bring to that process. It is clear from these comments that the designers who were interviewed were sensitive to human-factors issues but were largely unaware of resources they could tap to resolve them. Thus they were dependent primarily on their own logical analyses of the tasks that would be performed with the systems they were designing, and on their intuitions about users' needs and capabilities.

That designers differ considerably in their intuitions about what is needed in any particular situation is easily illustrated by a comparison of different systems that were intended to serve the same purpose for the same community. Roberts and Moran (1983) note, for example, that among nine text editors that they evaluated there were six different ways of invoking commands: "(1) type all or part of an English verb, (2) type a one-letter mnemonic for the command name, (3) hold down a control key while typing a one-letter mnemonic, (4) type a one-letter mnemonic on a chord set, (5) press a special function key, (6) select a command from a menu on the display" (269).

Some investigators have taken the approach of testing certain system features via simulation before the system has been developed. Thomas and Gould (1975) tested IBM's Query-by-

Example system (Zloof 1977) by having people attempt to compose queries that could be used to obtain answers to questions that had been given to them in English. Gould, Conti, and Hovanyecz (1983) experimented with a speech-to-text system ("listening typewriter") by interposing a human typist between the speech input and the text display. A similar approach was used in the development of the Lunar natural-language system (Woods 1973).

Nickerson and Pew (1977) caution that users who get involved in system development may not be representative of the average prospective user, simply by virtue of their involvement in the development process. To the degree that such prospective users become knowledgeable about the workings of a system, they may unwittingly accommodate to its design flaws and become effective users in spite of them. It is not safe to infer from the fact that such users find a system design adequate that inexperienced users will be able to use it effectively as well.

Norman (1983a) suggests that there are no simple answers to interface design questions, only tradeoffs. In a menu-based system, for example, menu size must be weighed against display time: the larger the menu, the longer it takes to generate it. Norman advocates attempting to quantify such tradeoffs by determining how user satisfaction depends on each of the trading variables independently and then plotting one satisfaction function against another. The resulting tradeoff functions are often concave upward, suggesting that satisfaction—if it is a linear combination of the satisfaction with respect to both features—is maximized by an all-or-none solution. (When the tradeoff functions are concave downward, satisfaction is maximized by a point that is not at the extremum of either continuum.) This is somewhat counter to intuition. One might think that if users liked both large menus and fast display times and it were not possible to get both, they would prefer a compromise in which an intermediate menu size was coupled with an intermediate display time. If the assumptions underlying the tradeoff functions are correct and the measurements that produced the concave-upward curves are valid, intuition in this case is wrong. As examples of tradeoffs other than the one between menu size and display time, Norman mentions command languages versus menus, the use of lengthy descriptive names for commands and files versus short cryptic abbreviations, and hand-held computers versus workstations.

The notion of tradeoffs is compelling. The goal of quantifying them and using user-satisfaction functions to guide interface design is ambitious. User-satisfaction curves may well vary from user to user, as Norman points out. However, the conclusion that the tradeoffs inferred from such curves really represent user preferences for combinations of system features requires the assumption that preferences with respect to individual features combine in a simple linear fashion and that interactions can be ignored. If this assumption does not hold, user satisfaction with an overall system may not be inferable from tradeoffs involving pairs of system features. These caveats aside, the approach that Norman is advocating deserves further exploration. If it proves possible to determine user-satisfaction curves for large classes of users and for the more important system features, and to use these curves to predict accurately preferences among systems described in terms of these features, the approach could provide very useful data to guide interface design. Whether the approach has this potential is an empirical question, and experimentation should give us the answer.

## Some Proposed Design Guidelines

The literature contains many general guidelines, rules of thumb, criteria, and recommendations for the design of interactive systems or parts thereof. Shneiderman (1980a), for example, presents in summary form design principles and guidelines for interactive systems suggested by several writers: Hansen (1971); Wasserman (1973); Kennedy (1974); Sterling (1974); Pew and Rollins (1975); Gaines and Facey (1975); Cheriton (1976); and Turoff, Whitescarver, and Hiltz (1978). Other relevant reports include Howell (1967); Howell and Getty (1968); Bennett (1972); Doherty, Thompson, and Boies (1972); Foley and Wallace (1974); Smith (1974); Sterling (1975); Chamberlain (1975); Engel and Granda (1975); Stewart (1976); Watson (1976); Nickerson and Pew (1977); Newman and Sproull (1979); Williges and Williges (1981); Bannon, Cypher, Greenspan, and Monty (1983); Gould and Lewis (1983); and Nakatani and Rohrlich (1983).

These documents differ considerably in scope and in the degree to which the recommendations are supported by empirical data. In general, more guideline information is available relating to the physical interface than to the cognitive interface. Ramsey and Atwood (1979) make the point this way:

In some well-established research areas, such as keyboard design and certain physical properties of displays, guidelines exist which are reasonably good and fairly detailed. Such guidelines may be quite helpful in the design of a console or other interface device for a system, or even in the selection of an appropriate off-the-shelf input-output device. As we progress toward the more central issues in interactive systems, such as their basic informational properties, user aids, and dialogue methods, available guidelines become sketchy and eventually nonexistent. The interactive-system designer is given little human-factors guidance with respect to the most basic design decisions. In fact, the areas in which existing guidelines concentrate are often not even under the control of the designer, who may have more freedom with respect to dialogue and problem-solving aids than with respect to terminal design or selection. (2)

Individual design recommendations range from the very general and abstract to the quite specific and concrete. There are some commonalities, at least in gist if not in detail, but the differences are also quite striking. An interesting exercise would be to have each of several people attempt independently to integrate the various lists of recommendations that have been produced into one master list, with elements arranged according to importance and suitably paraphrased—a list that would represent the combined wisdom of the writers as a group. One suspects that the master lists produced would differ considerably from each other. The field has yet to converge on a set of design principles that is widely recognized as the appropriate set or something close to it.

While most of the more general recommendations would probably be accepted by system developers as desirable objectives, merely stating them provides limited help to the developer. Moreover, there is little evidence that even those principles that appear to be most obviously sound and straightforward are really followed. Gould and Lewis (1983) argue that four principles for developing a system design must be applied if systems are to be produced that are easy to learn, useful, easy to use, and pleasant to use:

- Designers must understand who the users will be.
- A panel of expected users (e.g., secretaries) should work closely with the design team during the early formulation stages.

- Early in the development process intended users should actually use simulations and prototypes to carry out real work, and their performance and reactions should be measured.
- When problems are found in user testing, as they will be, they must be fixed. This means design must be iterative: there must be a cycle of design, test and measure, and redesign, repeated as often as necessary. (50)

Gould and Lewis claim that while these principles may appear to be obvious and designers may claim to believe them, they are not generally used in the design process. As evidence, they summarize the results of six surveys in which over 400 system designers and developers were asked to write down the five or so major steps one should go through in developing and evaluating a computer system. In spite of a liberal scoring policy in which credit was given for even the simplest mention relating to any one of the four principles, no matter how impoverished or incomplete the thought, only 2 percent of the respondents mentioned all four principles, 35 percent mentioned only one, and 26 percent mentioned none of them. Further, as Gould and Lewis point out, the fact that designers mention a principle as being important when answering a survey question is not conclusive evidence that they would follow it when designing a system.

Gould and Lewis offer several speculations as to why their "commonsense" principles are not as obvious or compelling to designers as they seem and why they are seldom used. Among the reasons they suggest are underestimation of user diversity (because designers have insufficient contact with users, they assume greater uniformity and ability than is warranted); too much faith in the ability to design adequately from rational analysis of how a task should be done; the fact that users often do not know what they need; and fear of lengthening the development process.

A major problem with most general objectives is figuring out *how* to meet them. Moreover, general guidelines tend to be vague. Further, intuitions, even those that are widely shared, about how to design an interactive system do not alway prove to be valid. Walther and O'Neil (1974) have shown, for example, that flexibility at the interface, defined as interface alternatives available to the user, is not uniformly beneficial or appreciated by all users.

There are a number of intuitions that appear to be sufficiently widely shared to assume the status of dogma. Here are a few examples:

- It would be a good thing if people could talk to computers in unconstrained natural language.
- Computer-generated speech should sound as human-like as possible.
- If an algorithm for a task can be written, it should be, and the task should be performed by a computer.
- The human's task in person-machine systems should be made as simple as possible.

While these and many other intuitions may prove correct, it is also possible that some of them may not. It does not seem prudent to accept them uncritically and without qualification. Moreover, it is not always apparent precisely what some of the intuitions mean. Everybody espouses such objectives as simplicity, versatility, and friendliness, but seldom are these terms defined. It is apparent that maximum simplicity, taken literally, is not a reasonable goal. A maximally simple display would be one with nothing on it.

The more concrete recommendations lack cohesion as a group, inasmuch as they are scattered throughout the literature and have not been related to any integrating conceptual framework. Fragmentary as they are, however, these specific ideas represent the collective wisdom of builders and users of systems on the question of making those systems more effective and convenient to use. Here are just a few of the many ideas one finds in the literature, or embodied in systems, about specific capabilities or characteristics that interactive systems should have.

- Unambiguous feedback to the user about what the system has interpreted particular inputs to be.
- Required confirmation of commands with potentially catastrophic effects (such as file deletion).
- "Undo," "backup," or "nullify" commands permitting one to retract a command that has been given.
- The possibility for the user to take control of the dialog at any time.
- The use of names for variables, commands, objects, files, and so

forth that are suggestive, insofar as possible, or what they represent.

- The possibility of extending the command language, by defining procedures, for example.
- The possibility of user-defined command-language abbreviations and equivalences.
- "What you see is what you get" text editing.

Even these guidelines, specific and concrete as they are, may not be as easy to follow as it would appear. Consider the second one, for example. Clearly it is a good idea to help protect the user from inadvertently invoking a command that will have a catastrophic effect. But how does one do that? Shneiderman, Wexelblat, and Jende (1980) suggest that the execution of potentially disastrous actions should require considerable mental force. Norman (1983b) points out that simply making the user confirm a particular action sequence probably does not satisfy this principle, because if the confirmation is easy, in time it will be invoked automatically as part of the command sequence.

The last suggestion in the list is necessitated in part by the fact that advances in hardware often outrun advances in software, and software that was developed for one set of hardware when transferred to more powerful hardware fails to exploit the full potential of the latter. Thus text-editing software that was developed for use with paper terminals is sometimes used with graphics terminals, and when this happens the versatility of the graphics terminals is underutilized. Among the most primitive and useful editing functions is the ability to delete the letter, word, or line that one has just typed. Unfortunately, there is no way literally to remove what has already been typed from a page of paper (at least this is true for the vast majority of paper-output terminals that have been widely used). Various conventions have been devised to permit the user to delete items from the electronic copy of a document and to represent these deletions on the displayed page by means of codes. For example, deletion of the preceding character might be effected by striking a specific control key and indicated on one's output in the following way: *the fa/aollowing way.* The slash represents the deletion of a letter and the letter following the slash represents the letter that was deleted. Similarly, deletion of a word or letter string back to the first interword space would be effected by another control key and represented on the output as follows: *the smalles<slargest file*. In this case the < indicates deletion of the preceding letter string back to the last space, and the letter following the < indicates the first letter in the deleted string.

When such conventions are carried over into an editing program that uses a graphics terminal, they are making poor use of the terminal's flexibility from a human-engineering point of view. What one would like to see on the scope face is the actual change that has been made in the document. When a letter is deleted, one would like to have that letter disappear; and the same is true, of course, for words, lines, or other segments of one's text. The general principle on which one would like to work is that "what you see is what you get." What hinders the realization of this objective is that some software must be relatively terminal-independent—for example, an electronic mail system that is to serve users with a variety of paper and scope terminals.

A large number of interactive computer systems have been developed to date, and presumably much has been learned about the design of such systems as a result of this activity. By and large, what has been learned has not been communicated effectively, however. It exists in the heads of individual system developers and, in fragmented form, in numerous reports in the literature. Collection and organization of the various proposed design guidelines—only a few samples of which I have recorded here—would be a service to the field, fostering constructive criticism and studies aimed at empirical validation.

### Formal Specifications as Design Tools

Many system designers and developers advocate the development of some formal specification as a first step in designing an interface. Formalisms that have been used for this purpose include state-transition diagrams, Petri diagrams, and command-language grammars (Blesser and Foley 1982; Feldman and Rogers 1982; Jacob 1983a, b; Parnas 1969; Reisner 1981, 1982; Roach and Nickson 1983; Wasserman and Shewmake 1982). Formal specifications may be developed at different levels of detail or for different aspects of a system. Foley and his colleagues (Foley and van Dam 1982; Foley and Wallace 1974) distinguish three levels of specification of an interactive interface: semantic, syntactic, and lexical. Jacob (1983a) describes these three levels as follows:

> The *semantic* level describes the functions performed by the system. It tells what information is needed to perform each function and the result of performing it. The *syntactic* level describes the sequences of inputs and outputs. For the input, this means the rules by which sequences or words (*tokens*) in the language are formed into proper (but not necessarily semantically meaningful) sentences. The *lexical* level determines how input and output tokens are actually formed from the primitive hardware operations (*lexemes*). (31)

Formal specifications can help the designer identify and correct problems in a design before it is implemented in software or hardware. Thus refinements can be made at a point where they are relatively easy and inexpensive. Formal specifications can also provide guidance for the system builders when it is time to implement a design. They are not trivially easy to develop, however, and do not appeal to all system designers and builders. The extent to which they are used is to some degree a matter of cognitive style.

### Guided Evolution as an Approach to Design

System development is often thought of as requiring two qualitatively different tasks: design and implementation. When the functions the system is to serve are well structured and clearly understood, these tasks can be performed seriatim, and it may make sense to insist that the design be finished before the implementation begins. As the complexity of the system increases, however, it is more and more often the case that what is required of it cannot be entirely specified in advance but is best determined by a process in which developer and user together explore various possibilities in an effort to converge on something useful. Factors that can prohibit the predesign of a system include the following: some complex tasks are poorly understood even by people who perform them; prospective users of an innovative system often have very limited understanding of what such a system might be expected to do and consequently are unable to plan effectively how best to apply it to their jobs; the introduction of new technology to a job situation often not only changes the way a job is done but fundamentally modifies the nature of the job.

In support of the exploratory approach, Sheil (1983) argues as follows:

Any attempt to obtain an exact specification from the client is bound to fail because, as we have seen, the client does not know and cannot anticipate exactly what is required. Indeed, the most striking thing about these examples [which Sheil has just given] is that the clients' statements of their problems are really aspirations, rather than specifications. And since the client has no experience on which to ground these aspirations, it is only by exploring the properties of some putative solutions that the client will find out what is really needed. No amount of interrogation of the client or paper exercises will answer these questions; one just has to try some designs to see what works. (132)

Sheil goes on to note the importance of developing computing environments that facilitate exploratory programming and to describe some of the features such environments should have.

Eason (1982) too asserts that, given the current state of computer technology, the evolutionary design and development of information systems is not only a possibility but the preferred approach.

The nature of modern computer technology means this (evolutionary) approach is now a practicable proposition and it means users have the time and opportunity to gather experience before they have to specify the systems they regard as desirable. In working with organizations we have tried to promote learning for design decisions by creating temporary design bodies, implementing pilot studies specifically for learning purposes, developing user design exercises, instituting regular evaluating procedures, and creating social structures for user support and system evolution. (212)

Evolutionary design in Eason's terms means building gradually, moving from small configurations that contain some of the desired functions to larger ones that contain more, and maintaining flexibility for change, insofar as that is possible, throughout the process.

The value of guided evolution as an approach to system development more generally seems to have been fairly widely recognized (C. M. Brown 1984; Carroll and Rosson 1984; Cushman 1983; Draper and Norman 1984). With respect to military command and control systems, Cushman (1983) asserts, "These themes of evolutionary development with intimate user involvement are so reinforced in studies and the literature on command and control systems as to leave no doubt as to their validity" (50).

One evolutionary approach to the design of interfaces is that of having users design them indirectly, by indicating through their attempts to use a command language with which they are not familiar what they think the system ought to be able to understand. This approach is illustrated by a system development effort of Wixon, Whiteside, Good, and Jones (1983). Using two principles to guide their effort, "(1) an interface should be built based on the behavior of actual users, and (2) an interface should be evolved iteratively based on continued testing" (24), they attempted to build an interface for an electronic mail system that would permit a novice user to perform useful work during the first hour without the aid of menus, documentation, or instruction. "The interface was to be so natural that a novice would not have to be trained to use it" (24).

The investigators had users attempt to perform a specified electronic mail task on a computer. Users were given no instruction on how to perform the task, so they structured their commands to the system in accordance with their intuitions. Commands that were not interpretable by the system were intercepted by a human operator, who translated them into commands the system could recognize. Presumably the users had the illusion that they were working directly with the computer and that the computer was recognizing all of their inputs. On the basis of inputs recorded during the sessions, changes were made periodically in the system software so as to increase its ability to process the user-generated commands. The initial version of the system software was able to handle about 7 percent of all the commands generated over the six-month period of the experiment. The final version was able to interpret correctly about 76 percent of them. Wixon and his colleagues describe the resulting command language as a "user-defined interface."

What are the implications for human-factors research of the idea that the best approach to the development of some systems is an exploratory one in which design and implementation evolve together? If conventional task and requirements analyses are of limited use here, what role should engineering psychologists play, and what special tools and methods do they bring to the problem? If guided evolution is the preferred way of developing computer-based systems, better methods are needed for providing the guidance. This means the development of techniques for making formative evaluations of systems, or aspects of systems, in such a way that the results can be useful in guiding the further development of those systems. It also means keeping options open, to the extent possible, when it is not clear which of a set of design possibilities is to be preferred.

### The Problem of Standardization

Lack of standardization has been a continuing problem in the computer industry, and it is felt in various ways. The keyboards on computer terminals differ enough that proficiency in the use of one of them is almost a guarantee of difficulty in using another. Seldom will a peripheral device designed for one mainframe function on a different one. Software that runs on one computer is unlikely to run on another, even if written in a language that is available on both machines, because compilers and interpreters are often less machine-independent than they are claimed to be. (For a delightful discussion of the effects of lack of standardization with respect to whether bit strings in bytes or words of computer codes are interpreted left to right or right to left, see Cohen 1981.)

The lack of standardization causes severe problems for system builders. Consider, for example, the task of the designer of an electronic mail system that must be able to function with a wide assortment of computer terminals, both hardcopy and video. An output format that is appropriate for some subset of the available terminals is very likely to be inappropriate for some other subset. The same type of problem is encountered in the design of information services of the Videotex variety (see chapter 9). These systems too must be accessible by a variety of terminals, including some that have not yet been designed.

Through hindsight, perhaps, one begins to realize the problems of introducing a Videotex system which is suitable for today's technology and yet allows for future expansion. One needs to describe images at the central data banks in such a way that they are completely independent of the data access arrangements at the central computer, the characteristics of the communication medium, and, what may be most important, completely independent of the display terminal construction and resolution capabilities. (Bown, O'Brien, Sawchuk, and Storey 1978, 2)

The approach that Bown and colleagues propose to take to the problem is an interesting one. They do not lobby for the adoption of standard terminal equipment, but rather propose the development of a way of describing images that is independent of the characteristics of the terminals on which these images will eventually be displayed. The technique they promote involves analyzing images into certain primitive components (points, lines, arcs, polygons) and transmitting instructions regarding how to construct specific pictures with these primitives (Bown, O'Brien, Lum, Sawchuk, and Storey 1979). This assumes, of course, that there is resident at the terminal enough computing power and the appropriate software to interpret and execute such instructions.

In the absence of standards, the user is forced to make choices involving difficult tradeoffs. To get a personal computer with the hardware features one wants, for example, one may have to forego getting one with a better program library. In short, the lack of standards inhibits the sharing of equipment and programs, assures much duplication of effort, and compels users to make unsatisfactory decisions.

Although the detrimental effects of lack of standardization are widely recognized, the development of standards has proved exceedingly difficult. Standardization necessitates compromise; and selection of one approach as a standard requires rejection of alternatives. Developers of equipment and programs typically believe that what they have developed is better than the alternatives; otherwise they would not have developed what they have. In other words, there is general recognition of the desirability of standardization but not much agreement on what the standards should be.

Within the military, standardization and interoperability were the key objectives in the introduction by former Secretary of Defense McNamara of the World-Wide Military Command and Control System (WWMCCS). The WWMCCS was built around a single off-the-shelf computer, the Honeywell 6000. As of 1981, 83 such machines had been procured and were deployed at 26 sites; the number of workstations was over 1,500. The development of the Ada programming language and its adoption by the Department of Defense as its standard language represent another response to the standardization and interoperability problem. The problem is far from solved, however, and the Standard Computer Resource Interface in Management Plan (SCRIMP) issued by the Center for Tactical Computer Systems (CENTACS 1980) cites the lack of a standard product line of terminals as responsible for "a growing proliferation which is impeding the attainment of interoperability, continuity of operations, security, reliability, availability, and maintainability" (11).

Here is one quite remarkable example of how the lack of standardization of computer equipment manifests itself in military systems. The function keypads for the TACFIRE system have 64 possible codes; 47 of these codes are common to the keypads that are used at the division and battalion levels, but only 19 of them are represented the same way on both pads (Synectics Corporation 1980). Such an arrangement ensures that a person who has become adept with the keypad that is used at one echelon will have difficulty with the one that is used at the other.

The U.S. Department of Defense recently issued a request for a proposal to develop a network query language for its Intelligence Information System. The problem is that the system includes about 40 data-processing and telecommunications sites that do not have common hardware or software. Approximately 20 different data management systems are used across these sites. This heterogeneity makes it impossible, or at least extremely difficult, for analysts at different sites to share databases. Thus the purpose of the intended procurement is to develop a database inquiry capability that will provide access to all systems by users who do not have detailed operating knowledge of those systems. All users will access the systems through the same network query language; it will be the job of the software to translate between that language and the database systems used at the various sites. This, like the approach of Bown and colleagues to the problem of designing Videotex systems that can service various types of terminals, demonstrates the hope of living with diversity by developing tools that can cope with it so the user does not have to. In this case the tool is a piece of software that can interface to diverse databases and provide users with a common access path to all of them.

A great many "higher-level" general-purpose computer languages have been developed. A relatively small number of these languages (including their various dialects) account for a very large percentage of all programming that is done. These include FORTRAN (FORmula TRANslator), COBOL (COmmon Business-Oriented Language), BASIC (Beginners' All-purpose Symbolic Instruction Code), LISP (List Processing Language), and Pascal (named for the French mathematician Blaise Pascal). It is probably not possible to show convincingly that any of these languages is better than the others in a general sense. Each has its advantages and its limitations, and each has its community of users. Different languages were developed with different applications in mind, and, not surprisingly, a programmer's choice of language tends to be somewhat applications-dependent.

The recent development of the Ada language, named after Augusta Ada, Countess of Lovelace (admirer and chronicler of Charles Babbage's prescient work), was motivated by concern over the proliferation of languages and by the desire of the U.S. Department of Defense to standardize on a single one. The design followed an intensive study of existing languages and represents an effort to produce something more versatile and effective than any of them. To the degree that other widely used languages tailored to specific application areas are especially well suited to their uses, standardization on a more general common language will necessarily involve a tradeoff. Whether the devotees of more specialized languages will find this tradeoff acceptable remains to be seen.

The idea of "layered protocols" (Cole, Higginson, Lloyd, and Moulton 1983; Denning 1985; Sherman and Gable 1983) appears to represent a promising approach to standardization of the communication protocols followed by the users of computer networks. Simply stated, the idea is that systems may be described at various levels (the level of the physical interface, the level of language syntax, the level of semantics, the level of system application). The model of network interconnections adopted by the International Standards Organization (ISO) recognizes seven layers of interconnections. Standardization can occur at any level but may be much more important at some than at others. A protocol is a set of rules that governs the way information is transferred between components of a communication system. Protocols can exist at any of these levels, and those that govern behavior at a particular level may be thought of as independent of those that govern behavior at another level. How much standardization should be attempted at each level and what the standards should be are questions for research. At present there are four major data communication protocols: Xerox's Network Systems (XNS) protocols; the U.S. Defense Department's Transmission Control Protocols/Internet Protocols (TCP/IP); IBM's System Network Architecture (SNA) protocols; and the International Standards Organization's Open System Interconnection (OSI) protocols (Raunch-Hindin 1984).

How can we standardize without stifling innovation? The history of language development illustrates the dilemma represented by the need for innovation on the one hand and standardization on the other. A great deal has been learned about how to design languages, and programming environments more generally, that make the power that is latent in computing machinery accessible in a psychological sense to the sophisticated user. This progress has come about because many capable people have enthusiastically taken up the challenge of developing languages that are better in some respects (more powerful, more convenient to use, better suited to specific applications) than those that already exist. This has also produced, however, a babel of languages and has greatly inhibited the sharing of results of programming efforts. We have noted that the Defense Department has responded to this unmanageable diversity by adopting in principle one language, Ada, for all of its future operations. There is already some evidence, though, that too strict an enforcement of this constraint could make inaccessible to the Defense Department some of the tools that it will undoubtedly want to have.

In general, new languages and operating systems are developed because people are dissatisfied with those that already exist. There can be no question that many of the languages and operating systems that exist today are much more powerful and easy to use than those of a few years ago. Moreover, it would probably not be possible to control the proliferation of languages and systems even if it were considered desirable to do so. The question is how to have the best of both worlds. How can we encourage progress in language development and operating-system design on the one hand, while fostering cooperation and reducing the need for duplication of effort on the other?

•••••••••••••••••••••••••••••••••••••••••••••••

Designing information systems is still more an art than a science. How designing is accomplished—whether it is accomplished—depends very much on the system developer. In many cases it is impossible to identify a design process that is independent of the work of actually building the system.

Design guidelines have not been articulated in a well-organized, generally accessible form, although there exist in the literature many suggestions that will doubtless be incorporated in guideline compendia when such things are produced. Most of the guidelines that have been suggested, however, lack empirical validation and are therefore open to challenge.

One point on which there seems to be fairly wide agreement is the merit of guided evolution as an approach to design. Potential users often cannot say what they want a system to be able to do or how they would like to be able to interact with it, because they do not have an adequate understanding of the possibilities. Also, the introduction of information technology often changes jobs qualitatively in unanticipated ways. When potential users try to imagine how information technology might be used in their work situations, they tend to think about how it might help them do more efficiently what they are in the habit of doing. They are unlikely to consider the possibility that with the new technology they may be doing rather different jobs.

The issue of standards is a difficult one in a rapidly evolving technology. Standards serve the objectives of increasing efficiency, decreasing duplication of effort, and ensuring compatibility among systems and software packages. They simplify the life of the user by reducing his need to make difficult decisions involving tradeoffs among computer hardware, compatibility of peripheral devices, and availability of desired software. On the other hand, standardization imposed too early or too aggressively can inhibit the development of improved methods and devices. The search for an optimal compromise is a continuing one.

# 13

## Some User Issues

### Attitude and Motivation

There may be many people who do not know a lot about computers, but there must be few in industrialized countries who have not heard a lot about them. And it is apparent that many people have fairly strong attitudes, both positive and negative, about them. Zoltan and Chapanis (1982) administered a questionnaire to some 500 professionals and found that computers were generally perceived by their subject group as "efficient, precise, reliable, dependable, effective, and fast" on the one hand but also "dehumanizing, depersonalizing, impersonal, cold, and unforgiving" on the other. Coupled with Walther and O'Neil's (1974) finding that people who had negative attitudes toward computers learned how to use them on specific tasks less efficiently than did those with positive attitudes, Zoltan and Chapanis's results suggest that attitudes may stand in the way of many people's becoming effective users of this technology.

An especially important attitudinal issue is the threat, real or perceived, that information technology represents to the job security of some people. When we think of this problem we tend to think first of lost jobs due to automation, but the threat may take more subtle forms. Skills and knowledge are valued in the workplace in direct proportion to their scarcity and their consequence to a company's operation. Scarcity presumably correlates highly with difficulty of acquisition: the longer it takes to become highly knowledgeable in some area or highly skilled at performing some task, the smaller the number of people who are likely to have that knowledge or skill. One of the results of introducing new technology in the workplace is that some of the old hard-won skills and knowledge are rendered obsolete. Consider a person who, by virtue of knowledge or skill acquired over many years, has become recognized by a company as a critical resource. It should not be surprising if that person fails to perceive as an unalloyed blessing a technological innovation that will make his job readily doable by a person with far less experience or, worse yet, by a machine. If the new tool is well designed ergonomically and easy to use, that will not necessarily be perceived as a plus. Indeed, it may exacerbate the problem. Not only have the individual's skills been devalued, but the job requirements have been simplified to the degree that even his demonstrated ability to acquire complex skills is a questionable asset.

The assumption that it is a good idea to provide people with more powerful tools and to make those tools easier for people to use seems a reasonable one. Certainly it is an assumption that underlies and motivates much human-factors work. However, innovation that is perceived as progress from one point of view may be perceived quite differently from another; in particular, the individual whose own job is involved is very likely to have a view that is strongly influenced by the implications of the innovation for his vocational status. It is a view that should not be ignored.

Other, more specific attitudinal problems have been reported. The finding by the National Institute of Occupational Safety and Health (NIOSH) (Murray, Moss, Parr, Cox, Smith, Cohen, Stammerjohn, and Happ 1981) that operators of visual display terminals (VDTs) report a greater number of health problems, especially emotional and gastrointestinal problems, than do people performing comparable jobs without video display terminals is a case in point. It is particularly interesting in view of the fact that the NIOSH study ruled out radiation from CRTs as a health hazard. The results of the study suggest that there is something about the operation of VDTs, or about the work that is typically done on them, that makes these jobs more stressful than they should be. (For an extensive review of health problems and issues relating to the operation of computer terminals and in particular VDTs, see Oestberg 1975).

Attitudinal issues represent challenges for research in several ways. There is the problem of assessing the attitudes themselves. There is the problem of determining the effects of attitudes on performance. There is the problem of determining when negative attitudes are justified and when they are not.

And there are the twin problems of changing negative attitudes when they are not justified and changing the conditions that have produced those attitudes when they are.

An area where motivational issues matter considerably is that of training. Carroll (1982a) urges that we try to understand better why it is that people appear to master complex, cognitively demanding computer games, such as Adventure, more easily than they master the use of cognitively demanding computer-based tools, such as text editors. Why do certain characteristics of the situation seem to be stimulants to learning in the one case and obstacles in the other? He suggests a variety of ways in which the situations facing learners of Adventure and learners of a text editor are similar and a variety of ways in which they differ. Among the differences that might be significant he notes the following:

- In Adventure the player cannot issue a command that does not elicit some message from the system. This is not true of the text-editing systems Carroll studied.
- In Adventure it is impossible for the player to do anything that aborts the game or leaves the player unable to get a response from the system. With the text editors it was possible to get hung up and unable to continue.
- Adventure is supposed to be fun; whereas the use of a text editor is not.

Carroll questions whether it really is "easier" to learn to play a game like Adventure than to learn to use software tools such as text editors. Perhaps the difference is primarily a motivational one: the learner is more willing to invest the required cognitive effort in the one case than in the other. Carroll makes the interesting observation that in our culture there has traditionally been a sharp distinction between work and recreation. Work is not expected to be fun. Undoubtedly, in the view of at least some people, work not only is not expected to be fun but *should not* be fun. To have fun while making a living is perceived as cheating—having one's cake and eating it too.

### User Acceptance of Interactive Systems

One might assume that the acceptability of a system to a user would depend on whether the system meets a bona fide user need and is convenient to use. Apparently there is more to it than that. Sometimes acceptance or rejection can hinge on other kinds of factors. Castleman, Whitehead, Sher, Hantman, and Massey (1974) have reported, for example, that physicians who were given an opportunity to use an automated medical-history-taking system were reluctant to do so until they had affected the system design in some way. "*In every practice*, once the physician had modified the original questionnaire in some way, often quite minor, it was much more highly regarded by the physician and was used with greater frequency" (17). The authors note "the necessity [for the intended users] to modify any such application program in some way, however slight, to customize it to their own particular circumstances, before the program was found really acceptable" (39).

In keeping with this finding, Igersheim (1976) has reported survey data that indicate that a system is more likely to be acceptable to people who have been involved in its implementation than to people who did not have that involvement. Eason (1981) has also noted that one of the major advantages of involving potential users in system design is that such involvement increases the likelihood of their acceptance of the final system. More generally, several writers have argued for the involvement of intended users in a system's design for a variety of reasons (King and Cleland 1975; Kriebel 1970; Martin and Parker 1971). Eason points out, however, that there are disadvantages as well as advantages in this involvement and makes some suggestions about how to minimize the disadvantages.

These observations are undoubtedly related to the finding by Monty, Perlmuter, and their colleagues that people do better on tasks they believe they have participated in choosing than on those they believe to have been imposed on them (Monty, Geller, Savage, and Perlmuter 1979; Perlmuter, Scharff, Karsh, and Monty 1980).

This is not to suggest that user involvement in system development is the only, or even the primary, factor determining whether a system will be accepted by intended users, but only that it is an important one. The most compelling reason for rejection of a system is the failure of the system to offer users what they want or need. It is certainly also the case, however, that systems that do provide useful functionality are sometimes not used by people who presumably could benefit from their use. I have tried elsewhere to identify some of the reasons why time-sharing systems were not used by some people who might have benefited from them (Nickerson 1981). The reasons include inadequate accessibility, unacceptable system dynamics (response time), inadequate documentation and training aids, and inconvenient command languages.

Sometimes what had been assumed to be an advantageous aspect of a computer-based system has turned out not to be. One finding in Castleman and colleagues' experiment with an automated medical-history-taking system (1974), mentioned at the beginning of this section, illustrates the point. The intent was that patients would interact directly with the system; however, when the investigators found that patients took an excessively long time to type answers to the history-taking questionnaire at the terminal, they then had each patient answer questions listed on the questionnaire to a clerk, who keyed them into the computer in about 5 percent of the time the patient would have taken to do so.

W.I. Card and colleagues (1970, 1974) have also studied the effectiveness of an automated medical-history-taking system. This system required no typing by the patient but only push-button responses to indicate "yes," "no," or "?". Most of the patients found the method acceptable, but the error rate in the completed interview forms was nearly twice as great as when the interviews were performed by doctors.

### How to Tell How Hard One Is Working

The measurement of user workload has presented methodological and conceptual problems in several contexts. The problems are no less severe in the context of person-computer interaction. The workload in this case is seldom apparent from the user's overt activity. How to measure how hard one is working, or the extent to which one's capacity is being strained, in any particular interactive situation is likely to pose a methodological challenge for some time to come.

Although the "information-processing load" imposed on the user has been proposed as an appropriate criterion for evaluating systems (Finkelman 1976), there is no general agreement as to how load should be measured. There have been some specific suggestions—for example, Treu (1975) has suggested that the time the user of a system spends thinking, immediately before issuing a command, might make one useful indicator of mental workload—but no single measure or set of measures has been widely accepted as adequate to the task. As Hammond, Long, Clark, Barnard, and Morton (1980) point out, protocol data collected at the terminal may fail to reveal many of the cognitive aspects of the user's behavior, such as goal structuring and higher-order planning; and such data do not record the user's activity during pauses in the interaction. In particular, they do not reveal whether, or how deeply, the user is thinking about the problem.

## Users as Sources of Input Errors

The user as a source of error in interactive data-processing systems remains a cause of concern for developers of systems— especially those systems that are to be used in situations in which errors could prove to be consequential if not disastrous. Input errors originating with users have been seen as a major problem in the military, and considerable attention has been given to the development of a better understanding of their types, causes, and impact on system operation (Mace, Harrison, and Seguin 1979; Nawrocki, Strub, and Cecil 1973). Several investigators have studied the errors that people make with specific systems. Typically, the approach has been to attempt to identify a variety of errors or error classes, to determine the relative frequency with which each is made, and, in some cases, to identify the underlying causes of the errors (Black and Sebrechts 1981; Galambos, Wikler, Black, and Sebrechts 1983; Mosteller 1981; Thomas 1976).

One obvious approach to the problem of input errors is to attempt to give the system itself the capability of detecting and flagging, if not correcting, errors (Obermayer 1977). Some of the errors that users make (such as misspellings, the use of synonyms) are detectable to some degree automatically, and in those cases this approach is feasible. Not all user errors are detectable within the current state of the art, however, and perhaps not even in principle (Nawrocki, Strub, and Cecil 1973). Other ways of addressing the problem include more adequate training, better human engineering of interfaces, and the use of verification procedures that force the user to think twice about potentially disastrous commands. Certain types of errors that beginners are likely to make can be prevented by the use of specially designed "training interfaces" that limit the commands and system functions to which the user has access (Carroll and Carrithers 1984).

Norman (1983b) has argued that an analysis of the types of errors humans make in a variety of situations is one reasonable way to begin to develop interface design principles. He suggests that preventable errors in human-computer interaction come primarily from a small number of sources and that many of them could be prevented or made easily correctable through the use of the following design principles:

- Feedback: The state of the system should be clearly available to the user, ideally in a form that is unambiguous and that makes the set of options readily available so as to avoid mode errors.

- Dissimilarity of response sequences: Different classes of actions should have quite dissimilar command sequences (or menu patterns) so as to avoid capture and description errors.

- Revocability of commands: Actions should be reversible (as much as possible), and where both irreversible and of relatively high consequence, they should be difficult to do, thereby preventing unintentional performance.

- Consistency of the system: The system should be consistent in its structure and design of commands so as to minimize memory problems in retrieving the operations.

### The Use of Metaphors in Training New Users

We tend to think of primitive people as living in a world they did not understand, but one can make a case that the average person today comes into contact daily with more things he does not understand than did his ancient, and not so ancient, predecessors. How many people who use it regularly understand electricity, for example? Or radio, or television, or microwave ovens? The fact that we are familiar with these things, know how to use them to our advantage, and are not frightened by them does not mean that we understand them. People do, however, develop models, or myths, of how these things work. These models are almost certain to be inaccurate in various respects, but unless they lead to actions whose results clearly and dramatically expose those inaccuracies, they are very likely to persist. A common model of a thermostat has it working like a valve in a water pipe (Kempton, in press). According to this model, the higher one sets the thermostat above the current temperature, the faster the heat will rise; so if the temperature is at 60 and you want to bring it to 70, you will get it there faster by setting the thermostat at 80 than you would by setting it at 70. The model is incorrect, but not so egregiously that in acting upon it one will necessarily discover its incorrectness. The valve metaphor for a thermostat is a reasonably functional one. In contrast, the belief that one raises the temperature by lowering the thermostat would be unlikely to persist for very long, because the behavior it would produce would yield compelling evidence that it was wrong.

The computer is certainly among the devices that people will be encountering and using more and more frequently without acquiring a very profound understanding of how they function. We may assume that people will construct their own theories about how these machines work. Unfortunately, when new users interact with a complex software system, they see it only in glimpses. They look at it through a series of small windows, as it were. Only if they work with the system over a fairly long period are they likely to develop a cognitive model of the system as a whole. Even then the model may be incomplete and, perhaps, inaccurate. A question of some practical significance, then, is that of what potential users of computers should be taught about their structure and operation. One possibility is to treat them as "black boxes" whose inner workings are irrelevant. According to this view, the user only needs to know enough about the input-output relationships to get the machine to do what he wants it to do. But if one assumes that people are likely to invent theories about what is going on inside the black box, there may be some practical benefit—not to mention intellectual value—in trying to assure that those theories are not totally at odds with reality.

At the opposite extreme from the black-box approach is that of "full disclosure." Why not simply teach users what is going on inside the black box in some detail? The problem is that not everyone is sufficiently interested—although one suspects that more people might be interested if they understood that the basic operations computers perform are really quite simple.

A compromise is the use of metaphors as models of how computer systems work (Brown 1977; Winston 1981; Young 1983). The assumption seems to be that the presentation of a conceptual model—a simple way of thinking about the system—at the beginning can help one acquire an accurate, if schematized, understanding sooner and can assure a reasonable correspondence between a user's model of the system and the real thing. The question is, what concepts, metaphors, or analogies should be used to characterize a system and one's interaction with it? To what extent should the representation of a system encourage users to think in terms of objects (documents, file folders, file drawers, desk tops) and actions (writing, editing, addressing, filing) with which they are already familiar? More generally, how should users think about computers—about what they do and about the data they hold? What kinds of conceptual models should be provided for beginning computer users who know little about how computers work technically and have neither the need nor the desire to learn?

Kay (1977) has suggested that the development of programs for complex systems might be facilitated by replacing concepts such as *data* and *procedures* with the more general idea of *activities*. Activities are

computer-like entities that exhibit behavior when they are sent an appropriate message. . . . Every transaction, description, and control process is thought of as sending messages to and receiving messages from activities in the system. Moreover, each activity belongs to a family of similar activities, all of which have the ability to recognize and reply to messages directed to them and to perform specific acts such as drawing pictures, making sounds, or adding numbers. New families are created by combining and enriching "traits," or properties inherited from existing families. (238)

The question of how to use metaphors effectively and what metaphors to use in teaching people about computer systems has also been addressed by Carroll and Thomas (Carroll 1984; Carroll and Thomas 1982; Thomas and Carroll 1982). They point out that people are likely to use metaphors spontaneously in learning about computer systems, just as they use them when learning about anything new (Folley and Williges 1982; Gentner 1980; Ortony 1979; Petrie 1979), and that an effort should be made to assure that the metaphors beginners use are effective ones and not misleading. If the metaphor one selects in a particular case is an appropriate one, learning will be facilitated; if it is an inappropriate one, learning will be inhibited. Even metaphors or analogies that seem quite straightforward can be misleading in some particulars, however, so care should be taken to acknowledge the boundaries of those that are used (Douglas and Moran 1983; Halasz and Moran 1982).

Carroll and Thomas make general recommendations regarding the use of metaphors to teach about computer systems, among which are the following cautionary ones:

- When introducing a metaphor, explicitly point out to the user that it is not a perfect representation of the underlying system and point toward the limits of the metaphor. (113)
- Keep in mind from the beginning that any metaphors presented to the user are to give an overview of the system and that there may be a time, at least for the continual user, that the metaphor is no longer useful. (114)

One metaphor that Carroll and Thomas believe has caused difficulties is the "communication" metaphor. They point out that the phrase "man-computer communication" may mean one thing to an engineer who thinks of communication as the point-to-point transmission of information (in the Shannon sense), and quite another thing to a lay person who thinks of communication as something that takes place between human beings and involves understanding of what is being communicated. They note that use of the term invites the naive user to impute to the computer the ability to engage in communication in the same way human beings do. This expectation can only result in frustration when the user begins to discover the limitations the computer really has in this regard.

In one effort to teach beginners how to use a text editor, Foss, Rosson, and Smith (1982) gave their subjects an "advance organizer" before having them study the system manual. The organizer used the metaphor of storing information in conventional file folders and file cabinets to explain how the computer could create, store, and retrieve files. Subjects who were given the advance organizer took slightly less time and used fewer commands to complete tasks than did subjects who studied the manual straightaway.

Metaphors that are helpful in getting a new user started may become a hindrance to further learning. Carroll and Thomas (1982) note that filing documents in a computer is not exactly like filing them in a file cabinet; when the file cabinet metaphor has outlived its usefulness, one must select a new metaphor or abandon the metaphorical approach in favor of a more literal one. This observation raises the question of how to determine when a metaphor should be discarded and how to wean the user from dependence on it. Highlighting the provisional nature of a metaphor at the time of initial learning, the authors suggest, may make it easier for the user eventually to acquire a fuller understanding of the system.

Carroll and Mack (1982b) also note the utility of metaphors in teaching beginners how to use computer systems. But they are critical of theoretical accounts of why metaphors work. In particular, they criticize theories (Bott 1979; Gentner 1980; Ortony 1979) that emphasize the importance of structural relationships between the metaphor and its target domain. Metaphors, in Carroll and Mack's view, are open-ended, incomplete, and inconsistently valid representations of the metaphorized domain. In these respects they differ from models, which, at least ideally, are explicit, comprehensive, and valid. Metaphors play the role of facilitating active learning by providing clues that can help the learner construct a model of the target domain: "While models are designed to represent some target domain, metaphors are chosen or designed to invite comparisons and implications which are *not literally true.* Metaphors are not 'right' or 'wrong' descriptions, as models are: rather they are 'stimulating' (or unstimulating) invitations to see a target domain in a new light" (13).

According to this view, a metaphor is a temporary expedient. The learner's objective is to construct an accurate model of the object of study. A useful metaphor is one that facilitates the development of that model. Once the model has been developed, the metaphor is no longer necessary. Carroll and Mack note that the salient dissimilarities between the metaphor and the metaphorized domain can facilitate learning by stimulating thought, as can the salient similarities.

### Manuals and On-Line User Aids

Manuals for users of computer systems often leave something to be desired (Chapanis 1982; Howard 1981). It is not clear that computer-system manuals are worse than those for other kinds of systems or devices, but that is of little comfort to the novice user who is struggling to extract enough information from a manual to permit him to get through his first work session without giving up in frustration.

Wright (1983) distinguishes between the problem of finding desired information in a document and that of understanding information once it is found. A user guide that is usable, as well as readable, must be designed so that people can locate information easily. Liberal use of section headings and subheadings can help in this regard, as can the provision of an extensive index.

Carroll and Mack (1982a) find that whereas most user guides assume passive learners, what can actually be expected of users is an engagement in *active* learning. When given the task of learning to use a computer system, the authors claim, people seldom simply attempt to absorb the information in the user's manual passively:

Put succinctly, our subjects seem to learn actively—not by passively reading or following rote exercises. They try things out according to self-generated agendas of needs and goals. They construct theories on the fly to explain what the system does. And they spontaneously anchor much of this active reasoning in their prior knowledge (e.g., of typewriters). This approach to the learning tasks is all the more striking when one views it in light of the fact that state-of-the-art systems and their training materials presuppose *passive* learners. Manuals tend to consist of rather extensive exposition augmented by prescribed exercises and practice drills. (9)

The desirabilty of on-line aids for users of interactive systems is apparent. It has long been realized that interactive systems should be able to train their users, and several systems have been built that do that, with varying degrees of effectiveness (Caruso 1970; Goodwin 1974; Kennedy 1975; Mayer 1967; Morrill 1967). A friendly and effective interface is nowhere more important than in the user-aid component of a system; there are few things quite so frustrating to a naive computer user as typing "help" and receiving in reply a cryptic bit of jargon that he cannot understand.

Perhaps the most important user aid is the error message. Virtually all systems notify users of various types of detectable errors when they are made. The errors that can be detected and the ways in which the users are informed of their occurrence differ considerably from system to system. Several writers have commented on the need for care in the design of error messages generated by compilers, interpreters, or operating systems and have offered suggestions for making them more helpful than they typically are (P. J. Brown 1982, 1983; Horning 1974; Shneiderman 1982). On the basis of some experimentation with error messages given to unskilled programmers attempting to compile COBOL programs, Shneiderman (1982) has proposed some guidelines. Error messages, he suggests, should have a positive tone, indicating what must be done rather than condemning the user for the error; they should be specific and address the problem in user terms, avoiding the vague "syntax error" or obscure internal codes and using variable names and concepts known to the user; and they should place the user in control of the situation by providing adequate information in a neat, consistent, and comprehensible format. Thomas and Carroll (1981) make the following recommendation regarding system-to-user messages: "Make clear to the user not only the content of the message, but also how it is to be taken. In other words—by spatial convention or otherwise—clarify whether a message is meant to inform of error, inform of state, prompt for action, or give feedback" (256).

Workers at IBM in San Jose, California, have developed a program that helps evaluate error messages. The program presents both erroneous code and appropriate error messages to users, who are asked to correct the error and to rate and comment upon the error messages. Users evaluate the error messages again after they have been modified in light of evaluative comments by a number of subjects. Data are obtained to determine whether in fact the messages have been improved (Isa, Boyle, Neal, and Simons 1983).

A feature that shows the sensitivity of designers to human-factors issues—and one that required a high degree of sophistication to implement—is the DWIM ("Do what I mean") feature of Interlisp (Teitelman 1972; Teitelman and Masinter 1981). The DWIM feature represents an effort to provide Interlisp with the ability not only to detect errors but also to figure out what the programmer intended. How much it can do along these lines is limited, of course, but the user is protected against faulty guesses on the part of the system, in that the system indicates what it thinks the user intended to say and the user is then free to accept or reject that interpretation.

In addition to providing error messages to users, many interactive systems have descriptive and tutorial information on line that can be consulted by the user on request. Such aids, though they can be extremely helpful, often require some special knowledge on the part of the user if they are to be used effectively. Developing on-line aids is, at the moment, more art than science. Research may eventually yield some helpful guidelines for designers of such aids; it might also be useful to attempt to synthesize what has already been learned by developers and users of "help" capabilities.

A few investigators have begun exploring ways in which interactive computer systems can function as monitors, critics, advice givers, or coaches for their users on problem-solving tasks (Baldwin and Siklossy 1977; Brown, Burton, Bell, and Bobrow 1974). Systems with these capabilities have the potential for use either as instructors or as problem-solving assistants. The opportunities for research and development here are great.

One intriguing possibility for performance aiding has been advanced by Carroll and Thomas (1982). They suggest that a computer-based system could "reframe" a user's task and thereby make it more interesting and motivating. An intrinsically dull task, they suggest, could be transformed into a competitive, gamelike situation. The "virtual task," the task as seen by the user, could be quite different from the actual task that had to be performed. What would be required is that effective performance of the former be translatable automatically into effective performance of the latter. In the absence of efforts to test the notion in specific cases, it is difficult to judge its merits. It does seem, however, to be an idea worth exploring.

### Meeting the Needs of Users with Different Levels of Skill

In the evolution of an interactive system, a conflict often arises between the desire for simplicity and the desire for functionality. The system is designed with an emphasis on simplicity, but as users get extensive experience with it they begin to demand the addition of greater functionality. The addition of functionality seems invariably to require an increase in complexity. Consequently, the system becomes increasingly powerful in the hands of an experienced user but also increasingly difficult for the novice. The problem for research is how to provide the functionality that the experienced user wants without sacrificing the simplicity that the inexperienced user needs.

More generally, one would like any system to be able to accommodate users with different levels of experience and skill with that system. This is not an easy desire to realize, because novices and experts have quite different needs and preferences. Novices, for example, need much instruction and detailed guidance from the system. They need error messages that not only alert them when they have made an error, but give them detailed information about what they have done wrong and how to do correctly what they intended to do. Experts, on the other hand, require little in the way of guidance or "hand holding." The simple flagging of errors usually suffices, and long explanations of the nature of those errors may be more of a hindrance than a help. Novices are likely to be more tolerant than experts of system delays or lack of responsiveness. Novices are likely to be satisfied with limited functionality, whereas to the expert, functionality is of the utmost importance. Indeed, no matter what a system can do, an expert user will think of other capabilities he would like it to have, and he may simply add them if he is sufficiently expert to do so.

One approach to the problem of meeting the needs of users with different levels of expertise has been to have two versions of a system concurrently available, one designed for the novice and the other for the skilled user. A system may, for example, have both a menu-driven and a command-driven interface and permit the user to select between them (Richards and Boies 1981). There seems to be some agreement that menu-driven interfaces are easier for novices to use than are command languages (Branscomb and Thomas, undated; Norman 1983a). Norman argues that menu-based systems are convenient for novice or infrequent users but too slow, tedious, and inflexible for experts. Command languages, on the other hand, are better for experts because of their flexibility and speed, but they are difficult to learn and provide no on-line reminders of action alternatives.

The foregoing comments suggest a dichotomous view in which users of a given system can be neatly partitioned into novices and experts. While much of the literature that addresses this issue seems to reflect this view, expertness is a continuum and people are experts to varying degrees. Moreover, any given individual may be expert with respect to some aspects of a system and quite inexpert with respect to others. Having two versions of a system, one for experts and one for novices, is likely to be a less than satisfactory approach for the many users who are neither completely in the one category nor completely in the other. Providing the system with features that can be selectively enabled or disabled depending on the user's experience level is somewhat less all-or-none, but still requires that users be partitioned in an oversimplified way.

There are some specific things that have been done to increase a system's flexibility with regard to the degree of expertise required of its users. Giving the user the ability to terminate lengthy and noninformative messages from the system is one simple step. Having the computer output brief coded messages but be able to provide full explanations of those messages in response to requests from the user is another. A third approach is to include within the system software tutorials explaining advanced system capabilities that are presented at the user's request (Novell 1967). In taking this approach, however, one must be concerned with the possibility that the inexperienced user may fail to learn enough about the system to know how to use the tutorials effectively.

This possibility points up another issue with which system developers should be concerned. In addition to wanting a system to be able to interact effectively with users at different skill levels, one would like a system to be able to provide novice users with the kind of feedback and assistance that will facilitate their becoming experts. There are some systems being used today by people who are exploiting only a small fraction of the capabilities of those systems. What is particularly unfortunate about this is that it is possible for people to use a powerful system in a suboptimal way indefinitely, without ever discovering how powerful the system really is and how much more efficiently they might be able to accomplish the same tasks if they only knew how to exploit the system's capabilities.

But how does one build into a system the capability to help the novice user become an expert? Clearly this involves more than simply giving it the ability to provide help on request. Novices lack the knowledge they need to ask the questions they must in order to gain the knowledge that will advance them toward expert status. Ideally, a system should notice when a user is doing something in an inefficient way or failing to take advantage of a system feature that would simplify the task, and should be able to volunteer the needed information without being asked for it, much as a human tutor would.

The usual approach to simplifying life for the beginning computer user is to try to design systems, languages, and interfaces in such a way as to minimize what the user will have to learn. Another approach is that of providing the beginner with learning strategies that will facilitate acquisition of whatever knowledge he needs (Coombs, Gibson, and Atty 1982). This approach has received relatively little attention, but it seems a reasonable one to develop. In the context of a discussion of the training of users of business telephone systems, Dooling and Klemmer (1982) point out the merits of focusing training on the use of performance aids. The idea is to provide a good system of aids to which the user can turn while using the system, and to concentrate during an introductory training program on how to use those aids effectively. This too seems an approach that could be used to advantage with interactive computer systems. Although most systems provide user manuals and many have on-line help facilities, often these aids are not as helpful to a user as they should be because the user lacks the knowledge that is needed (and typically more knowledge is needed than the system designer realized) to use these resources effectively.

• • • • • • • • • • • • • • • • • • • • • • • • • • • • • • • • • • • • • •

Computers are being used by so many people for so many purposes that it would be impossible to catalog all the user-related issues that arise. Here I have mentioned only a few of those that strike me as especially noteworthy.

For the first couple of decades after computers appeared on the scene, most of the people who used them directly were technically trained and had a basic understanding of how they worked. That situation has been changing now for some time as computers and computer-based systems have been introduced into more and more work situations. A growing percentage of users of these systems are people who have not had technical training and who do not necessarily understand how these machines do what they do. That percentage is further increased with the use of computers outside the workplace for avocational, recreational, and other personal purposes. User-related issues multiply as the size and heterogeneity of the user population increase. The fact that a large percentage of this population is not technically oriented gives special significance to issues of attitude and training. These and many other issues relating to the problem of matching systems to the capabilities, limitations, and preferences of their users will remain challenges and opportunities for research.

# 14

## *Programming*

Human beings have been writing for perhaps 7,000 years. For the most part, what we write today is not very different from what people who could write wrote hundreds or even thousands of years ago: agreements of various sorts, including treaties and business contracts; treatises intended to describe or explain some aspect of the world for instructional purposes; personal letters and notes; records of events and transactions; poems, stories, plays. There have emerged in our own century, however, a new purpose for writing and a new form of writing. The new purpose is that of providing a machine with detailed instructions for performing some task, and the new form is the computer program.

The emergence of programming as a new type of intellectual activity has implications beyond the obvious practical one of providing a new class of job opportunities for people. It motivates the codification of procedural knowledge, and it provides a vehicle for representing and cumulating that knowledge, for making it accessible to future users. Further, it provides a new way of testing the depth or adequacy of our understanding of specific processes.

While the term *program* has a fairly precise meaning in the present context, namely, a set of instructions for a computer, it is vague by virtue of its inclusivity. In this respect it is a bit like the term *bridge*. While the notion of a structure that provides a path over an obstacle (body of water, railroad tracks) is clear enough, in fact real bridges include planks across ditches and engineering marvels such as New York's Verrazano Narrows Bridge and San Francisco's Golden Gate. Computer programs also vary in complexity, from very simple ones composed of a few statements to those containing hundreds of thousands of

lines of code and requiring hundreds of professional programmer years to construct.

People learn to program for a variety of reasons. Programming ability is a demonstrably useful and marketable skill;
the number of professional programmers in the world has
been growing rapidly for the last three decades and is likely
to continue to do so for the forseeable future. Programming
can be an intellectually stimulating and satisfying activity.
And the ability to program can be a source of status and social
reinforcement.

Although the potential social benefits of being able to program have not received much attention, they may be important
motivators for some people. In an interview study of six unusually accomplished programmers less than 14 years of age,
Kurland, Mawby, and Cahir (1984) noted the role that programming played in the social lives of these youngsters. These
programmers associated with other young people who also programmed. Their programming skill provided them status
among their peers, and their access to computer bulletin
boards provided them with a social network through which to
communicate.

### Programming as a Cognitively Demanding Task

There are at least two reasons why the activity of programming
should be of interest to psychologists. First, programming is a
career for a large and growing number of people; it would be
useful to know what the cognitive prerequisites are for learning
to do it well and how skill in programming can best be acquired.
Second, programming is an intellectually demanding task with
certain characteristics that make it a suitable vehicle for studying problem-solving behavior more generally.

Birnbaum (1982) refers to the computer as "probably man's
intellectually richest invention" and to programming as "the
most complex craft ever practiced" (763). Undoubtedly, one
could find practitioners of a variety of other crafts who might
take issue with the latter claim. Be that as it may, programming is certainly a complex activity and for some people a
totally engrossing one. Molzberger (1983) has noted the importance to expert programmers of being able to spend large
blocks of uninterrupted time at the task.

Probably most people who have programmed more than a

trivial amount are aware of how captivating an activity programming can be. I have heard programming referred to seriously by more than one practitioner as an addiction. Is there
such a thing as compulsive programming? Weizenbaum (1976)
believes there is:

> Wherever computer centers have become established, that is to say, in
> countless places in the United States, as well as in virtually all other
> industrial regions of the world, bright young men of disheveled ap-
> pearance, often with sunken glowing eyes, can be seen sitting at com-
> puter consoles, their arms tensed and waiting to fire their fingers,
> already poised to strike, at the buttons and keys on which their atten-
> tion seems to be as riveted as a gambler's on the rolling dice. When
> not so transfixed, they often sit at tables strewn with computer print-
> outs over which they pore like possessed students of a cabalistic text.
> They work until they nearly drop, twenty, thirty hours at a time.
> Their food, if they arrange it, is brought to them: coffee, Cokes,
> sandwiches. If possible, they sleep on cots near the computer. But
> only for a few hours—then back to the console or the printouts. Their
> rumpled clothes, their unwashed and unshaven faces, and their un-
> combed hair all testify that they are oblivious to their bodies and to
> the world in which they move. They exist, at least when so engaged,
> only through and for the computers. These are computer bums, com-
> pulsive programmers. They are an international phenomenon. (116)

Undoubtedly, many people, myself included, who would admit to having spent more than one marathon sleepless session
glued to a computer terminal, totally engrossed in developing
or debugging a program, would see Weizenbaum's description
as a pejorative one. People have been known to be absorbed by
work of various types long before computers appeared on the
scene, and the ability to persevere at a task has even, on occasion, been seen as a commendable thing. What may be different
about the phenomenon of the "computer bum" is the number
of people who might qualify for the appellation. It may well be
that programming has captured the imagination, and the time,
of a larger number of individuals—especially young people—
than has any other cognitively demanding activity. The interesting question, to which we do not know the answer, is what
effect a period of computer bummery is likely to have on one's
further development and what it is likely to contribute to one's
future success or failure in life. I know of no compelling evidence one way or the other on this question.

Programming is a multifaceted activity. From reports obtained from expert programmers, Kurland, Mawby, and Cahir (1984) estimate that the actual coding of programs consumes only about 20 to 25 percent of a programmer's time; the rest of that time is spent in planning, debugging, documenting, and program testing. Some writers have also pointed out that programming is an intensively and extensively knowledge-based activity (Atwood and Jeffries 1980; Pennington 1982). Accomplished programmers call upon a wealth of knowledge gained from years of experience regarding general approaches, heuristics, and procedures for accomplishing specific tasks.

Acquiring the ability to program requires much more than learning one or more programming languages and some specific techniques. It requires first an understanding of what it means to specify a procedure completely, quite independently of representing that procedure in a symbology that will permit it to be executed by a machine. This is true because writing programs for complex tasks seldom amounts to expressing in a computer language a procedure that has already been made explicit in natural language. More typically the first task a programmer faces is that of explicating the procedure that is to be programmed (Kowalski 1979). Or as Bacon (1982) puts it, much of the work involved in designing software for a particular task really amounts to systematizing and designing the task itself. Thus, while learning to program involves learning how to talk to a computer, a prerequisite is knowing how to talk precisely with oneself.

In other words, programming can be thought of as involving two quite different tasks. One is that of specifying completely and unambiguously a procedure for accomplishing some specific objective. The other involves representing this procedure in a particular programming language. The first of these tasks is by far the more creative, but it would appear to require less in the way of knowledge that is specific to programming. One might assume that people would learn to specify procedures in natural language just by virtue of years of experience in communicating with other people. What little evidence there is on the issue suggests, however, that nonprogrammers do not find it easy to produce complete and detailed procedure specifications in natural language (Miller and Becker 1974).

Programming is difficult because precision in communication is difficult. We are not accustomed to dealing with literalists. We are surprised at the difficulty of programming because of untenable models we have in our heads about how we communicate with each other. Interpersonal communication is a *cooperative* activity between intelligent beings. The listener/reader is as active in this process as the speaker/writer. One does not extract meaning from an utterance or a written passage so much as impose meaning upon it; and the ability of the listener/reader to do so is something that the speaker/writer takes for granted. Unfortunately, we tend to be unaware that we make such assumptions and consequently find it exceedingly easy to carry them over to the situation in which the communication is directed not to other people but to computers.

Programming also illustrates the necessity for organizing one's thinking about complex entities into manageable chunks. A program is completely described by a specific pattern of bits (typically represented by ones and zeros) in a computer's memory and by the decoding circuitry in the hardware. No one could think effectively of any but the most trivial of programs in this way, however. At the next level of abstraction one might represent sequences of bit patterns as individual instructions to perform logical operations on other bit patterns. At a still higher level one might represent these sequences as instructions in a language that has constructs with which we are familiar from other contexts, such as algebra. This type of representation is convenient for understanding small program segments, but even this is an inadequate vehicle for providing an overview of a large, complex program. Here it becomes necessary, or at least convenient, to think in terms of hierarchies of processes. One can then conceptualize a program at different levels of the hierarchy. To represent to oneself the operations of a given component in this hierarchy, one tends to think about the next-lower-level components of which it is composed, but not in terms of the details of those components or those of any lower-level components into which these could be further analyzed.

Programming has been the focus of some research attention (Brooks 1982; Dunsmore 1983; Jeffries, Turner, Polson, and Atwood 1981; Mayer 1981; McKeithen, Reitman, Rueter, and Hirtle 1981; Shneiderman 1980a; Soloway, Ehrlich, Bonar, and Greenspan 1982), but surprisingly little in view of its practical importance and the fact that it is such an intrinsically interesting activity. The studies of programming that have been done have focused on such issues as the effects of goals on programming (Weinberg 1971), the structures and characteristics of programs (Dahl, Dijkstra, and Hoar 1972; Knuth 1972; Saal and Weiss 1977), the errors that programs commonly contain (Boies and Gould 1974; Youngs 1974), and program debugging (Atwood and Ramsey 1978; Gould 1975; Gould and Drongowski 1974).

Weinberg's (1971) study illustrates the type of thinking that has been done in investigating the effects of goals on programming. He asked four programmers to work on a problem that was estimated to require about one-fifth of each individual's time over a period of ten weeks. The description of the problem was identical for all four programmers, but for two of them the instructions emphasized the desirability of completing the program as quickly as possible, whereas for the other two they emphasized the desirability of producing a program that was as efficient as possible. The programmers who were instructed to complete the program quickly took only about one-third as much programming time as did those who were instructed to produce an especially efficient program. The programs produced by the first pair of programmers took about ten times as long to execute, however, as did those produced by the programmers working for efficiency.

In a second experiment of the same type, problems were used that were assumed to provide less of an opportunity for a spectacular time savings. The results in this case were similar but not as dramatic. Programmers were asked also to estimate how long they thought their tasks would require. The programmers who had been instructed to finish as quickly as possible were more conservative in their time estimates than those who had been instructed to be as efficient as possible; and the first group, but not the second, did better than their estimates.

Weinberg concluded that much of the variance between programmers on a given job is attributable to different conceptions of their goals. When a goal is set explicitly, a programmer is likely to work toward that goal, possibly at the expense of other goals (which is presumably what one hopes in setting explicit goals); and, perhaps more interestingly, estimates relating to emphasized goals are likely to be more accurate than those relating to goals not emphasized. As Weinberg points out, the increased accuracy may come about because the programmer is motivated to make the estimate correct with respect to the emphasized goal.

Interpreting the results of programmer performance studies is complicated by the possibility that they are contingent on such factors as the language in which the programming was done, the specific programming problems used, and the experience and skill of the programmers. One approach that has been taken to the last problem has been to use subjects who have had no programming experience at all (Adam and Cohen 1969; Gold 1969; Miller 1974; Smith 1967). But this approach also produces results with limited generality. What proves to be true for novices may not be true for experienced programmers.

Debugging is an aspect of programming that deserves special attention because of its importance in the process—Boehm (1973) estimates that debugging accounts for 25 to 50 percent of the time required to produce a new program—and because it is prototypical of diagnostic problem solving in general. Several studies of debugging activity have produced some surprising or counterintuitive results. Gould and Drongowski (1974), for example, planted bugs in some FORTRAN statistical library programs and had experienced programmers try to find them (off-line) with and without the use of certain debugging aids (the input-output of the buggy program, the input-output of the buggy program plus the output that would have resulted had the program been free of bugs, identification of the class of bug the program contained, specification of the line of code that contained the bug). The aids proved not to be helpful in reducing debugging time. Yasukama (1974: reported in Shneiderman and McKay 1976) found that high-level comments on a FORTRAN program did not help subjects locate bugs; and Shneiderman, Mayer, McKay, and Heller (1977) got a similar negative result with respect to the helpfulness of flow charts as debugging aids.

Bugs can be classified in a variety of ways, but one major distinction is that between what might be referred to as syntactic or surface-structure bugs and semantic or deep-structure bugs. A syntactic or surface-structure bug is a violation of the language's rules of grammar. Such bugs are relatively easy to find and in many cases can be detected automatically by software. Semantic or deep-structure bugs are errors in logic or conception and usually cannot be detected except by someone who understands what the program is intended to do. It is much more difficult, and in many cases impossible, to write software that will detect such bugs.

Discussions of program bugs and debugging sometimes promote the idea that programs can be neatly partitioned into two groups, those that contain bugs and those that do not. To be sure, there are certain types of bugs that will guarantee that a program will fail to do what it is intended to do, and it is important that programs be free of bugs of that type. But there are also many types of program shortcomings with less consistently catastrophic effects that nevertheless might qualify as bugs. These are program features that will permit the program to fail under unusual conditions or that make the program less efficient or more complex than it needs to be. Debugging in the most general sense includes not only eliminating the more catastrophic bugs but improving a program with respect to more subtle inefficiencies or inelegancies as well.

The opportunities for research on debugging and on programming more generally are great. How should programmers be trained? (Read "programmers" here to include users of personal computers, because as Kay [1977] points out, although personal computers come equipped with applications software, users will probably have to do some programming if their computers are going to be useful in a more than superficial way.) What kinds of programming and debugging aids are worth including in systems? How does programming performance depend on the characteristics of the language and the computing environment being used?

The growing interest in the development of multiprocessor machines poses another major challenge to psychologists; which is to explore ways to decompose complex problems so that they can be solved by parallel processes. In the future, computers will be built that have many processors (tens, hundreds, thousands), all running in parallel. It is not yet clear how such machines will be programmed, because very little is known about how to solve problems in parallel mode. The increasing availability of multiprocessor machines will provide an incentive to begin thinking about some problems in ways that are qualitatively different from how we have thought about them in the past. We have tended to partition problems into sequential steps, both because we work sequentially ourselves and because the computers at our disposal have been serial-processing machines. This is not to suggest that there are no models for parallel approaches to complex tasks. Perhaps the most obvious examples are tasks that involve many people—erecting a building, publishing a newspaper, running a government. Efficient use of resources in such cases requires careful planning, task decomposition, scheduling, and monitoring. There is much to be learned about how to structure programming tasks so that they can exploit the parallelism of multiprocessor machines.

## Cognitive Prerequisites for Programming

What kinds of intellectual capabilities and knowledge are essential or important to the learning of how to program? Kurland, Clement, Mawby, and Pea (1984) addressed this question in a study in which they looked for correlations between certain evidences of two specific skills—procedural reasoning and the ability to decenter—and the progress made by students in learning to write LOGO programs during a six-week summer course (roughly 90 hours of relatively unguided programming). Modest positive correlations were found between procedural reasoning and decentering and certain tests of programming proficiency administered after the course. The results are difficult to interpret, however, inasmuch as none of the subjects learned to program very well.

Kurland, Mawby, and Cahir (1984) found that the main thing that one group of accomplished programmers had in common was lots of practice and enthusiasm. Estimates of the amount of time they spent at a terminal while learning ranged from 20 to 35 hours per week. Reports of lengthy continuous work sessions of 30 hours or more and 60- to 100-hour weeks were not unusual.

There is some speculation that programming is particularly attractive to people with certain cognitive styles or predilections for logically structured approaches to problem solving. Interestingly, it has been noted that many expert programmers are also accomplished musicians (Kurland, Mawby, and Cahir 1984).

Contrary to one popular view, computer programming does not require a great deal of formal training in mathematics. While there is no reason to suppose that a deep knowledge of mathematics would be a disadvantage, there is little evidence that it is a great help; and that it is not essential is demonstrated by the fact that many highly competent programmers are not mathematically trained. Programming is, however, a knowledge-based skill. One must have a knowledge of the grammar of the language in which a program is to be expressed; and the more tricks and procedures with which one is familiar, the more effective one is likely to be at writing programs that work and do interesting things. Unfortunately, the knowledge that highly skilled programmers possess is only beginning to be organized and codified so that it can be acquired through formal training.

Weizenbaum (1976) has argued that programming is a relatively easy craft to learn, that "almost anyone with a reasonably orderly mind can become a fairly good programmer with just a little instruction and practice" (277). He warns, however, that just because it is easy to learn the rudiments of programming, and because one can very quickly get to the point of seeing tangible results from one's efforts, programming can be very seductive for the beginner.

Moreover, it appeals most precisely to those who do not yet have sufficient maturity to tolerate long delays between an effort to achieve something and the appearance of concrete evidence of success. Immature students are therefore easily misled into believing that they have truly mastered a craft of immense power and of great importance when, in fact, they have learned only its rudiments and nothing substantive at all. A student's quick climb from a state of complete ignorance about computers to what appears to be a mastery of programming, but is in reality on a very minor plateau, may leave him with a euphoric sense of achievement and conviction that he has discovered his true calling. . . . for the student this may well be a trap. He may so thoroughly commit himself to what he naively perceives to be computer science, that is, to the mere polishing of his programming skills, that he may effectively preclude studying anything substantive. (277)

While I believe the concern expressed by Weizenbaum has some legitimacy, I take issue with the categorical way in which it is expressed. I know of no empirical evidence that programming appeals more to immature students than to more mature ones. Moreover, the immediate feedback that programming provides could have very considerable educational value. The importance of intellectual engagement to learning is widely recognized and a continuing challenge to teachers, who must attempt to structure situations that ensure the active participation of students in the learning process. Programming cannot be done passively; it requires an intensive expenditure of intellectual energy. As an activity that has proved highly motivating and rewarding to many people, it invites exploitation for educational purposes.

## Cognitive Consequences of Programming

Success in programming depends on a number of abilities—to plan, to use language precisely, to decompose complex problems into subproblems or steps, to think about a problem at different levels of detail, to generate and test hypotheses, to think inventively—that are essential to the successful performance of a variety of complex tasks. One reason, then, for psychologists to study programming is for what it can reveal about cognitively demanding tasks more generally. Conversely, if programming really is prototypical of many cognitively demanding tasks, it might be an effective vehicle for teaching generally useful thinking skills (Howe, O'Shea, and Pane 1979; Nickerson 1983).

Whether learning to program will spontaneously produce generally useful thinking skills and whether programming *could* be used as a vehicle for teaching such skills are two different questions. So far the evidence for the former is sparse (Dalbey and Linn 1984; Ehrlich, Abbott, Salter, and Soloway 1984). Linn and Fisher (1983) have argued that much programming instruction is not well suited to the development of such skills because the programs students write are so simple that they do not require the use of such strategies as planning and problem decomposition. Pea and Kurland (1984a, b) have concluded that if the teaching of programming is to be used as a vehicle for fostering the development of generally useful thinking skills, the targeted skills should be taught explicitly and not simply left to emerge spontaneously from the programming experience.

Some investigators have suggested that experience with programming should help one adopt a more procedure-oriented approach to problem solving and that this approach should transfer to situations not involving programming. It is well known that many students have trouble with algebra word problems (Carpenter, Corbitt, Kepner, Lindquist, and Reys 1980). The "reversed-equation" problem, which has been investigated by a group of researchers at the University of Massachusetts, illustrates the difficulty people sometimes have in finding mathematical expressions for relationships described in natural language, or in translating mathematical expressions into corresponding natural-language descriptions. Several studies have shown that when asked to write an equation to represent the relationship "There are six times as many students as professors," many people (37 percent of a group of 150 first-year engineering college students, in one case) are unable to do so. The most common error is production of the equation $6S = P$. Essentially the same finding has been obtained in several studies (Clement 1982; Clement, Lochhead, and Monk 1981; Rosnick and Clement 1980). The error has been interpreted as evidence of misconceptions at a fairly deep level about the nature of variables and equations. These misconceptions have proved quite resistant to correction by training (Rosnick and Clement 1980).

One method of rectifying difficulties of the reversed-equation type that did show some promise involved having people write a computer program that would produce the value of one of the variables, given the value of the other as input, rather than having them simply write an equation representing the relationship (Clement, Lochhead, and Soloway 1979; Ehrlich, Soloway, and Abbott 1982). The investigators took their results as supportive of the idea that the reversal problem has its roots in the tendency to view the relationship statically, whereas an equation should be seen as describing an equivalence that would hold if specified operations were performed. Presumably, thinking in terms of a computer program helped to make explicit the operation that is identified in the equation (multiplication in the professor-student example) because a program is by nature something that "runs," or performs operations. A question of some educational significance is whether problems of the reversed-equation type might be avoided altogether if children were introduced to equations via programs at the outset.

### Programmer Productivity

It has been true for some time that the cost of the software required to run a large and complex computer system is often greater than the cost of the system hardware. (The operating system for IBM's System/360 computers required more than 5,000 man-years to produce [Brooks 1975].) As hardware costs have come down dramatically, the relative costs of developing software have steadily risen, so the disparity between the costs of the two components has been increasing. According to Birnbaum (1982), the performance-to-cost ratio of computer hardware rose by a factor of 1,000,000 between 1955 and 1985 (two orders of magnitude every ten years), while programmer productivity increased only by a factor of about 3.6 during the same period. The main limitation to progress at present, he suggests, is our difficulty in dealing with complexity, and with the complexity of software in particular.

Comparisons between hardware costs and software costs should be interpreted cautiously. To be sure, whereas in the early days of computers the hardware for a system often accounted for the larger portion of the costs, today the reverse is true. On the other hand, the software that is being developed today is very much more complex and powerful than that of the early days. Nevertheless, the increasing *relative* cost of programming is of concern to the industry. It is not surprising, therefore, that the issue of programmer productivity has been receiving increasing attention (Scott and Simmons 1974; Walston and Felix 1977).

The concern about this issue is heightened by the fact that what it will cost to develop a given piece of software has typically proved to be very difficult to estimate, and estimates have almost invariably been on the low side. There seems to be a pervasive belief that *any* sizable software development project will cost more and take longer than most estimates would indicate. This attitude is captured in the language of the final report of a software acquisition and development working group commissioned by the U.S. Assistant Secretary of Defense For Communications, Command, Control, and Intelligence:

It is common knowledge that software development projects rarely meet cost-benefits originally projected, usually cost more than expected, and are usually late. In addition, the software delivered seldom meets user requirements, oftentimes is not usable, or requires extensive rework. [Our] analysis has shown that all facets of the software acquisition and development process need varying degrees of improvement. (Jones 1980, 103).

Given that the Department of Defense currently spends from $5 to $6 billion per year on software and that the annual expenditure has been predicted at over $30 billion by 1990 (Steier 1983), it might be economical to spend a fair amount of money on developing tools and techniques that would increase programmer productivity generally even by as little as, say, 10 percent.

In spite of the great interest in the topic of programmer productivity, no generally accepted method for measuring it has been developed, although some attempts have been made to quantify it (McCall, Richards, and Walters 1977; Perlis, Sayward, and Shaw 1981; Walston and Felix 1977). The quality of programming also remains largely a matter of judgment, in spite of some attention to the problem of measuring that as well (Boehm, Brown, and Lipow 1977; Gilb 1977; Shneiderman 1977). Shneiderman (1980a) has reviewed a variety of efforts to develop metrics for evaluating such aspects of programs as reliability, maintainability, complexity, and comprehensibility. He concludes that in spite of these efforts no acceptable set of metrics for evaluating programs has yet emerged. The difficulty of measuring programmer productivity notwithstanding, the interest in increasing productivity is substantial. A partial answer to this problem may come from new programming techniques (structured programming, object-oriented programming) and tools to facilitate program sharing (emulators, cross-compilers).

The tools that programmers have to work with are indeed increasing in scope and versatility, as well as in number. In the area of graphics, for example, programs exist that will take a description of an object to be displayed and automatically translate that description into display code. Some of these can also change the size or orientation of figures and can, in effect, rotate them in three-dimensional space. Some attempts have been made to develop intelligent program composition and editing aids—programmers' assistants or apprentices—that will produce code for accomplishing specified generic tasks (Rich 1984; Waters 1982).

The problem of programmer productivity is far from solved, however, and it is likely to become of even greater concern as the programs that must be developed become more and more complex. Bacon (1982) argues that the logical complexity of computer software often far exceeds that of its hardware; he notes that in a large machine that performs commercial applications, programs may have as many as $10^7$ lines of code. Programmer productivity is an example of those problems that may be subsumed under the general rubric of complexity management.

The need to invent more effective approaches to the management of complexity, in many contexts, is clear. It seems clear also that if such approaches are developed, they will make use of information technology in some way. One might hope that some of the techniques used to manage the complexity that is found within information technology would be applicable in other contexts as well. One thing that has been learned is that complex problems often differ from simple problems qualitatively, so it does not suffice just to scale up approaches that work in the latter case and assume they will work in the former. In particular, it is not necessarily true that the best way to solve a complex problem is to get many people to work on it; ten people working in concert may be able to lift ten times the weight that one could lift alone, but ten people are not necessarily ten times as smart as one, and the problem of bringing the talents of groups of people to bear effectively on intellectually complex problems is itself a problem in complexity management that we do not yet know how to solve.

Object-oriented programming is seen by some as one way to reduce the complexity of large software systems (Robson 1981). In an object-oriented system the basic software building block is the *object*, which is a segment of computer code that has a name, receives messages (commands to modify itself in specific ways), and does what these messages instruct it to do. An object might be the code for representing a window on a visual display, for example. Among the messages that such an object might receive would be *move, enlarge, overlap, delete*. The coded representation of such an object would include representation of the procedures for carrying out such commands. Users of the object need not understand how the object carries out a particular command; they need to know only the fact that it is able to do so and the appropriate message to send so as to evoke the desired action. An important concept in object-oriented programming is that of *inheritance*. Objects inherit the properties of the class to which they belong. Class properties are also inherited by subclasses contained within them.

Discussions of programmer productivity have typically

focused on the question of the productivity of the individual programmer. At least equally important, however, is the question of the productivity of programming as a corporate human enterprise. It must be the case that programs to accomplish certain specific tasks have been written countless times by as many different programmers. It seems a shame that such multiplication of effort is necessary because of the lack of effective mechanisms for sharing procedures and code.

• • • • • • • • • • • • • • • • • • • • • • • • • • • • • • • • • • • • • • • • • •

Programming defines a new relationship between people and machines. Before computers appeared on the scene, people controlled the machines they used primarily by interacting with them manually, pushing and pulling levers, turning knobs, setting dials, throwing switches. To control a computer, one typically does none of these things; rather, one writes down a set of instructions that the computer then executes on its own time scale. For some reason that I have not been able to determine, sets of instructions for computers came to be known as programs, and in the absence of an obviously suitable verb for denoting the process of generating programs, the noun was used for that purpose. So now we have "programming" as well. Programming has come to be a very broad term and now covers a range of activities that vary greatly in complexity, but it does retain its basic meaning of producing instructions that a computer is to follow.

People program for a variety of reasons. One, of course, is that they get paid to do so. Programming skills are highly marketable today and have been for several years. People also learn to program, however, for reasons unrelated to employment, and these are not yet well understood.

Programming is a cognitively demanding task. It should be studied not only for the purpose of finding out how to teach programming skills to people who wish to acquire them but also because it is prototypical of many problem-solving tasks. It involves planning, problem decomposition, precision of expression, and a number of other activities of general interest. Debugging, which constitutes a critical aspect of programming inasmuch as programs are seldom free of errors from the beginning, is an interesting intellectual activity in its own right. It involves hypothesis generation and testing and is a pattern for many problems of diagnosis and troubleshooting.

The study of programming is complicated by the diversity of the activity. The specifics of the programming task, the programming language used, the experience or skill level of the programmer, are all germane to the understanding of programming performance. Moreover, programming cannot be studied in the abstract. One must deal with specific programmers, specific languages, and specific programming tasks. This is not to suggest that nothing of general interest can be learned from the investigation of specific situations, but only that the details of the situations should be carefully factored into the interpretations of the results of these studies. Programming of parallel-architecture machines is a largely unexplored domain but will take on increasing significance as the number of such machines in use grows.

The cognitive prerequisites to learning to be an effective programmer have only begun to be studied. It is not certain that there are any prerequisites beyond general intelligence and, perhaps, what might be thought of as a cognitive style. But programming is a knowledge-based activity, and one aspect of programming expertise is having at one's command a large number of algorithms, heuristics, and tricks for accomplishing specific objectives. This knowledge is only beginning to be codified in such a way that much of it can be learned by means of books and formal training, as opposed to experience.

The cognitive consequences of programming are also not clear. There has been some speculation, but little solid evidence, that what one learns in learning to program generalizes to other contexts. If it were established that people who program are more effective problem solvers in general than are people who do not—and it has not been established—we would be left with the question of whether learning to program improves problem-solving skills or whether people with generally good problem-solving skills tend to become programmers. Whether the teaching of programming can be used as a vehicle for teaching generally useful thinking skills and whether in learning to program one automatically acquires those skills are two different questions, the answers to which are also not yet known; both deserve more attention from researchers than they have received.

Programmer productivity is a major practical concern, because the production of software for computer systems is expensive, and as the costs of hardware continue to decline, the costs of software are accounting for an ever greater fraction of the total costs of an operating system. Productivity is difficult to measure, however. There are as yet no measurement techniques that are widely recognized as adequate. There is a general impression, though, that productivity is not what it could be and that methods for improving it are much needed. An important aspect of the problem as it relates to programming activities in the aggregate, as opposed to the productivity of individual programmers, is the duplication of effort. In the absence of effective methods for facilitating the sharing of software, we must assume that many procedures are programmed numerous times.

# 15

## Artificial Intelligence and Expert Systems

Artificial intelligence (AI) has been an area of research interest for a small community of scientists in universities and research organizations for almost thirty years. Until very recently, few people outside that small community had much interest in this research or even knew it was going on. Suddenly, artificial intelligence has become fashionable. It has been written up in every major news magazine. It has become a golden gleam in the eye of the investment community. It is discussed openly in mixed company of all types. What has caused this dramatic change of status? According to Reitman (1984b),

If AI applications are here to stay, it is not primarily because of any radical new achievements in AI itself. All the hard old problems are still there. Instead, the conditions have changed. The universities now have produced sufficient numbers of skilled, experienced AI researchers to make business investment in human AI resources feasible. Senior people in the field have become deeply involved in the marketing and funding of AI applications. Radically decreased computing costs now make AI applications economically viable. And finally, . . . we are seeing what in the language of the seventies would be termed a general raising of consciousness in the MIS [management information systems], business, and financial communities about the potential benefits of more intelligent hardware and software. (1)

As evidence of the growing interest among researchers in finding ways to apply artificial intelligence, a number of books on the subject have appeared, including Szolovits's (1982) *Artificial Intelligence in Medicine,* Reitman's (1984a) *Artificial Intelligence Applications for Business,* and Winston and Prendergast's (1984) *The AI Business.* (Each of these three is a collection of essays by participants in a symposium or conference on the present state and future possibilities of the indicated field.)

Just what is artificial intelligence? Given that psychologists have found it impossible to agree on a definition of (real?) intelligence, we ought not be greatly surprised by the lack of complete consensus among AI researchers as to what it is that they are studying or attempting to develop. One can find definitions or characterizations, however. Here are three examples, two of which refer to AI as an activity or field of inquiry and the other as an object of study:

- the study of methods for enabling computers to do the things that make people seem intelligent (Winston 1977);
- the study of inherently ill-defined problems, with the aim of transforming complex, ill-defined problems into defined ones (J. S. Brown 1984, 83);
- stuff that is interesting that we do not know how to do yet (Kay 1984a, 277).

Kay's definition makes AI a function of our ignorance. According to it, after we have learned how to do something—or, more precisely, how to program a computer to do it—we can no longer consider that activity to require intelligence, even though we viewed it as intelligent behavior before we knew how to program it. This definition is an allusion to the view that behavior that can be understood cannot be intelligent. AI researchers have observed on several occasions that the criterion for what constitutes thinking or intelligent behavior has changed along with the accomplishments of the AI community: $x$ may be among the set of activities considered to be examples of intelligent behavior, until someone manages to program a computer to do $x$, at which time it is removed from the set. The definition is, of course, self-contradictory: it rules out the possibility of developing artificial intelligence. Programs that we do not know how to produce do not exist; and once we have produced them, they fail to meet the criterion for inclusion in the set of interest. But this is the point. Kay is suggesting, I believe, that the debate about whether or not a particular activity requires intelligence is not very useful and maybe not even very interesting, inasmuch as it can so easily become a matter of semantics. The interesting question is whether there are limits to how far we can go in the direction of giving computers capabilities that human beings have and, if so, what those limits are. The history of artificial intelligence, as an area of work, has been a gradual chipping away at the set of things that people can do and computers cannot. That set is not likely to become empty any time soon, if indeed it ever will; but progress has been steady.

## What Motivates Artificial-Intelligence Research?

There are at least three reasons why one might want to build an intelligent machine: (1) to learn more about human intelligence, (2) to extend the Industrial Revolution to the realm of the intellect, and (3) to meet an intellectual challenge. With respect to the first, simulation with artifacts has been found to be an effective way to study natural processes; another, of course, is to observe them as they occur in nature. Few scientists would argue that either strategy should be used to the exclusion of the other; however, there is no denying the power of the simulation approach. As its proponents have pointed out, in spite of centuries of observation of birds, not much was learned about aerodynamics and the principles of flight until serious efforts were made to build machines that could fly.

Several of the pioneer workers in the field of machine intelligence have been motivated by a desire to understand human thought processes. To them machine intelligence provides a vehicle for the study of processes that underlie intelligence and thinking as they are found in human beings. The goal is to develop computer programs that mimic human behavior. The philosopher's injunction to "know thyself" finds an operational paraphrase in our efforts to duplicate our own intellectual capabilities in machines.

One of the things that make this approach attractive is that the attempt to simulate often forces one to face up to critical issues that might otherwise escape notice. As we shall note presently, for example, the attempt to program a machine to understand natural language brings one up against many problems that might be glossed over or not even recognized as problems if one were content to observe the processes of language acquisition and use as they occur in a child.

A caveat is in order here. Sometimes a successful simulation of some aspect of intelligent behavior is viewed as an explanation of that behavior as it is found in human beings. But this is not a justifiable view. A computer program that successfully simulates some aspect of behavior represents a specification of a set of operations that is *sufficient* for producing that behavior.

It is an existence proof that the behavior can be reduced to simple mechanistic operations. It does not establish that the behavior as it occurs in human beings is based on the same operations as those embodied in the program.

The second motivation for building intelligent machines is explicitly practical. As the Industrial Revolution extended our muscles, so now there appears to be the possiblity of extending our minds. Some see in the computer and in machine intelligence an unparalleled opportunity to enlarge our control over the environment, increase the production of material goods while decreasing the necessity of human involvement in the process of production, augment our problem-solving skills, and generally improve the quality of life. The notion of a population of robots of varied sizes, shapes, and capabilities, dedicated to the service of humankind, which only a short time ago would have been considered utterly incredible by most people, is now the vision of some serious and technically knowledgeable persons.

With respect to this goal, it does not matter whether intelligent systems invoke the same types of processes as do human beings when performing intellectually demanding tasks. The only question of practical significance is whether the processes the systems use accomplish the desired objectives and do so in a cost-effective way. Of course, even system builders who have no interest in human problem solving per se might attempt to produce programs that mimic human approaches if they are unable to invent more effective ones.

The third motivation for building intelligent machines is perhaps the most intrinsically human of all. If one reads the biographies of the outstanding scientists of the past, one discovers a truly heterogeneous collection. Some were physically robust, some were sickly; some were highly educated, others had little or no formal schooling; some were humble, others were insufferably arrogant; they represented every conceivable political, philosophical, and religious persuasion. There is hardly a trait or dimension with respect to which we could make a comparison that would not produce representatives of opposite extremes. The exception is this: to a person, great scientists have loved to think. They have gotten deep satisfaction out of meeting and mastering intellectual challenges. (Not, of course, that only scientists have had this trait.)

What one thinks about, however, must depend to some extent on the times and circumstances in which one lives. All scientists build upon the work of their predecessors. One wonders what Newton would have thought about had he lived a thousand years ago—or today. In any case, the question of the possibility of intelligent machines is one of the most intellectually challenging questions to present itself to our age. By "present itself," I mean that science and technology have developed to the point where the question forces itself upon us. Given the insatiable inquisitiveness of the human mind, it is inconceivable that the challenge would go unanswered.

The detached and unemotional scientist, who observes nature and develops theories in a disinterested and objective way, is a myth. Scientists are personally and passionately involved in a creative activity. Scientific theories are not discovered; they are produced. Newtonian mechanics is as much the creation of Newton as the Mona Lisa is that of Leonardo. Moreover, nothing has stimulated the creative impulses of scientists and artists alike more than has humanity itself. It is claimed that upon completing a sculpture of a man, the early-Renaissance sculptor Donatello would say to it, "Why do you not speak?" Had Donatello lived in the twentieth century instead of the fifteenth, it is possible that his creative instincts might have found expression in the building of an artifact that could carry on a fairly respectable conversation. Whether one would consider the latter achievement as beautiful as Donatello's statues would depend, of course, on one's aesthetic values and tastes. In any case, whatever the motivation, the possibility of developing intelligent machines has captured the imagination not only of a few scientists, but now of a growing number of industrialists and investors as well.

## The Hidden Complexity of Common Abilities

"Despite all the marvelous things that computers can do today," begins a recent article in *Science* on artificial intelligence (Kolata 1982), "they simply lack many of the qualities that are present in human intelligence—they don't even have common sense." Significant here is the implication that common sense is a minimum requirement of anything that is to be considered intelligent. One of the surprises that have come out of the work on artificial intelligence is the realization that how difficult a task appears to be for a human being is not a good indication of how difficult it will be to get a machine to perform that task. Paradoxically, it has proved far easier to program some tasks that can be performed effectively by relatively few highly trained people, such as diagnosing a medical problem or configuring a computer system, than to get a computer to understand natural language or recognize common visual scenes, tasks that the vast majority of children perform with ease.

Consider, for example, the problem of getting a computer to understand speech. To a first approximation speech is composed of sequences of acoustic units known as phonemes. The word *pat,* for example, comprises three of these units (*p, a,* and *t*), as does the word *thatch* (*th, a,* and *tch*). Speech scientists have identified about forty such phonemes in English. The phonemes themselves are sometimes thought of as distinctive combinations of even more primitive entities called features. This being the case, one might think that the problem of speech recognition could be solved in a straightforward way: the computer would simply apply to the speech signal a set of feature tests to identify each phoneme in turn, and having identified the phonemes, would find word recognition a relatively trivial task—except in cases of homonyms (*meet, meat, mete*) and homophonous phrases (*I scream, ice cream*), in which the same string of phonemes can have more than one meaning. Things are not that simple. Even if phonemes and words were easily identified, there would still be some difficult problems to be solved before one could say that one had accomplished the objective of getting the computer to understand speech. And as it happens, even at the level of phoneme and word recognition, the task is far more difficult than these comments suggest. Among the reasons for this are the problems of variability, segmentation, and noise.

People differ greatly from one another in many aspects of their speech. Words are emitted by different speakers at different rates. Voices differ widely in their fundamental frequency, or pitch, as well as in the timbre or quality of their sound. Some people talk in a monotone, some use a great deal of inflection. Some articulate words distinctly, others have what speech teachers refer to despairingly as lazy speech. Add to these and other individual differences the effects of regional dialects, foreign accents, and speech impediments, and it is easy to understand how the acoustic signal for a phoneme produced by one speaker may have little in common with the signal for the same phoneme produced by a different speaker. Moreover, the characteristics of a given phoneme can vary considerably not only from speaker to speaker but from utterance to utterance for the same speaker. This is in part because the acoustic features of phonemes are somewhat context-dependent: the way a given phoneme is articulated can be influenced by what immediately precedes or follows it. Indeed, the variability is so great that some speech scientists question the utility of the very concept of the phoneme. Most do not go that far, but all would acknowledge the difficulty of establishing a set of context-free feature tests that will invariably be able to identify phonemes correctly as they occur in running speech.

The problem of segmentation—identifying the boundaries between adjacent phonemes or words—is no less severe. Not only may phonemes overlap in the speech signal, but sometimes a single nonsegmentable feature of a speech sound will carry information about two successive phonemes. Establishing word boundaries is perhaps even more difficult, and more critical to the task of speech recognition, than delimiting phonemes. The problem is not only that the acoustic representation of a word spoken in continuous speech tends to merge with the representations of neighboring words, although this would be bad enough. Matters are further complicated in that the acoustic characteristics of words, like those of phonemes, depend substantially on the contexts in which they occur. One way to make the boundary problem tractable is to limit one's vocabulary to words whose beginnings and endings are most readily identified, and this is an approach that has sometimes been taken. For example, inasmuch as *the big* is more easily segmented than is *the large* because voicing is momentarily interrupted at the initiation of the *b* in *big*, one can simplify the segmentation problem by permitting the use of only the former term. This may seem like cheating, but it is not unreasonable to attempt to solve the easier problem before facing the more difficult one. Trying to get a system to function reasonably well under highly favorable conditions is an appropriate first step toward making one that would be more versatile.

The problem of noise is one that is encountered in any pattern-recognition task, and it has two aspects: the presence of unwanted data and the absence of data that would be useful if they were available. The unwanted data in the speech signal come from background noise, such as voices of other people and environmental sounds, and from the speaker himself—"ums," coughs, clearing of the throat, sighs, and other non-speech sounds. The speech recognizer must be able to tell the difference between the genuine speech sounds that are produced by the speaker to whom it is attending and the various other sounds that may be mixed in with that speech.

The problem of missing data is probably more bothersome than that of the presence of extraneous sounds. It is natural to assume that the acoustic waveform that represents speech normally contains all the information that is necessary to recognize what a speaker is saying. This assumption is demonstrably false. In fact, what a speaker is saying, or, more precisely, what a listener hears, often is *not* all contained in the speech signal. Much is inferred by the listener on the basis of what he knows about grammar, the phonological rules of the language, the topic of discussion, the speaker, and so forth. To be successful at speech recognition, a computer must be able to do the same; it must be able to fill in gaps and resolve ambiguities by the application of phonological, syntactic, and semantic information that it has stored away in its memory.

So much for the problems involved in identifying phonemes and words. The point was made earlier that even if these could be solved we would not yet be justified in saying that the computer had been given the ability to understand speech. Having determined what sequence of words has been uttered by a speaker, the machine still must determine from those words the message that the speaker intended to convey. Here one discovers a host of thorny problems that linguists discuss under the rubric of *semantics*.

The human's ability to extract meaning from speech is indeed remarkable, and not well understood. There are a few hundreds of thousands of words in the English language, by some counts, although between five and ten thousand seem to account for about 95 percent of usage. Even with such a modest number of words, however, the number of meaningful sentences that can be formed is for all practical purposes infinite. Not only is it possible to say something that has never been said before, but each of us probably does so all the time. Certainly it must be that many, if not the vast majority, of the sentences that any given individual encounters he has never encountered before and never will encounter again; they are literally once-in-a-lifetime events. It is this richness of language that makes the human's ability to use it—to produce it and to understand it—so impressive.

Of the numerous problems that relate to meaning and its implications for speech recognition by computer, we will consider briefly only the following two: the fact that words often have multiple meanings and the fact that information can be carried not only by the words that are spoken but by the way they are said as well. One can readily appreciate the problem of multiple meanings by simply leafing through an unabridged dictionary. The word *grip* has eighteen definitions in my dictionary, among them "an energetic grasp," "a handle," "a piece of luggage," and "a contagious viral disease." I count over forty meanings for *run* as a transitive verb, and over forty more as an intransitive verb. I did not bother to count its definitions as a noun. Such a proliferation of meanings poses a problem of some magnitude for the programmer who would teach a computer to understand speech. He must give it the capability of distinguishing, when it learns that something is running, whether the thing is going by moving the legs rapidly, contending in a race, becoming a candidate, playing a musical passage quickly, flowing rapidly, or melting. The machine must realize that colors can run, as can time, or a theme, or an eye, or an advertisement—but each in a different way. A man who runs a business can also run his car on gasoline—on the road that runs by the river that runs into the sea, if he wants to! It is not enough that the computer recognize that the word is admissible in each of these contexts; it must be able to distinguish the differences in meanings, some sharp, some subtle, that are involved.

The fact that information is carried not only by what is said but also by the way it is said is another problem that must be solved by any generally effective speech-recognition procedure. The problem is encountered both at the level of individual words and at that of phrases or sentences. The meanings of some words (for instance, *object, attribute, permit, invalid*) depend on which syllable is accented. The question, "Does he speak French or German?" has one interpretation if the pitch of the voice rises on the last syllable and another if it falls. In the former case the speaker is asking a question that can be answered by "yes" or "no"; in the latter he expects as an answer either "French" or "German." Moreover, the problem is not always simply that of deciding which of two possible meanings to attach to an utterance. A subtle change in inflection or stress can alter the message in a sentence in one of many ways. One need not be a speech scientist to recognize that one can convey a variety of messages by the way one says "Good bye." And consider how many inflectional embellishments one can put on the question, "Isn't that nice?"

These comments have not touched all of the aspects of the problem of speech understanding by computer, but perhaps they suffice to establish the complexity of the task. Speech understanding as it occurs in human beings is among the many phenomena that we take for granted only because they are so common. It it were not for the fact that people *do* communicate by talking, one might well conclude on theoretical grounds that the development of such a capability would be too much to expect of such an organism. Efforts to give the computer the capability of understanding speech and natural language are certain at least to further our understanding of the process as it occurs in humans, even if it is a long time before it can be effectively duplicated in the machine.

What *are* the prospects of developing a respectable speech-understanding system in the near future? The answer depends on the specific aspirations implied by "respectable." In 1971 Newell and other members of a study group listed several level-of-aspiration questions that are germane to the issue of estab-lishing reasonable goals for speech recognition projects. Among the questions they raised are the following: What sort of speech (isolated words, continuous discourse) should the sys-tem be able to accept as input? How many speakers (one, small set, open population) should it be able to accommodate? What sort of speakers (cooperative, casual, playful; male, female, child) should it recognize? In what sort of auditory environ-ment (quiet room, computer room, public place) should it be able to function? With what sort of communication system (high-quality microphone, telephone) should it be able to cope? How much training (few sentences, paragraphs, full vocabu-lary) should the system require? How much training should be required of the users? How large and free a vocabulary should the system be able to use? What sorts of syntactic constraints (fixed phrases, artificial language, free English) should be nec-essary? What sorts of tasks (highly constrained, open) should the system be able to perform? What information is the system to have concerning the user? What kinds of error rates are to be tolerated? How fast must the system be?

We are still a long way from a general solution to the speech-understanding problem. That is to say, the ability of the best speech-understanding programs is not close to that of human beings and is not likely to be for some time to come. The technology has advanced considerably in the past few years, however, and is now at a stage of development where it is beginning to be used in applied situations—systems that are capable, after some tuning to the speaker, of isolated word recognition with limited vocabularies (a few tens or possibly hundreds of words) have been in use for some time. We are beginning also to see some limited recognition of connected speech. While it is not clear that a system that can recognize connected speech as well as a human being can will ever be developed, it is clear that speech technology will be good enough—is nearly good enough now—to permit the use of speech input in a variety of operational situations.

## The Problem of Knowledge Representation

The question of how to represent knowledge in a computer so that the machine can readily access and use it is a central, if not *the* central, question in artificial intelligence today. This marks a significant change in thinking among AI researchers. Much of the early work in this area focused on the question of how to give a machine the ability to learn. The general idea was that the way to produce an intelligent machine was to build one that had little, if any, knowledge to begin with but had the ability to acquire information from its environment and to adapt its behavior on the basis of preprogrammed learning rules.

Although numerous machines were built that could be said to learn in some rudimentary sense, it took only a few years of these efforts to convince researchers that if one wanted a machine to perform relatively complex intellectual tasks, one had to program into it a considerable amount of knowledge at the outset. This includes linguistic knowledge, some general world knowledge, and knowledge specific to the domain in which the machine is to function. How to represent such knowledge so that it would be accessible soon became a major problem for the field. A variety of new formalisms have been developed for representing knowledge, including production (*if-then*) rules

(Davis, Buchanan, and Shortliffe 1977; Davis and King 1977; McDermott 1982a; Shortliffe 1976), frames (Pauker, Gorry, Kassirer, and Schwartz 1976), scripts (Cullingford 1978; Schank and Abelson 1977), logic (Kowalski 1979; Moore 1982), and networks (Duda, Gaschnig, and Hart 1979).

The issue of representation is a generic one that relates to intelligence, both natural and artificial, deeply and in a variety of ways. The many forms of spoken and written language are representational systems. We have learned to supplement what might be referred to as "common language" with many symbol systems to facilitate the representation and communication of special types of concepts. These include the elegant and powerful Hindu-Arabic number system; the notations of algebra, calculus, and other forms of mathematics; and the notational systems that serve physics, chemistry, and music. A primary heuristic for problem solving is to find a way to represent the problem one wishes to solve with a picture or diagram. A variety of pictorial or diagrammatical representations have proven useful in a wide assortment of contexts. These include graphs, contingency tables, trees, flow charts, and circuit diagrams. There is a continuing need, however, for new, more powerful representational schemes. The general question of how best to represent information—knowledge, problems, processes within computer systems—is one that is likely to be around for a very long time.

### The Commercialization of AI

Artificial intelligence has recently captured the fancy of industry, venture capitalists, and the general public. A great deal of money is being poured into efforts to turn this technology to practical and profitable use. It is not hard to find claims like this one: "The new AI systems are so radically different from conventional computer programs and systems that their commercial application may cause changes in business and society which will be more dramatic than any technological innovation conceived by man" (Brown 1983, 57).

The U.S. market for AI products and services has been estimated at $40 million to $75 million in 1983, growing possibly to over $8 billion by 1993 (Manual and Evanczuk 1983; Rauch-Hindin 1983). Among the U.S. companies investing heavily in artificial-intelligence research and development are Dupont,

Schlumberger, DEC, IBM, General Electric, Texas Instruments, Allied Corporation, Xerox, Fairchild, and Tektronix. Although workers in this field are finding the sudden attention gratifying, if not somewhat intoxicating, there is growing concern that expectations may have risen to unrealistic heights and that a serious backlash may occur if those expectations fail to be realized in accordance with the assumed timetable (Waldrop 1984a; Winston 1984). For a sampling of what several members of the AI research and investment communities think about the attention AI is now receiving from business and industry, see Winston and Prendergast (1984).

## Expert Systems

One of several terms relating to information technology, and particularly to artificial intelligence, that have recently become quite popular is *expert systems*. Like most such terms this one means different things to different people. It does invariably refer, however, to a computer system or program that knows a lot—or, if one prefers, that contains a considerable amount of information—about a particular topic.

Chandrasekaran (1984) points out that it is hard to be precise about what qualifies a system to be called an expert one. He identifies several features or "dimensions" that have been suggested as important, including expertise, a search capability, uncertainty in data, the use of symbolic knowledge structures, and the ability to explain its reasoning; but he considers none of these to be a truly defining property. McDermott (1984) sees the ability to bring large amounts of domain knowledge to bear on a problem as that which sets expert systems apart from other programs. The main problem in developing an expert system, he suggests, is figuring out how to represent that knowledge so that it will be available when needed.

Hayes-Roth (1984, 264) gives the following list of what expert systems do:

- They solve very difficult problems as well as or better than human experts.
- They reason heuristically, using what experts consider effective rules of thumb.
- They interact with humans in appropriate ways, including the use of natural language.

- They manipulate and reason about symbolic descriptions.
- They function with erroneous data and uncertain judgmental rules.
- They contemplate multiple competing hypotheses simultaneously.

Sometimes these systems are intended to replace their human counterparts. More often they are intended to function as assistants or advisers to people in the performance of specific tasks. The distinction between expert systems, decision support systems, and software assistants, associates, advisers, or consultants is not very sharp.

The reasons for trying to develop expert systems include all those for doing research in artificial intelligence more generally. However, much of the support for this work derives from the assumption that the systems being developed will have great practical utility. One hope is that expert systems will bring to specific tasks capabilities that are brought by (often scarce) human experts, without some of their liabilities. The following comment (reproduced here with apologies to its unknown author; the attribution is missing from my notes) succinctly captures this hope: "Unlike some human experts, an expert software system does not get tired or sick, demand more compensations from its employer, rub colleagues the wrong way, carry away trade secrets to a competitor, or quit and start its own company."

There are very few expert systems in operational use, but many are under development or in use on an experimental basis. The number of experimental systems in Europe has been estimated at about 200 (Tate 1984). It is difficult to know what to make of such an estimate, however, as the term *expert system* has such broad meaning. Application areas for which expert systems are being used or developed include computer system configuration, locomotive maintenance, oil exploration, biological research, medical diagnosis, business information management, and education. A sampling of these systems follows.

- XCON (also known as R1): A system developed at Carnegie Mellon University and used by the Digital Equipment Corporation to configure VAX computer systems in accordance with the needs and wishes of individual customers (Kraft 1984). Instead of marketing a small number of preconfigured systems, DEC offers a variety of system components (over 1,000 options) from which buyers can customize systems to their tastes. Not all components are compatible with each other, however, and configurations must be designed with the knowledge of the constraints. XCON is claimed to be "the largest expert system in daily use in an industrial environment anywhere in the world" (Abramson 1984; McDermott 1982a). It has more than 2,500 if-then rules in its data base. It is a batch processing system and has no interactive capability.

- XSEL: An interactive expert system developed by DEC to assist its salespersons in estimating prices on various VAX and PDP-11 configurations. It is a rule-based system and contains about 300 rules (McDermott, 1982b, 1984).

- PTRANS: A manufacturing management assistant, developed as a companion to XCON and XSEL, that is designed to help anticipate and prevent problems arising in the manufacturing process (Haley, Kowalski, McDermott, and McWhorter 1983).

- DELTA/CATS-1 (Diesel-Electronic Locomotive Troubleshooting Aids/Computer-Aided Troubleshooting System): Developed by General Electric to help diagnose problems with railroad locomotives and to facilitate maintaining them. It contains over 500 rules. Incorporation of a videodisk player permits the system to provide the user with drawings, photos, and movies as appropriate (Artificial Intelligence Report 1984).

- Dipmeter Advisor: Developed by Schlumberger for analysis of oil well drilling data, the Dipmeter Advisor gets its name from the fact that one objective of the system is to determine the angular displacement, or "dip" from the horizontal, of subsurface mineral strata. Its purpose is to help geologists interpret data obtained from a variety of probes inserted into drill holes. It operates on about 90 rules (Baker 1984; Davis, Austin, Carlbom, Frawley, Pruchnik, Sneiderman, and Gilreath 1981; Gershman 1982; Schlumberger 1981).

- Drilling Advisor: The Drilling Advisor was developed jointly by Teknowledge Inc. and the French national oil company National Elf Aquitaine. Its purpose is to provide consultation to the supervisor of an oil rig regarding the problem of "sticking," which is often encountered in the drilling of production oil wells. Sticking refers to a situation in which it is impossible either to continue drilling or to raise the down-hole equipment to the surface. The Drilling Advisor is intended to help diagnose the most likely causes of such problems and to recommend actions aimed at alleviating or avoiding them. Its knowledge base contains about 250 if-then rules.

In diagnosing a problem, the Drilling Advisor attempts to identify the most likely of six possible causes of sticking. It qualifies each hypothesized diagnosis with a probability reflecting its degree of certainty. Diagnoses are accompanied by explanations of the reasoning on which they are based. Prescribed treatments are also selected from a relatively small set of possibilities. In diagnosing, the system requests information from the user regarding the well, constituent rock types, type of activity immediately preceding the sticking, depth of drill bit, and so on. When it has proceeded far enough to form a tentative hypothesis, the specific questions it asks are contingent on that hypothesis (Hollander, Iwasaki, Courteille, and Fabre, undated).

- Prospector: Developed by SRI International, this system also assists in the exploration of mineral deposits (Duda, Gaschnig, and Hart 1979; Duda, Hart, Nilsson, and Sutherland 1978; Hart, Duda, and Einaudi 1978).

- Dendral: The first expert system to be used very widely, Dendral was developed by Stanford University and SRI International and is used (along with its successor, Genoa) by chemists to aid in the study of molecular structures. It helps determine the structure of organic molecules from various inputs, including mass spectra and nuclear magnetic resonance data (Buchanan, Duffield, and Robertson 1971; Buchanan and Feigenbaum 1978; Feigenbaum 1983; Lindsay, Buchanan, Feigenbaum, and Lederberg 1980; Michie and Buchanan 1974). As of 1983 Dendral had helped generate about 50 publications in the chemistry literature (Duda and Shortliffe 1983).

- MOLGEN (Molecular Genetics): Developed at Stanford University, MOLGEN assists molecular geneticists in planning gene cloning experiments. It uses a "constraint-posting" approach to the problem of dealing with interactions among variables that are important to a planning or design process. Constraint posting is a way of narrowing the search space by applying constraints as they are discovered without prematurely focusing on specific hypotheses that may later prove to be wrong. If, for example, the problem on which MOLGEN is working involves a selection of a bacterium and a vector (a self-replicating DNA molecule) for a cloning experiment, one constraint is that the bacterium and the vector be biologically compatible. Rather than select a specific bacterium-vector pair early in the planning process, MOLGEN "posts" the compatibility constraint to ensure that this is taken into account when a tentative selection is eventually made. Constraints are propagated from problem to problem. Problem solving consists, to a large degree, in finding objects in the knowledge base that satisfy the various constraints that have been formulated. MOLGEN is not yet considered by its developers to be a useful computational aid for geneticists. Its knowledge base is considered to be too narrow, and even in the restricted class of experiments that it is able to help plan, there are laboratory techniques beyond its knowledge (Martin, Friedland, King, and Stefik 1977). Interestingly, the implementation of MOLGEN makes an explicit distinction between planning and metaplanning: planning involves designing an experiment; metaplanning involves strategizing about the design process (Stefik 1981a, b).

- Genesis: Also assists in the planning and simulation of gene-splicing experiments (Engelmore and Nii 1977).
- Mycin: Also developed at Stanford University, Mycin assists in the diagnosis and treatment of infectious diseases and in the selection of antibiotics appropriate to their treatment (Davis 1984; Davis, Buchanan, and Shortliffe 1977; Feigenbaum 1983; Shortliffe 1976; Shortliffe and Buchanan 1975). Mycin's database contains about 500 if-then rules. In attempting a diagnosis, Mycin tests the various rules in its database against information that has been provided about the patient. If, upon attempting to apply a rule, it discovers that necessary patient data are not at hand, it can request additional information from the user, who is probably the attending physician.

  Mycin has the ability to explain to the user at least some aspects of its reasoning. If, for example, the user types *why* in response to a request from the program for additional information, the system responds with an explanation of why it wants the information requested. The explanation reveals the rule that it is currently working on and why it is working on that rule. By typing *why* repeatedly, the user can back the system up through its entire chain of inferences. This feature adds to the usefulness of the system for purposes of training.

  Mycin has been shown to do about as well as trained physicians in the diagnosis of certain bacterial infections and meningitis (Yu, Buchanan, Shortliffe, Wraith, Davis, Scott, and Cohen 1979; Yu, Fagan, Wraith, Clancey, Scott, Hannigan, Blum, Buchanan, and Cohen 1979). It has been successful enough that efforts have been made to generalize its procedures and apply them in other contexts (Van Melle, Scott, Bennett, and Peairs 1981; Aikins, Kunz, Shortliffe, and Fallat 1982).

- EMycin (for Essential Mycin, or Empty Mycin, or Engine Mycin): EMycin is Mycin stripped of its original knowledge base regarding infectious disease; it is, in a sense, Mycin's inference engine (Van Melle 1979; Feigenbaum 1983).

- Puff-VM: Developed by Stanford University and the Pacific Medical Center, Puff is a rule-based system for helping to diagnose lung disorders. It was developed by providing EMycin with a set of rules for pulmonary function diagnosis (Feigenbaum 1983; Osborn, Fagan, Fallat, McClung, and Mitchell 1979). It takes a patient's history and a variety of measurements and test results as inputs and produces a diagnosis, which is added to the patient's records and is checked by a physician. Puff's interpretations of test results (pulmonary function tests) have been compared with those of clinicians and found to agree in 80 to 90 percent of the cases (Aikins, Kunz, Shortliffe, and Fallat 1982; Basil and Edwards 1984).

- SACON (for Structural Analysis Consultant): This system was also built by providing EMycin with a new knowledge base. It was intended to be used with a finite-element analysis package in the design of airplane wings (Bennet and Engelmore 1979).

- Neomycin: A tutoring system based on a restructuring of Mycin (Clancey 1981; Clancey and Letsinger 1981).

- Teiresias: A system designed at Stanford University to help build a knowledge base that can be used by an advice-giving system such as Mycin. It functions as a bridge between an expert and the program the expert is trying to "educate" (Davies 1982).

- Guidon: Also developed at Stanford University, Guidon is intended to serve as a bridge between a system such as Mycin and a student who wishes to use it as a vehicle for learning about its domain of expertise (Clancey 1979). Guidon becomes, as it were, a tutor that has access to the expert system's knowledge base. The methodology used in developing Teiresias, Guidon, and other Mycin derivatives is considered sufficiently general to be applicable to domains other than medicine, and several efforts have been made to apply them in domains as disparate as the repair of electromechanical devices and the design of intelligent computer terminals (Davis 1982).

- Internist-1: Developed at the University of Pittsburgh, Internist-1 assists diagnosis in internal medicine (Pople 1977, 1982, 1984a). Its diagnostic capability was intended to be broader than that of previously developed systems and to apply to the diagnoses of multiple and complex disorders (Miller, Pople, and Myers 1982). The inferential methods it uses to arrive at a set of possible diagnoses and to select the most appropriate alternative from among that set were modeled after those that are believed to be used by physicians when confronted with similar diagnostic problems.

The knowledge base of Internist-1 represents 15 person years of work and contains over 500 disease profiles, approximately 3,550 disease manifestations (symptoms), and about 6,500 relations among manifestations (information about how the presence or absence of a given manifestation may influence the presence or absence of other manifestations). Associated with each manifestation in a disease profile are an evoking strength (the degree to which that disease explains that manifestation) and a frequency (the frequency with which patients with that disease have that manifestation); also associated with each manifestation is a disease-independent import (the extent to which the manifestation requires an explanation). Diagnoses are produced by application of a scoring procedure involving assigning numerical values to evoking strengths, frequencies, and imports and combining these values in accordance with a set of ad hoc heuristics.

Internist-1 has been evaluated by comparing its diagnoses with those of attending clinicians and case discussants for several case records of the Massachusetts General Hospital published in the *New England Journal of Medicine*. Of 42 potentially appropriate cases reviewed, 23 were excluded from the study because their diagnoses were not yet represented in Internist-1. This fact illustrates the ambitiousness of the task of developing systems that can function as experts in such knowledge-rich domains as internal medicine. With respect to the 19 suitable

trial cases, the performance of Internist-1 was roughly comparable to that of the clinicians who originally dealt with the cases
and that of the case discussants. The system is viewed by its
originators as still a research tool, and much of the current
work is focused on identifying its specific shortcomings and
limitations for the purpose of paving the way to the development of more effective systems (Miller, Pople, and Myers
1982).

- Caduceus: The successor to Internist-1, Caduceus also performs medical diagnosis and provides consultation to physicians on internal medicine. It currently contains data on more
than 600 diseases and 4,000 disease manifestations (Pople
1982, 1984b).

- CASNET (Causal-Associational Network): A system developed
at Rutgers University to provide consultation on glaucoma
(Kulikowski and Weiss 1982; Weiss, Kulikowski, and Safir
1978; Weiss, Kulikowski, Amarel, and Safir 1978), CASNET
presents more than one diagnostic hypothesis and qualifies
each with a degree-of-certainty judgment expressed non-
numerically (possible, probable, almost definite, and so on). It
also identifies possible causes of the reported conditions, if a
single cause cannot be determined with certainty. CASNET
makes heavy use of causal relationships in its reasoning (hence
its name). It uses pattern-recognition techniques, probabilistic
scoring rules for evaluating hypotheses, and a semantic network for representing causal relationships between disease
states and symptoms. Evaluation of CASNET has primarily
involved judgments by ophthalmologists of its clinical
proficiency, its applicability to glaucoma research, and its importance to health care. The large majority of clinicians making
such judgments have rated the system as at least very competent and moderately applicable or important (Kulikowski and
Weiss 1982).

- Expert: Also developed at Rutgers University, Expert is a successor to CASNET (Weiss, Kern, Kulikowski, and Safir 1976).
The intent in developing it was to provide it with a generalized
and extended knowledge base and the inferential capability of
CASNET. This system has been applied in the domains of
thyroid problems, rheumatology, and neuro-ophthalmology.
Such applications require the development of specialized
databases, of course, but the system provides the formalisms

and representational schemes into which the appropriate information can be put. Expert is really a framework that is intended to facilitate the development of databases that are appropriate to a variety of disease categories.

- ONCOCIN (Oncology Chemotherapy Consultation): A Stanford-based system for managing patients with certain types of cancer, including Hodgkin's disease and oat-cell lung carcinoma. The rule-based system makes recommendations regarding patient treatment to the attending physician.

- PIP (Present Illness Program): Developed at MIT and the Tufts New England Medical Center, PIP is a program for providing consultation on kidney disease (Pauker, Gorry, Kassirer, and Schwartz 1976).

- Sophie: A knowledge-based computer-assisted instruction system, developed at Bolt Beranek and Newman, for teaching electronic troubleshooting at an expert-grade level. It can simulate an electronic circuit in such a way as to permit the student to perform experiments by making different kinds of measurements and altering the circuit's design. It is also able to monitor and critique a student's reasoning about a faulty circuit, as that reasoning is revealed in the system-student dialog (Brown, Burton, Bell, and Bobrow 1974; Brown, Burton, and de Kleer 1982).

- Steamer: Also developed at Bolt Beranek and Newman, Steamer is a graphics-oriented system for training operators of a steam propulsion plant. The system contains a model from which it can generate graphical representations of the plant, or components thereof, at different levels of detail. It can also represent graphically the flow of water or steam through the system and the consequences of specific malfunctions. It permits structured tutoring in which it presents problems to the student and guides the session, and also exploratory learning whereby the student can perform *what if* experiments and thus discover the consequences of various operator actions (Stevens, Roberts, Stead, Forbus, Steinberg, and Smith 1981; Williams, Hollan, and Stevens 1981).

Probably more efforts have been made to develop expert systems in medicine than in any other area. This work can be seen as a continuation of a wide assortment of efforts to apply computer technology to medical decision making during the 1960s and early 1970s. The following catalog of that earlier work is taken from Nickerson and Feehrer (1975, 178; the works cited may be found in the reference list at the end of this volume):

Several experimental computer-based systems have been developed for the purpose of facilitating various aspects of decision making in the medical context. Applications that have been explored include initial patient interviewing and symptom identification (Griest, Klein, and Van Cura 1973; Whitehead and Castleman 1974); analysis, organization, and presentation of the results of laboratory tests (Button and Gambino 1973); personality analysis (Kleinmutz 1968; Lusted 1965); storage and retrieval of individual patient data (Collen 1970; Greene 1969); on-demand provision to practitioners of clinical information (Siegel and Strom 1972); automated and computer-aided diagnosis of medical problems (Cumberbatch and Heaps 1973; Fischer, Fox, and Newman 1973; Fleiss, Spitzer, Cohen, and Endicott 1972; Gledhill, Mathews, and Mackay 1972; Horrocks and de Dombal 1973; Jacques 1972; Lodwick 1965; Lusted 1965; McGirr 1969; Yeh, Betyar, and Hon 1972); management and graphical representations of data to aid research in pharmacology and medicinal chemistry (Castleman, Russell, Webb, Hollister, Siegel, Zdonik, and Fram 1974); modeling of physiological systems and explorations via simulation of effects of alternative courses of treatment (Siegel and Farrell 1973); and training (Feurzeig 1964; Feurzeig, Munter, Swets, and Breen 1964).

The fundamental components of an expert system are usually considered to be a *knowledge base* and an *inference engine* (Robinson 1983). The knowledge base is the corpus of information contained in the system that relates to its domain of expertise. The inference engine is the collection of rules and procedures used by the system in the application of its knowledge to specific problems. While both the knowledge base and the inference engine are considered essential, increasing emphasis has been placed recently on the importance of the knowledge base. Feigenbaum (1983) says flatly, "the power does not reside in the inference procedure. The power resides in the specific knowledge of the problem domain. The most powerful systems will be those which contain the most knowledge" (38).

While the performance of many expert-system programs is remarkably good on some subset of the problems of interest, many of them exhibit what has been called the "plateau and cliff effect": their performance tends to degrade ungracefully when they encounter problems that are on the edge of the domain of interest or that involve complex interactions among the variables with which they have to deal (Szolovits 1982).

One of the current problems of interest to researchers is that of coupling expert systems to existing decision-support systems such as management information systems (MIS) and database management systems (DBMS) (Jarke and Vassiliou 1984). The objective here is to provide the user with assistance in getting the needed information out of the MIS or DBMS and in applying it effectively to his decision problems.

People who have worked on expert systems have addressed a variety of questions, including how to identify the knowledge such a system should contain; how to organize and represent that knowledge; how to extract or infer from a knowledge base the information one needs to solve problems and make decisions; how to give a system the capability of explaining to a user the basis of its conclusions, decisions, or advice; how to ensure effective communication between the system and its users; and how to give a system an understanding, in a nontrivial sense, of its domain. All of these problems are difficult ones, and while there has been considerable progress on most of them, much remains to be done on all of them. The most challenging of these problems, and the one that is in some sense a key to all the others, is the last one, the problem of understanding. It is also the one on which the least progress has been made to date. A major part of the difficulty here is that it is not clear what it means for a person to understand a domain, let alone how to mechanize understanding.

One of the most interesting aspects of efforts to develop expert systems is what these efforts reveal about the nature of the reasoning involved in solving difficult cognitive problems. An interesting distinction that has been made, for example, is the distinction between probabilistic and categorical reasoning (Szolovits and Pauker 1978). Probabilistic reasoning may be represented in the abstract by an assertion of the type "If condition $x$ holds, then usually (almost always, often, never) $y$ also holds." Categorical reasoning would involve a statement such as "If $x$ is observed, do $y$." Both forms of reasoning may be found in expert system implementations.

Here is a reasoning problem of the kind that work on diagnostic systems brings to focus: How does one decide when one has enough information to accept a diagnosis as sufficiently probable to warrant acting as though it were correct? Consider the following two decision rules: (1) select whichever diagnosis first acquires enough corroborating evidence to exceed some preset threshold; (2) accept a diagnosis as true when the judged probability of its being true exceeds the probability of the next most likely candidate by a preestablished amount. The first decision rule is based on an absolute threshold; the second, on a differential threshold. Both types of rules have been used in systems, and there is some evidence that in at least some instances the differential-threshold rule tends to produce slightly more accurate diagnoses than the absolute-threshold rule (Sherman 1981, cited in Pople 1982).

As is apparent from the foregoing examples, most of the expert systems that have been developed to date are intended for use by professionals in work-related situations. If such systems proliferate as some writers expect them to, they could eventually give the average person access to expert advice and assistance of many types. Very little effort has been given to developing expert systems for the average person, but a beginning has been made in this direction. Kbol, for example, is an experimental system for answering natural-English questions about some of the products supplied by a garden store (Walker 1984; Walker and Porto 1983).

### The Codification of Expertise

The idea of an expert system prompts several questions. What is an expert? What constitutes expertise? How might the expertise of a human expert be codified? How does one determine how expert an expert system is? How does one compare the degree of expertness of an expert (whether human or machine) in internal medicine, say, with that of one in paleontology, or English literature? Does it make sense to try to do that? Such questions are of considerable interest to psychologists. They are of increasing interest to developers and users of computer systems that are intended to play the role of expert in one way or another.

What is expertise? One answer is that expertise in a given domain is what the people who are acknowledged to be experts in that domain define it to be. This would be neither surprising nor disconcerting were it not the case that experts in the same subject area quite often disagree. It is not unusual, for example, for the two parties in a court proceeding to obtain diametrically contradictory testimony from accredited experts in the same domain.

Some writers have suggested that expertise is a combination of a large amount of domain-specific knowledge and good reasoning ability. An expert probably not only knows a lot about his area of expertise but has ready access to that knowledge. A nonexpert might have some of the same knowledge but be unable to access it without searching through irrelevancies at some significant effort. There is undoubtedly the possibility of a tradeoff between knowledge and reasoning ability, but it is limited: an individual who has little or no knowledge of a domain is not likely to be able to solve problems effectively in that domain, no matter how clever he is.

As we have already noted, it should not be assumed that the method used by an expert system to solve a given problem is the same as the method used by a human expert to solve that problem. There is no logical necessity that they be the same (exhaustive searches are sometimes feasible for computer systems, but they seldom are for humans), and, in fact, in at least some cases they are not. On the other hand, many of the methods that have been programmed into expert systems are indeed patterned after methods that human experts use. There are two reasons why this is so. First, in many cases system builders have been unable, or have not tried, to invent methods that work better than those that human experts use. The first order of business is to find methods that work at all, and studying the behavior of human experts who demonstrably can do what is wanted seems like a reasonable way to begin. Second, although a major reason for building expert systems is the practical one of making expert assistance more widely available, there is also the interest in learning more about human expertise through efforts to simulate it. In the latter case, the goal is to build a system that not only looks like a human expert from the outside but derives its expertise from the same underlying principles and approaches as well. The locus of interest here is not only expert performance but a better understanding of the bases of human expertise.

If one wants an expert system to have a knowledge base that is the same as, or similar to, that of a human expert in a particular area, one must find a way to transfer the human expert's knowledge to the computer. It will not do simply to ask the human expert to tell a computer all he knows about the domain. Neither experts nor run-of-the-mill humans can report all they know about a given subject on demand, unless the subject is one they know next to nothing about. The fact that one knows something is not evidence that one can retrieve what one knows under any circumstances, nor is the fact that one is unable to retrieve some knowledge on demand evidence that one does not have it. Asking an expert to report all he knows on a particular subject is more or less like asking a person who knows a great many people to report the names of all the people he knows. The names would be available when needed but would not necessarily be accessible for such an unlikely purpose as listing all of them on request.

McDermott (1984) points out that the expert systems that have been developed to date have acquired their knowledge over a long period of time. Often they begin life with a small subset of the knowledge they will eventually have. This is provided by the original system designer, who has extracted what he can from a domain expert. The original knowledge base is then extended, sometimes over a period of years, whenever the need becomes obvious for new knowledge to deal with unanticipated constraints (McDermott and Steele 1981).

In developing the knowledge base for an expert system, one discovers that a very large fraction of the information the system will need is to be used only in low-frequency cases. This has been described as the "20 percent/80 percent effect": 20 percent of the total effort will take care of 80 percent of the cases, and 80 percent of the effort will go into providing for the last 20 percent of the cases. In other words, the core knowledge of a domain may be represented by relatively few rules, but the rule set must become very large in order to cover the whole domain (McDermott 1982a; Chandrasekaran 1984). There is also the problem, as Sridharan (1985) points out, that because what is known about a topic changes in time, sometimes rapidly, expertise is subject to obsolescence. This fact has not received much attention yet from the builders of expert systems.

Given that it is not a straighforward task to find out all that an expert knows, it seems reasonable to ask whether it would be possible to finesse the problem by giving the computer the ability to accumulate information over time and to organize this information in a model of expertise in the domain of interest. Having provided a computer system with the ability to acquire knowledge, one might then let it observe a human expert in a variety of situations in which he uses his expertise. The system would serve an apprenticeship, as it were. Indeed it might serve such an apprenticeship with several human experts; its model of expertise could then be a composite or amalgam of the knowledge of its several tutors.

We have already noted that giving the computer the ability to learn is not a new idea. One of the first AI programs ever written, Samuel's (1963) original checker-playing program, had the ability to improve its game and eventually became good enough to beat its inventor. However, giving the computer the ability to acquire expertise in a substantive domain such as medical diagnosis or oil drilling is not something anyone yet knows how to do. One is left with no alternative but to attempt to determine what experts know about an area of interest and what they would do under a wide variety of conditions, and then to codify this information—often as a set of if-then statements or production rules—in such a way that it can be represented in a computer program.

The task of discovering and codifying the knowledge of human experts is sometimes referred to as *knowledge engineering* (Browndi 1983). Davis (1984) has referred to it as intellectual cloning. Both aspects of it—determining what experts know and do, and getting the computer to mimic them—are challenging work, although quite possibly not as uniquely so as is sometimes claimed. Basil and Edwards (1984) refer to the writing of programs for expert systems as "the most complex creative process ever undertaken." Without denying the complexity of the process, one must wonder what Goethe, who took forty years to produce his *Faust,* would think of such hyperbole, or Delaunay, who took twenty years to write and check the two-volume equation that describes the exact position of the moon as a function of time (Pavelle, Rothstein, and Fitch 1981). This quibble notwithstanding, the task is certainly a difficult one, and few people know how to do it well.

Brown (1983) has suggested that "because of their virtually unlimited potential applications, the commercial market for knowledge-based expert systems may become the fastest growing area in the computer systems industry of the next decade" (58). Of special relevance to the question of the role that psychologists and human-factors engineers may play in the development of information technology is Brown's claim that "the major near-term restraint on developing new [expert] systems is a shortage of 'knowledge engineers,' persons trained to formulate the knowledge and reasoning processes of human experts into a computer program" (58).

### Evaluation of Expert Systems

If a system is alleged to be an expert, it seems reasonable to ask *how* expert it is. Not surprisingly, inasmuch as the field is young, relatively little attention has been given to the question of how to evaluate these systems. Possibly it is premature to put a great deal of emphasis on evaluation, but it is an issue that must be faced before long. Funding agencies and investors will want to have performance goals and criteria by which to judge, beforehand, whether the targeted capabilities of a system are worth the cost of developing it, and, after the fact, whether the objectives have been realized. Prospective users will want some reasonable way of comparing the capabilities of competing systems in the same area of expertise. The field as a whole would benefit from a theory of expertise that would serve as a vehicle for accumulating and organizing what is being learned. The articulation of such a theory should be facilitated by efforts to develop concepts and methods for comparing systems, even— perhaps especially—across domains of expertise.

Ideally, one wants the evaluation of an expert system to yield more than a figure of merit on its performance. One wants to know not only at what level of expertise it performs but *why* it performs at that level. That is to say, the evaluation should include a diagnostic function, whose purpose is to help pinpoint specific strengths and weaknesses and provide the kind of information that will lead to improvements in that system or in its successors.

An important distinction to be maintained in this process is that between general intelligence and expertise. As these terms are generally used, they connote related but quite different things. *Intelligence* has to do with ability to learn; *expertise* connotes knowledge that one has already acquired. *Intelligence* refers to general intellectual competence; *expertise* connotes indepth understanding of a specific and, typically, narrow domain. Intelligence rests on a set of cognitive abilities for abstraction, classification, generalization, drawing inferences and analogies, and so forth; expertise is the ability to access and apply information about a given topic on demand. The implication of this distinction for the evaluation of expert systems is that the criteria for assessing general intelligence probably should differ in many respects from those that would be appropriate for assessing expertise. There is agreement among people working in artificial intelligence that an expert system is easier to build than a system that is intelligent in a general sense. Commonsense aspects of intelligence—those with which we are least impressed when we see them in human beings— are proving the most difficult of all to program into a machine.

A second useful distinction is that between intelligence and the *appearance* of it. It clearly is possible to make an individual appear, at least for a while, to be more intelligent than he is (say by providing him with answers to the items on an intelligence test before he takes the test). Similarly, it is possible to make computer programs appear to be more intelligent than they are. (Weizenbaum's Eliza [1966, 1967] is the classic illustration of this fact.) This has important implications for the design of "user-friendly" systems: the appearance of intelligence can lead to unrealistic expectations and hence to frustration on the part of users. With respect to the problem of evaluation, the fact that a machine can be made to appear more intelligent than it is represents a source of measurement noise.

A third important distinction is the distinction between assessing the intelligence or expertise of systems and evaluating them in terms of a broader set of criteria including cost, maintainability, usability, speed, accuracy, and general effectiveness in achieving their purposes. The general effectiveness of a system can be determined only relative to its intended function and its community of users. It is not at all difficult to imagine a situation in which the less intelligent or less expert of two systems would be considered the more effective one for a specific application.

One of the most striking differences between the knowledge contained in knowledge-based systems and the knowledge that human beings have is the insular nature of the former. Machine knowledge tends to exist in specialized packets that are self-contained and dissociated from anything else. One would not expect to find a human being who could diagnose illnesses but was unable to comprehend natural language and knew nothing about the world outside the realm of medicine. One might find two experts who could not converse with each other about their respective areas of expertise, but they would be expected to be able to carry on a conversation about any number of topics that draw upon their common knowledge of the world.

None of the expert systems that have been developed to date has the ability to conduct a meaningful dialog with a user on a topic outside its narrow area of expertise. Even with regard to topics that relate closely to its area of expertise, a system is likely to be completely uninformed. As Minsky (1984, 244) puts it, "The medical systems are wonderful, but they do not know what a person is. We have a magnificent kidney program at MIT that in its area is probably as good as the best medical specialist, but it does not know that the kidney is in the body in any sense. Nor does it know what a body is. Nor does it know that if you yell at someone, he will get depressed and act sick."

This is not a criticism of expert systems; their purpose, after all, is to provide expertise in narrowly defined domains. In the words of Duda and Shortliffe (1983), "the goal of expert systems research is to provide tools that exploit new ways to encode and use knowledge to solve problems, not to duplicate intelligent human behavior in all its aspects" (266). Duda and Shortliffe go on to suggest that one of the ways to help make sure that realistic solutions to problems are forthcoming is to constrain the problems that are addressed. Chandrasekaran (1984) notes that the apparently paradoxical situation in which machines can engage in high-level problem solving while being less than ignorant about the most obvious things results from the decoupling of common sense and general-purpose reasoning from domain-specific expertise, and that given our current state of understanding of human knowledge, this decoupling is probably necessary. The assumption on which much of the work on expert systems is based is that "once the body of expertise is built up, expert reasoning can proceed without any need to invoke the general world and common sense knowledge structures. If such a decomposition were not in principle possible, then the development of expert systems would have to await the solution of the more general problem of common sense reasoning and general world knowledge structures" (Chandrasekaran 1984, 46).

One implication of these differences between human and

electronic experts is that a first-order evaluation of the expertise of an expert system should be limited to the domain of alleged expertise. Does it make sense also to try to compare systems whose expertise is in different domains? Might one reasonably hope to come up with some way of scaling *level of expertness* independently of domain? A possibility that comes to mind is to grade the performance of systems in specific domains relative to what would be expected of human experts in those domains. Whether or not the attempt to develop a methodology for comparing expert systems from different domains were successful, it could have some useful results. One such result could be a better understanding of the determinants of expertness.

We can assume that there are certain techniques or principles that are generally useful in the implementation of expert systems and that some of them are more powerful or more widely applicable than others. An effort to compare systems with respect to level of expertness across domains might facilitate the identification and scaling of such generally useful techniques or principles. The methods used by systems judged to be more expert, independently of domain, would be of great interest because of the possibility of their usefulness in other domains as well.

### The Need for a Theory of Expertise

A challenge of psychologists that relates to the problem of evaluating expert systems is that of developing a domain-independent theory of human expertise. It would be useful to have a way of characterizing what an expert is, what an expert should be expect to know or to be able to do, that is independent of the field of expertise. The following might be included among things one expects an expert to be able to do:

- Assimilate new information and revise or enlarge one's knowledge base in one's area of expertise.
- Tell the difference between other experts and nonexperts in the area.
- Discriminate among levels of certainty with respect to elements in one's knowledge base; qualify one's answers to questions with reliable judgments of their dependability.

- Use information inferentially to answer questions the answers to which one has stored only implicitly.
- Recognize contradictions within one's own knowledge base or between one's knowledge base and new information.
- Perform adequately under conditions of uncertain or incomplete information.
- Recognize when one needs additional information in order to solve a particular problem, and, usually, know whether that information is obtainable.
- Know whether a problem one has been asked to solve is sufficiently unambiguous and well formed to be approachable.
- Explain what one is doing and why.
- Understand the limits of one's own expertise.

To amplify on the last item: one very important type of knowledge that an expert has is knowledge of his own knowledge vis-a-vis his area of expertise. He knows not only what he knows but, in a sense, what he does not know. Further, he can distinguish, within limits, between what *he* does not know and what is not known. The expert has a model of his area of expertise, of his own knowledge, and of how the one relates to the other. For example, with respect to questions he cannot answer, he should be able to distinguish among (a) nonsensical questions, (b) questions that do not relate to the domain, (c) meaningful questions the answers to which are likely to be known by other experts within the domain, and (d) meaningful questions the answers to which are not (yet) known by anyone—that is, questions that exceed the knowledge base of the domain. In short, an expert should be able to make judgments of meaningfulness, relevance, difficulty, and answerability.

In addition to a theory of expertise, it would be helpful to have a theory of problems that can account for the difficulty of any problem in terms of its cognitive demands and how those demands relate to human capabilities and limitations. The identification of potential components of a theory of problems would be facilitated by analyses of problems of different types and different levels of judged difficulty. Conversely, tests of a developing theory would involve *inferring* problem difficulty on the basis of the hypothesized determinants of difficulty and checking the accuracy of the predictions against more direct indications such as ratings by human judges or the performance of human problem solvers.

## Assessing the General Intelligence of Machines

The problem of assessing the intelligence of systems in a broad sense is more complicated than that of evaluating domain-specific knowledge or expertise. All the systems that have yet been developed have relatively restricted knowledge bases; no one has attempted to build a system that contains a knowledge base as broad as the one acquired by the normally developing child. Probably the systems that come the closest to this are those that have been built for the purpose of extending computer understanding of natural language. Many investigators expect that in time systems will be developed that have not only the broad world knowledge of human beings but the full spectrum of their linguistic and reasoning capabilities as well. Whether or not that expectation is ever realized, there is no evidence that current systems are up against any fundamental limitations, so we will surely see the development of systems with more extensive capabilities than those of the systems that now exist.

As the capabilities of intelligent systems broaden, the idea of assessing those capabilities by means of something analogous to a general intelligence test becomes more meaningful. Can we learn anything from the history of the assessment of human intelligence that can be applied to this problem? Certainly one lesson is that a consensus on what would constitute an appropriate methodology is likely to be a long time in coming. On the other hand, an effort to develop an assessment methodology could help identify areas of disagreement and perhaps bring greater coherence to the field.

It has often been observed that after more than a hundred years of trying to define human intelligence, the best we can do is say that it is whatever intelligence tests measure. From one point of view this is not a bad operational definition: if one wants to know what psychologists believe intelligence to be, or better, what they think an individual with a certain level of intelligence ought to be able to do, one need only examine the tests to find out. And, in fact, different tests are rather similar in content, and results across tests are fairly highly correlated. When one is asked what constitutes machine intelligence, on the other, one cannot resort to a what-the-tests-test definition. Perhaps it would be useful if one could. If one had a set of tests that served the same function with respect to computer systems that intelligence or general-abilities tests serve with respect to human beings, one would have an operational definition of machine intelligence.

It would be surprising indeed if the development of such a battery of tests did not prove to be a difficult task and the objective a controversial one. The effort should bring the question of what constitutes machine intelligence into focus, however, and proposed tests would represent at least a tentative and partial answer to that question. Any proposed battery could be expected to generate discussion and debate, which should help assure that the critical issues are articulated. This in turn should help to refine the concept of machine intelligence and should lead to the development of more effective ways of measuring it. A serious effort to develop tests of computer intelligence would undoubtedly have some effect on our conception of human intelligence as well.

### Looking Ahead

Interest in artificial intelligence and expert systems is presently widespread and strong. Many government agencies have initiated projects to investigate the applicability of work in these areas to their local problems. The U.S. Air Force has recently established the Automated Information Management Technology (AIMTECH) Program to investigate the potential applicability of artificial intelligence technology to the design of controls and displays for air crews and for command, control, and communication systems. The U.S. Department of Agriculture has funded a project to evaluate the role of artificial intelligence and expert systems in food and agriculture science and education over the next 25 years. Numerous other agencies as well as private businesses are carrying out similar investigations.

Some of the researchers who are most knowledgeable in the area of artificial intelligence have expressed concern that the short-term expectations from this technology may be unrealistically high. If they are, there is the risk of disenchantment that could work against continued smooth progress in the field (Pople 1984a; Waldrop 1984a). Critics of AI have reacted to what they see as self-serving touting, by some workers in the field, of the importance of the work that has been done to date and unwarranted extrapolation or generalization of limited results and findings (Dreyfus 1972; Weizenbaum 1976).

Martins (1984) claims that expert-system technology has been greatly oversold, that programs do not work as advertised in some popular accounts, and that the profession suffers from a certain "glamorizing the familiar with pretentious new terminology." Alexander (1984) concludes, on the basis of a survey of several expert systems and interviews with AI researchers and expert-system developers, that

despite their origins in artificial-intelligence research, expert systems have no more special claim to "intelligence" than many conventional computer programs. In fact, expert systems display fewer of the attributes classically associated with intelligence, such as the ability to learn or discern patterns amid confusion, than code-breaking and decision-aiding programs that first emerged during World War II. These older efforts never achieved anything like the glamour of expert systems, because they were shrouded in secrecy or wore unexciting labels like "operations research." (106)

Tate (1984) suggests there are indications that developers of expert systems are setting somewhat more modest goals now than they did even a few years ago:

Grandiose projects begun in the seventies suffered from a lack of understanding of both the technology and the tasks at hand. The resulting high cost, delays, and overzealous predictions of potential benefits by expert-systems proponents have tempered user companies' enthusiasm for the technology. (86)

The situation is being changed, Tate observes, by current efforts to develop simpler systems with more limited but attainable capabilities.

What if the efforts to apply AI fail? What constitutes failure is not always clear and may depend on one's perspective. The story of the efforts to develop automatic language translation systems in the 1950s is well known: initial optimism gave way to resignation to the idea that the task was many times more difficult than had been anticipated. Thirty years later, there still does not exist a system that can perform automatic translation with anything like the facility of a human translator. So in a sense, one might consider this effort to have failed. On the other hand, much was learned about natural language as a consequence of this activity; moreover, today there are systems that can aid human translators considerably, and the hope of eventually developing systems that will do translations as well as human translators is not dead.

Similar points can be made about work on natural language and speech. While there has yet to be developed a system that can understand speech or natural language as well as an average five-year old child, the efforts to do so have added very considerably to our knowledge of how humans understand natural language and speech. In particular, they have yielded a new appreciation of the complexity of the processes involved. They have made us more aware, for example, of how context-dependent is the meaning of an utterance; they have made clear the importance of knowledge (both about language and about the world in general) to language understanding; and they have taught us that everyday language is far more metaphorical and less literal than we had realized.

In addition to a few expert systems with practical utility and a better understanding of certain fundamental processes, several other positive results have come out of the work on artificial intelligence to date. These include some useful languages—notably LISP and PROLOG—and programming environments, a variety of programming tools, and some powerful methods for representing knowledge.

One of the effects that the further development of expert-system technology could have is to make specialized knowledge and expertise more readily available to the average person. If this happens on a large scale, it could have some interesting and profound implications for various professions and professionals. What distinguishes a professional is specialized knowledge. We are reminded frequently that what one pays for when one engages a doctor, a lawyer, an accountant, or other professional is not so much what the individual does as what he knows. It is assumed, presumably with justification, that professionals have a wealth of knowledge that they can apply to specific problems by virtue of years of study and experience in their fields. But suppose the knowledge that defines a professional field were codified in an expert system and made readily and inexpensively available to anyone who wanted to use it. Suppose, for example, that when one needed legal advice, one could get it at minimal cost from a computer system and could be assured that the advice one was getting represented the best that the legal profession, as a whole, had to offer. What now would be the function of lawyers in society? One function might be to maintain and extend the expert system's knowledge base. Even that might diminish in time, however, as techniques are developed that permit systems to learn and extend their expertise on their own.

Being the possessor of important knowledge that is shared by relatively few people in a society has given individuals privileged status for as long as there have been societies. It is not my purpose to claim that it is unreasonable that that should be the case. People who have special knowledge usually have acquired it at some considerable effort and expense. The fact that special knowledge confers special standing has served as an effective means of motivating people to seek education, and the services rendered to society by the custodians of special knowledge have undoubtedly been valuable and important services. But the question of interest is what the role of the holders of special knowledge will be when the knowledge that they now hold is readily available to the average person via computer-based systems that also hold that knowledge.

One of the most significant consequences of the work on artificial intelligence over the past two decades has been a greatly sharpened awareness that a prerequisite to intelligent behavior is a large amount of knowledge, particularly the kind of knowledge about the world that everyone acquires in the normal course of life. Precisely because everyone acquires such knowledge, its importance was not recognized until efforts were made to build intelligent machines. Initially the approach was to start with a tabula rasa as a computer memory and give it the capability of acquiring even first principles by interacting with the environment. This approach did not get very far. The second approach—to give the machine as much knowledge as possible, explicitly—has led farther, but here too there is a limit, because it is impossible for programmers or system developers to anticipate all the knowledge that the machine is likely to need.

It seems reasonably clear at this point that there is a need or both approaches. Systems that are to perform knowledge-based tasks will have to be given a considerable amount of knowledge when they are implemented. It will be very difficult, perhaps impossible, however, to preprogram all the knowledge that truly sophisticated systems will require. Provision will have to be made for adding to the knowledge base as the need becomes apparent. There is the option, of course, of having this knowledge added by human beings. There is also the possibility of giving systems the ability to acquire at least some types of knowledge in the course of being used. There is the further possibility of giving a system the ability to recognize the limits of its own knowledge base and to indicate explicitly what additional knowledge would be useful in performing specific tasks. The question of how best to grow knowledge bases so as to make them more and more adequate to their purposes is a key one in the development and use of expert systems.

There are two ways to get an answer to a question: look it up of infer (compute) it. Sometimes it is possible to do either, but one approach is more practical than the other. What is practical depends on one's tools. Before the days of pocket calculators the easiest way for most people to get the value of any of a variety of mathematical functions (a logarithm, say) was to look it up in a book of tables. Now it is often simpler to have the value computed by a pocket calculator. It would be possible, of course, to store function tables in a computer's memory and retrieve them by having the computer do the lookup. In most cases, however, it is more cost-effective to have the computer calculate the number when it is needed. It is cheaper to do the computation each time a value is wanted than to tie up the memory that would be required to store an extensive set of tables.

One might argue that there are few items of information as readily computable as mathematical functions and that for most information one might wish to obtain from a database, a computation is not a viable option. On the other hand, it is clear that much of the information that we human beings are able to obtain from our individual stores of knowledge is probably not represented in those knowledge stores explicitly, but is inferred (computed?) from other information that is there. I never learned explicitly, for example, that my great-great-grandfather from whom I get my surname was once six years old. But if asked if he was, I would say yes without hesitation. I know from the fact that he left progeny that he lived beyond puberty; and from my understanding that one is highly unlikely to do that in less than six years, I can infer that he must once have been six years old.

In order to infer the answer to a question, one must of course have access to information from which the inference can be made, and one must have the capacity to make the inference.

Developers of intelligent systems are constantly having to decide what information to represent explicitly in the system's knowledge store and how to provide the system with the ability both to access that explicit information when needed and to use that information as a basis for inferring what is there only implicitly. Knowledge and inferencing are the two cornerstones of intellectual competence in people or machines. While some tradeoff may be possible whereby especially good inferential capability may compensate to some degree for a limited knowledge store, or an especially extensive knowledge store may make up partially for a less-than-adequate inferential capability, truly outstanding intellectual performance requires both.

The discovery of the importance of domain-specific knowledge, and lots of it, to expertise also helped cause the pendulum to swing from an emphasis on general problem-solving heuristics (Newell and Simon 1972) to an emphasis on knowledge-based procedures. "Expert systems must be knowledge-rich even if they are methods-poor," argues Feigenbaum (1983, 47), among others. This, he suggests, has only recently become well understood by AI researchers: "for a long time AI focused its attention almost exclusively on the development of clever inference methods. But the power of its systems does not reside in the inference methods; almost any inference method will do. The power resides in the knowledge" (47). J. S. Brown (1984) has cautioned, however, that expertise depends on more than knowledge-base rules and that if an expert system is not to collapse when it gets slightly outside the knowledge explicitly represented in its rule set, it must have effective inference mechanisms that can operate on deep conceptual models.

It is somewhat ironic that while the AI community, and especially those members of it who have been developing expert systems, have discovered the importance of domain-specific knowledge, there has been a growing interest among educators and educational researchers in the teaching of problem-solving strategies and generally useful thinking skills in the classroom (Costa 1981; Covington, Crutchfield, Davies, and Olton 1974; Hayes 1981; Karplus 1974; Nickerson, Perkins, and Smith 1985; Rubenstein 1975, 1980; Schoenfeld 1980). This interest appears to stem from the assumption that nearly exclusive emphasis by an educational system on the acquisition of domain-specific knowledge produces people with an underdeveloped ability to think and to apply that knowledge effectively outside the classroom. It is too early to tell where the thinking-skills movement will eventually go and what effect it will have on educational philosophy and practice.

The evidence is fairly compelling that human experts differ from novices *both* in their breadth of knowledge of their fields *and* in their ability to use problem-solving heuristics that are applicable across a variety of domains. Further investigations of intelligence and expertise, and efforts to enhance the ability of either people or machines to accomplish intellectually challenging tasks, will need to give due weight to both content and process.

# 16

## Some Research Challenges

Since computers and computer-based systems first appeared on the scene, experimental and engineering psychologists have had an interest in their design and use. These machines and the systems they make possible must be used by human beings, and, not surprisingly, many human-factors questions arise in their implementation and utilization. Because the technology is new and the resulting systems are unique in many ways, these questions often cannot be answered by consulting the existing human-factors literature but require research.

Effective research in the area has proved difficult, in part because of the rapidity with which the field has evolved. The results of work focused too narrowly on specific systems have had a good chance of being obsolete before being published, because of the speed with which systems are replaced with improved successors. Much research on the human factors of computer systems has been done, however, and progress has been made. Questions are being asked, experimentally, whose answers apply to a broad range of systems and tell us something fundamental about human capabilities, limitations, and preferences. These questions have to do with information representation and presentation, language (both natural and artificial), expertise, function allocation, training, communication, job satisfaction, and a host of other topics.

We need not deny that progress is being made on understanding the human factors of computer systems when we acknowledge that, at the same time, there is much room for improvement in systems design from a human-factors point of view. Undoubtedly, more is known, or hypothesized, about how to build "friendly" systems than is universally applied; it is not readily available to system builders because it is scattered throughout the literature or, in some cases, exists only in people's heads. But even if all that is known or believed were readily available, there would be a need for research. Many, perhaps most, of the design guidelines that have been proposed are in need of empirical support. Moreover, there are many questions to which different people give conflicting answers and some for which no answers have been proposed.

The following are a few of the psychological research questions relating to information systems that seem to me worthy of attention. I suggest them in the spirit of provoking the generation of a better list. Elsewhere (Nickerson 1982) I have suggested that there are two broad categories of psychological problems relating to information technology: usefulness and usability issues, and quality-of-life concerns. The first category has to do with designing systems so they match well the needs, capabilities, and preferences of their users. The second relates to the problem of assuring that this technology will improve the quality of the lives of its users and of people in general. My list is not partitioned along these lines, but both types of questions are represented in it. The order of the items on the list carries no significance.

- What will be the impact of information technology on office work? On job qualifications? On training requirements? On worker satisfaction?

- How can jobs involving the use of (interaction with) computer-based tools be designed so as to present an acceptable level of challenge to the user? (An acceptable level of challenge is assumed to be one that is neither prohibitively great nor so small as to make the job boring or demeaning.)

- How might information technology be used to personalize and enrich jobs? Surely it would be unwise to assume that personalization and enrichment will occur if no explicit efforts to ensure those outcomes are made.

- What can be done to help ensure a graceful accommodation to the rapid changes that information technology is likely to cause?

- How will the use of computer-based message systems affect the ways in which users communicate with each other and the effectiveness of the communication in general?

- How will document preparation tools, such as composition and editing aids, affect the writing (quantity and quality) of people who use them consistently?

- Computer-based communication systems have certainly facilitated collaboration among some scientists in different locations. What additional capabilities do such systems need for them to become significantly more effective?

- Toong and Gupta (1982) have speculated that the proliferation of personal computers may change not only how business is conducted and how people organize their personal affairs but even perhaps how they think. How might this conjecture be explored?

- From experience to date, what can designers learn about maximizing the utility of Videotex, Teletext, and other information services?

- The number of computer terminals is increasing rapidly, as is the number of personal computers that can also serve as terminals. How does one design an electronic mail system that will exploit the capabilities of the high-end terminal while being able to run also on the low-end terminal?

- How does one determine what an expert in a particular area knows about that area, how does one characterize that knowledge, and how does one represent it in a computer?

- What kinds of tools should be developed to help people manage the enormous amounts of information that they will be able to store in private files with the introduction of new storage technologies?

- What can be done to realize the benefits of standardization without stifling creativity and innovation?

- How will the development of parallel-processing systems affect our thinking about problem solving?

- How might we anticipate the societal effects of new information technology and the things it makes possible, such as instantaneous national referenda, telecommuting, and electronic classrooms?

- The amount of information to which the average person has relatively direct and easy access will be very much greater in the future than it has been in the past. Are there ways to equip people to take advantage of this fact?

- How do we determine the potential positive and negative effects of anticipated technological innovations before they become realities? What can be done through controlled experimentation in this regard?

- What are the best kinds of laboratory facilities for experimental studies of user–information system interaction?

- How should information technology be brought to bear on job situations for which user requirements cannot be accurately specified in advance or at least cannot be specified by means of conventional methods?

- How might the collective knowledge of developers and users of interactive systems be codified and made accessible to designers of future systems?

- What new input-output methods might be developed that would increase the bandwidth of the communication channel between information systems and their users?

- What new methods of representing and displaying information might be developed that would capitalize on the computer's ability to generate dynamic and interactive displays and that would facilitate information assimilation by the user?

- What techniques might be developed that would help the user browse effectively through a very large database, such as the Library of Congress?

- What types of metaphors or mental models of information systems will help computer-naive users to make effective use of those systems?

- In what ways should the abilities of an information system resemble the abilities of its users and in what ways should they not?

- Under what conditions would speech be a preferred output medium for an information system? How should the characteristics of the speech be determined? How important is it that speech that originates with the computer (synthesized speech) differ noticeably from speech that originates with a human being?

- Under what conditions would speech be a preferred input mode for an information system? How well can people accommodate to the requirements of the limited speech recognition capabilities of specific systems?

- Is "friendliness" an appropriate concept to use in describing characteristics of computer-based systems, or does it invite inappropriately anthropomorphic notions of what these systems are?

- How can systems be designed so as to meet the needs of users with varying degrees of experience and expertise?

- What can be done to assure that people's attitudes toward computer-based systems are well founded and that their expectations regarding what these systems can and cannot do are accurate?

- How can programming skills be taught more effectively and efficiently?

- How can potential programming skill be recognized?

- What can the study of programming tell us about thinking and problem solving in general?

- To what extent can programming be used as a vehicle for teaching thinking and problem-solving skills?

- What methods can be developed to enhance programmer productivity?

- What can be done to facilitate the sharing of the results of programming efforts and to minimize duplication (or multiplication) of effort?

- What techniques might be developed that would facilitate the management of complexity, whether that complexity be found in computer programs, integrated circuits, multiperson-multimachine systems, or international politics?

- What principle(s) should be used to allocate functions to people and machines, especially those functions that both people and machines can perform?

- How can workload be measured when the worker is the user of an information system?

- What performance measures are appropriate for the purpose of assessing the effectiveness of an information system?

- What capabilities and properties would a teleconferencing system have to have for it to be preferred over face-to-face conferences even when travel time and cost were not issues?

- What can be done to maximize the chances that emerging information technologies will be beneficial to people with various types of handicaps?

• What might be done to increase the chances that this technology will significantly add to the job opportunities for handicapped people?

Perhaps the greatest challenge to research is to determine where the leverage for research really is. There are many questions that can be asked, some of which can be answered through experimentation. It is not clear that all the questions that can be asked, or even all those that can be answered, are worth asking. Certainly not all are worth equal effort. The fundamental problem is that of deciding where to put limited resources.

# 17

## Quality of Life: The Fundamental Issue

### On Weighing Potential Positive and Negative Effects

When people want to make the point that technological developments can have negative effects as well as positive, they often focus on the destructive potential of modern weaponry to make the case:

I would suggest that the success of modern technology, which has put each of the super powers in a position to destroy the other (and much of the rest of the world), presents a basic challenge, not only with respect to work but also with respect to all human values. It remains to be seen whether or not the potential of modern technology will turn out to be a blessing. (Ginzburg 1982, 75)

While one cannot deny the poignancy of this example, the evidence for the argument need not be limited to nuclear weapons:

• The automobile, for instance, though it has served extremely well the purpose of getting people from place to place, has also brought hundreds of thousands of people prematurely to their graves. It has contributed to noise and air pollution, has helped deplete natural resources, and may well have inhibited the development of more effective public transportation systems.

• Household detergents were an improvement over lye soap and, in conjunction with the electric washing machine, made possible a new standard of personal cleanliness. But even such an apparently innocuous invention was not completely harmless. Detergents that are made from petrochemicals are not broken down by bacteria in waste disposal systems as is soap that is made from organic materials. And noticeable damage to water supplies had been done before the problem got the attention it deserved and successful efforts were made to develop biodegradable detergents in the mid-1960s.

• Television has brought news and entertainment daily to millions of people around the world. In the process it has made us better informed and has raised our awareness of issues that concern us all. On the other hand, it has failed abysmally to realize its potential for education and intellectual stimulation, serving instead as a mind-numbing, time-stealing, universal soporific.

• The telephone has made it possible for us to talk with friends, relatives, and business associates at any time across any distance, but it has also made us constantly accessible to pollsters, interviewers, salesmen, and cranks.

In his inaugural address as the thirteenth president of MIT in October 1971, Jerome Wiesner referred to "the need for increased sensitivity to the dangers arising from the careless exploitation of new technology" (15). He pointed out that until very recently the relationship between technological change and our social, biological, and physical environments had rarely been examined; only the obvious benefits and immediate costs of technological innovations had been considered. Paradoxically, he noted, "science and technology have helped create our present predicament by extending to most of us options in modes of living and working that were previously reserved for a privileged few" (16). Unfortunately, our ability to guide changes for the good of mankind has not kept pace with our ability to effect change. On the positive side, there appears to be a general awakening to the importance of anticipating the risks associated with technological innovation; and perhaps for the first time in the history of the species considerable energy from many sources is being applied to that problem.

It is not clear how one should weigh the pros against the cons in order to decide whether any particular innovation is, on balance, a good or bad thing, or to indicate how good or bad it is. My own sense is that on balance the effects of technology have been very positive. Whether individuals are, on the average, truly happier in those societies that have been most affected by technology than were people in earlier times is a question that it is probably impossible to answer. Certainly we live longer and have much greater freedom of choice regarding what we do with our lives. I suspect that people who long for the good old days usually imagine themselves as part of the landed gentry, residing in comfort and good health amid pastoral tranquility. Few imagine themselves dying of smallpox or diphtheria, or performing the drudging labors of a scullery maid or a serf, or being ostracized from society and suspected of possession by demons because one happens to be an epileptic, or bringing into the world six children of whom three die before they are out of the cradle.

By attempting to anticipate both the good and the bad effects of each technological innovation, we can hope to maximize the former and minimize the latter; and the more powerful the technology, the greater the importance of doing so. The human factors of information technology clearly go far beyond considerations of usability and efficiency, as important as those may be. The anticipated developments are going to affect our lives profoundly. The potential that information technology has for effecting change for the better, or for the worse, is enormous.

Numerous writers have expressed concern over the potential social implications of computers. Data banks and privacy, and automation and employment, have been common themes (G.E. Brown 1982; Miller 1971; Tomeski and Lazarus 1975; Westin and Baker 1972). Wiener (1960) has pointed out that while it may be impossible in principle to build a machine whose individual actions we cannot comprehend, it is certainly possible to build one that is so fast that understanding of what it has done in any particular instance cannot be attained until long after the action was completed. Wiener cautions, "If we use, to achieve our purposes, a mechanical agency with whose operation we cannot interfere once we have started it, because the action is so fast and irrevocable that we have not the data to intervene before the action is complete, we had better be quite sure that the purpose put into the machine is the purpose that we really desire and not merely a colorful imitation of it" (1358).

These are important concerns and deserve at least as much attention as they have received. But information technology is likely to have many other social implications, more mundane perhaps, but important nonetheless. The increasing use of information technology in offices, for example, seems very likely to affect the ways in which people interact with each other and perhaps to affect organizational structures and management techniques as well. Computer-based message technology will increase greatly the accessibility of people to other people in the same organization, irrespective of their positions within the lines-of-authority structure. We do not know what effects this is likely to have on organizations and on the way people within them relate to each other. Perhaps psychologists should be attempting to anticipate the kinds of changes that will occur, in the interest of preparing people to accommodate to them without undue strain. The question of how to introduce new information technology into the workplace gracefully has received some, but not enough, attention (Helps 1970). It is important not only to recognize that the introduction of new technology can be threatening to people in the workplace (Igersheim 1976), but to work on the problem of reducing that threat.

The next few sections of this chapter focus on three worldwide problems that information technology should help us address. This is not to suggest that these are the only problems deserving attention, or even necessarily the most pressing ones, but they are serious problems whose global significance cannot be denied. They are problems that there is good reason to believe information technology, if thoughtfully and energetically applied, could help solve or ameliorate. And they are problems that could be aggravated if this technology is developed further and used with an insensitivity to their existence.

### Food and Nutrition

Among the most urgent problems facing the world today are hunger and malnutrition. These evils are both the causes and the effects of the others with which they tend to be associated, such as poverty and disease. Poverty ensures hunger and malnutrition, which promote disease and despair, which ensure greater impoverishment. About a quarter of the world's population is trapped today in this vicious cycle (Mahler 1980). The problems of hunger and malnutrition are not spread evenly throughout the world; they tend to be concentrated in the rural areas and urban slums of the underdeveloped countries. There are also large disparities across the world in the cost of food to the average consumer. In general, the proportion of one's income that is spent on food is much higher in underdeveloped countries than in developed countries. The average American consumer, for example, spends a smaller percentage of his disposable income on food (15.7 percent in 1971) than do his

European and Russian counterparts (26 and 30 percent respectively), and only about a quarter as much as the average consumer in developing countries (55 percent) (Heady 1976).

There are various approaches that can be taken to combat the problems of hunger and malnutrition. Some of them are fairly obvious; others are less so. One possibility is to increase the amount of land under cultivation. The total farmed land today is about 1.4 billion hectares, somewhat under half of what is assumed to be available: about 3.2 billion hectares of the 13 billion hectares of the earth's land surface that is free of ice is assumed to be cultivable (Revelle 1976).

A second possibility is to increase the productivity of cultivated land through more effective use of fertilizers and pesticides, double cropping, the development of plant species with higher yields or greater nutritional content, and the adoption of measures to reduce spoilage and waste. It has been estimated that enough food to feed the present world population (4 billion people) could be raised on 170 million hectares of land, or less than one eighth of the land currently under cultivation. Among the reasons for the discrepancy between the minimum land required and the actual amount used are the following: some of the cultivated land (about 10 percent) is devoted to the raising of non-food crops; some is devoted to the raising of foodstuff for livestock and poultry (which consume from 50 to 100 times as much food energy as is contained in their edible products); some of the food grown is destroyed by pests, disease, or weather, or lost to waste or spoilage at various stages of the distribution process; and much of the land under cultivation produces less than it is potentially able to yield (Revelle 1976).

A third approach involves the development of new sources of food. The energy captured by plants from sunlight and stored in chemical form exceeds human needs by roughly a factor of 200, even though plants capture only about one percent of the total incident energy (Janick, Noller, and Rhykerd 1976). There would seem to be much room for exploring ways of making a larger fraction of that stored energy available for human nutritional needs. About 80 percent of the world's food supply for humans currently comes from only eleven plant species (Loomis 1976), and about two thirds of that from only four species: wheat, rice, maize, and potatoes (Harlan 1976).

Ensuring the production of enough food to meet human

needs is only part of the solution to the problem of hunger and malnutrition; the other part is ensuring an equitable distribution of the food that is produced. Thus, increasing the effectiveness of food distribution is a fourth approach. An adequate overall food supply within a country is not a guarantee against undernourishment of any segment of its population (Scrimshaw and Taylor 1980). It is possible for a country to increase its food production and, as a consequence, decrease the availability of food for its own people. The production of higher-priced foods for export, for example, can have the effect of reducing the amount of land used to grow staple foods for domestic consumption.

What all these observations point to is the need for the development of a better understanding of the complex processes of food production and distribution. Controlled by many variables that interact in unpredictable ways, these processes are probably not fully understood by anyone. It is easy to find examples of explicit policy decisions that were made with the intention of solving a specific agricultural problem but that have had the effect of creating other problems, perhaps even more severe than the one originally addressed.

Recognizing the importance of financial incentive as a factor in determining what farmers will grow, governments have sometimes attempted to control agricultural production through the manipulation of prices. Evidence that the complex cause-effect relationships in food production and distribution are not well understood is seen in the fact that price fixing has sometimes produced effects quite different from those intended. In Egypt, for instance, note Scrimshaw and Taylor, "the government controls the procurement prices of cotton and wheat, but not the consumer price of meat. As a result crops have shifted from cotton and wheat toward clover, which is used in part to fatten animals for relatively well-to-do consumers in the towns" (1980, 83). As another illustration of the same point, the authors observe that high government-supported prices for grain have resulted in a drop in the production of legumes and consequently an increase in the cost of the diet of low-income groups.

A further example of unanticipated negative effects from efforts to alleviate hunger and malnutrition may be found in the distribution of food surpluses. What could be more clearly desirable than that countries with food surpluses should make those surpluses available at little or no cost to countries in need? But even this straightforward approach is not without its complications:

The food generosity of the industrial countries, whether in their own self-interest (disposing of food surpluses) or under the mantle of alleged distributive justice, has probably done more to sap the viability of agricultural development in the developing world than any other single factor. Food aid not only has dulled the political will to develop agriculture, but also, by augmenting domestic production with grain grown abroad, has kept local prices at levels that destroy incentives for indigenous farmers. (Hopper 1976, 203)

This observation obviously is not intended as an argument against providing food to meet shortage crises arising from such disasters as famine and drought; it is an indictment, however, of long-term food assistance policies that inhibit a country's development of its own resources.

Even in achieving solutions to the age-old problem of how to get enough water to growing crops, farmers have encountered unanticipated negative effects. Irrigation can be an effective means of increasing food production by increasing the amount of land under cultivation, by increasing the yields of individual crops, by making it possible to grow more than one crop per year on a given piece of land, and by making farming less risky to the individual farmer. On the other hand, it can damage soil through waterlogging and salinity buildup, and it can also contribute to the propagation of waterborne disease and parasites (Ambroggi 1980).

One could easily multiply the examples of well-intentioned actions that have had detrimental consequences owing to an incomplete understanding of the relationships among the variables that affect food production and distribution. The point is that a global effort to develop and share a better understanding of these processes is essential to any significant and lasting progress on the eradication of hunger and malnutrition. At the heart of the problem is the need for information and for methods that assure the universal availability of that information. We need

- more powerful predictive models of the type described by Leontief (1980; Leontief, Dunchin, and Szyld 1985), relating crop yields to the numerous variables that are known to affect them (the use of fertilizers and pesticides, crop rotation, soil conservation policies) and relating these variables to others by which they in turn are affected (for instance, the fact that the amount of fertilizer used per unit of land tends to vary inversely with the ratio of the cost of fertilizer to the farmer's proceeds from the final crop);

• better models relating food prices to various factors that can affect them: rises in the costs of energy, crop failures due to unfavorable weather conditions, crop diseases, damage done by insects and other pests, economic positive-feedback effects (unavailability of credit for farmers);

• a better ability to anticipate the potential effects of new developments in agriculture (such as aquaculture, the cultivation of microorganisms as food, and the synthesis of food from petroleum);

• more accurate weather forecasting, both short-term and long-term; greater ability to predict multiyear weather changes such as the current sub-Saharan drought (Kerr 1985); and greater knowledge among farmers and policy makers of how to take full advantage of weather information;

• better monitoring techniques, including, as suggested by Mayer (1976), a famine early-warning system that uses weather satellites, economic indices, and clinical indicators, such as charted weight and growth curves for children;

• systems that can facilitate worldwide coordination and planning of food raising, so as to avoid large surpluses of some foods and shortages of others;

• more effective means of distributing information—in easy-to-understand terms—about the nutritional requirements of individuals (depending on age, occupation, living environment, and so forth) and about alternative ways of meeting those requirements, for use by those who habitually plan and prepare meals.

There are numerous ways in which information technology can be applied to the meeting of these needs. I will focus here on one, namely, making information about agriculture and nutrition readily and continuously available to people who need it. One can imagine a system that could deliver such information via computer networks to users—planners and policy makers, farmers, homemakers—everywhere. To planners and policy makers it would deliver the kind of information that would permit efficient use of national resources and coordination with the agricultural policies of other countries. (This would require, of course, greater cooperation among countries than has been possible in the past; but political problems aside, providing the information necessary for such coordinative planning is a technical challenge that deserves considerable attention from information technologists.) To farmers such a system would deliver information about the requirements for growing various crops, the efficient use of specific parcels of land, the risks in specified decisions, the management of specified resources, appropriate responses to unforeseen problems at various phases of crop cycles, and so on. To homemakers it would provide information that would enable them to decide how best to satisfy the nutritional requirements of their families, given the constraints within which they are operating: local availability of foodstuffs, costs, family preferences, and special dietary needs.

Ideally, the information provided by such a system should be available in interactive question-answer mode. The system should also be able to offer advice that is not explicitly requested, when the need for it can be inferred. The idea of advice-giving systems leads naturally to the idea of the application of expert-system technology to this problem domain.

Prospective users of expert systems in agriculture and nutrition range from the individual who is interested in making optimal use of a small plot of land for a family garden, or the homemaker concerned about satisfying the nutritional needs of a family, to national and international policy makers concerned about issues of food production and distribution globally. An expert system that would be useful to the farmer or the backyard gardener would incorporate the kind of knowledge that is needed to make effective decisions pertaining to agriculture on a local scale: information that would permit one to plan the planting, tending, and harvesting of specific crops, taking into account such factors as soil composition, duration of the growing season, local pests and diseases, temperature, and rainfall. Systems serving local interests would be relatively easy to build, because there are many human experts with appropriate knowledge regarding specific crops and geographical areas.

A second type of expert system that could be useful in agriculture is one that understands food production and distribution on a global scale. Such a system is impossible at this point, because it is not clear that there are any human experts who understand thoroughly how the numerous variables that affect these processes relate. What appear to be needed are systems that incorporate and integrate the expertise of many human experts in such a way that they can deal effectively with situations that are beyond the scope of individual humans.

Such systems will have to develop in an evolutionary way. It will be necessary to begin with fairly crude approximations to what might now be considered a reasonable ultimate goal (but with the realization that the goal itself is almost certain to change in time). Even an initial version, however, will have to contain a knowledge base that represents a reasonable subset of the fundamental concepts, relationships, principles, and processes relating to agriculture. It should have models of food production and distribution at several levels of inclusivity (local, regional, national, global) and the ability to exercise such models to explore the effects of changes in various production and distribution parameters. And it must be structured in such a way that it can be easily modified for the purpose of increasing its scope and utility.

Agriculture scientists have exchanged information by means of networks for some time. Currently there are more than a hundred international agricultural networks in operation (Plucknett and Smith 1984). The kinds of research information exchanged through these networks include the results of testing of advanced breeding material, the transferability of agrotechnology within families, livestock diseases, and many others. Plucknett and Smith point out that this information exchange has surmounted ideological, ethnic, and language differences among participating countries. As evidence they note that more than 130 nations joined networks to test breeding material during the 1970s.

Networking among agricultural scientists has not depended on computer technology in the past and in many cases has involved simply an agreement to share information by conventional means. But the increasing use of modern computer and telecommunication technologies should advance collaboration, not only by increasing the speed with which information can be disseminated but by facilitating data processing and interaction among the scientists involved in the research. The types of systems envisioned in the preceding pages go considerably beyond the exchange of information among agricultural researchers. The ultimate goal should be to make information relating to food and nutrition readily accessible to everyone who has a part, large or small, in helping to assure a nutritionally adequate food supply for anyone.

## Education

Another area that represents both a challenge and an opportunity to information technology is education. While ignorance and illiteracy, like hunger and malnutrition, are worldwide problems of very considerable magnitude and consequence, here I shall concentrate on the situation in the United States.

America's public educational system is in serious trouble. Nationwide, Scholastic Aptitude Test (SAT) scores have declined steadily for several years and as of 1979 were at the lowest point in the fifty-year history of the administration of these tests (*College Bound Seniors* 1979). Many writers are claiming that a distressingly large fraction of students graduating from high school and entering college are not prepared to meet the intellectual demands that their college experience will place on them (Karplus 1974; Renner and Lawson 1973). And a growing body of data provides evidence that students frequently "get through" basic math and science courses without acquiring more than a superficial understanding of some of the concepts and relationships that are fundamental to the subjects they have studied (Carpenter, Corbitt, Kepner, Lindquist, and Reys 1980; McCloskey, Caramazza, and Green 1980; Nickerson 1985).

In August 1981 the National Commission on Excellence in Education was created by the Secretary of Education and directed to examine the quality of education in the United States. The commission was chaired by David P. Gardner, president of the University of Utah and president-elect of the University of California. Members included other university presidents or past presidents, superintendents of schools, school principals, and other people in positions to be knowledgeable about education in this country. The general tenor of the commission's report is reflected in its title, *A Nation at Risk*. In the words of the report, "the educational foundations of our society are presently being eroded by a rising tide of mediocrity that threatens our very future as a nation and a people. If an unfriendly foreign power had attempted to impose on America the mediocre educational performance that exists today, we might well have viewed it as an act of war" (Gardner et al. 1983, 5).

Strong words indeed, and, not surprisingly, they have not gone unchallenged. The authors of the report are careful to acknowledge that *the average citizen* today is better educated and more knowledgeable than was his counterpart of a generation ago. *The average graduate* of schools and colleges today, however, is not as well educated as the average graduate of a generation ago, they claim. A greater percentage of the citizenry attends school today, but the effects of that attendance are not as positive as they once were. Again in the words of the report, "more and more young people emerge from high school ready neither for college nor for work. This predicament becomes more acute as the knowledge base continues its rapid expansion, the number of transitional jobs shrinks, and new jobs demand greater sophistication and preparation" (12).

Disheartening statistics have been extensively reported in the media. Articles addressing the problem have appeared in scholarly journals and popular magazines and as special reports (Adler 1982; Goodlad 1983; Sizer 1983; Twentieth Century Fund 1983). Numerous seminars, symposia, and workshops have brought together teachers, researchers, school administrators, parents, politicians, agency representatives, and others to discuss the problem and what to do about it. So far there is little evidence of progress, and in the continued absence of improvement, concern threatens to give way to resignation or counterproductive measures born of frustration.

A particularly disturbing aspect of the educational picture in the United States is the problem of illiteracy among adults. According to a study by the University of Texas at Austin, approximately one in every five American adults is incompetent with respect to literacy or functions with difficulty, and one in two is merely functional and not proficient in necessary skills and knowledge (Adult Performance Level Project 1977). The National Assessment of Educational Progress (1981) reports that only about half of the country's 17-year-olds can write a wholly satisfactory piece of explanatory prose and only about 15 percent can defend a point of view effectively with a persuasive argument. A third study informs us that as of 1980, 28 percent of the U.S. Army recruits training at three centers read at or below the seventh-grade level (Fialka 1980).

While one may question the preciseness of the numbers in specific studies, one would have to discount a great deal of converging evidence in order to avoid concluding that public education in the United States is seriously deficient. The question of interest in the present context is whether information technology might help address this problem. Although the failure of computer-assisted instruction so far to have the impact on education that was predicted in the 1960s has left many people skeptical of its potential, some researchers and educators believe that the early expectations were too extravagant only with respect to timing and that eventually the impact will be very great indeed. Moreover, there seems to be a growing conviction that with the arrival of the relatively inexpensive personal computer, the stage is set for a greatly accelerated growth in classroom use.

Recently the U.S. Department of Education (Office of the Assistant Secretary for Educational Research and Improvements) reported the results of a conference on computers in education convened by the department in 1982. Participants included some forty computer scientists, psychologists, educational researchers, teachers, school administrators, and parents. Their basic conclusion was that "striking improvement in the quality and productivity of education through computer-based instructional systems is attainable, but only with a national investment that continues reliably for several years" (Lesgold and Reif 1983, 4). The potential role of computers in education and the need for the educational system to take action to assure computer literacy among its graduates have also been getting considerable attention from the U.S. legislature. Some twenty bills dealing with these topics were before the 98th Congress at one time, calling for, among other things, tax breaks for computer manufacturers who donate equipment to schools, support for teacher training, the mandating of courses in computer literacy, and the establishment of a national educational software corporation to stimulate and support the development of educational software (Watt 1984a).

We have noted in chapter 3 various evidences that computers are being more and more widely used in schools at all levels. Some observers believe it will not be very long before it is economically feasible for every student to have one. It is to be expected that the availability of computers will affect classroom instruction in both direct and indirect ways. Even in cases where computer use is not an explicit aspect of classroom teaching or out-of-class assignments, what teachers teach and what they expect from their students are very likely to be influenced by the knowledge that this technology exists and is readily accessible to students. Just as one would not assume that radio, television, and the telephone have influenced education only insofar as these instruments are used in the classroom, so one can expect that behavior in the classroom and everywhere else will be affected by computers and the various resources that become commonplace because of them.

Text editors and other computer-based office tools have been adapted for use in teaching children to write. The idea is to permit children to get experience with the more interesting and rewarding aspects of writing (planning, composing, and editing) before they have mastered spelling, grammar, and penmanship. Software tools that have had relatively widespread use in this context include the Bank Street Writer from Bank Street College, the Writer's Workbench from Bell Laboratories, Quill from Bolt Beranek and Newman, and the Talking Screen Text-Writing Program from Arizona State University (Watt 1984b; Rubin and Bruce 1983).

One reason for the great interest in the computer as an educational tool is the assumption that it is possible to design computer games that enable children to learn at least some aspects of mathematics, language usage, physics, biology, or whatever, while having fun. Unfortunately, there is little concrete evidence so far to bear out this assumption. While numerous programs have been developed and are being promoted as educationally effective, very few efforts have been made to substantiate their effectiveness with data from carefully controlled tests, and perusal of the programs that exist has led some observers to conclude that many of them are of dubious educational worth (Feurzeig, Horwitz, and Nickerson 1981; Holden 1984).

The problem of evaluating educational software is a serious and difficult one. A good evaluation methodology has not yet been developed. Much of the software is too new for a careful evaluation to have been done, even if such a methodology were available. Developers of the software typically are not motivated to put great energy into evaluations, nor is it clear that they should be. Qualitative assessments have been made, but from a scientific point of view these must be considered stopgap measures that should be replaced with more objective and replicable assessments.

Some educational programs simply provide a kind of reinforcement that is intended to keep a child working at drill and practice tasks. Monster Math, for example, displays arithmetic problems on a video screen, and as the player solves them the monster who also appears on the screen begins to disappear, piece by piece; the object is to solve problems fast enough to make the monster disappear entirely within a 60-second time limit. Some programs put the player in a fantasy situation in which progression through some maze (rooms in a castle, a network of connecting caves or dungeons) is contingent on the solution of a series of problems. Incorrect answers produce unhappy consequences in the fantasy world, whereas correct answers advance the player toward the goal. Most educational games are designed to take advantage of the presumed natural competitiveness of the players, and consequently they evaluate one's performance with a score of some sort. Some games permit competition between players; with others the idea is to play against oneself. On the whole, the educational programs that are available commercially today are limited more by the imagination of their developers than by the computational capabilities of the machines.

Although interest in using computers in the classroom is generally high and schools are acquiring them at a great rate, even the teachers who have welcomed them most enthusiastically have sometimes raised questions about possible risks. Bacon (1984), for example, ends a very positive article about his experience as a teacher using personal computers for classroom instruction with the following comments:

I often wonder, however, if there's another side. For all the current clamor about computer literacy, is it wise to occupy young, developing minds with computer applications? Not all students were enamoured with the prospect of spending 25 percent of class time isolated in front of a PC monitor. More than once, it became an issue that threatened to divide the group into two factions.

What of the concern that computers, along with other technological developments, have further alienated minorities and that computers have widened the gap between the sexes? What about an excess of reliance on technology that can create a social or political imbalance or that can deepen the rift between developed and developing nations? What of the computer's seductive nature, its uncanny ability to chain its user to compulsive use, the sedentary life style of some users, the neglect of social relationships, and the stress this fabulous tool places on the material world, perhaps at the expense of the spiritual? (299)

Nobody knows what the effect will be of making powerful computing facilities widely available to entire student bodies and faculties at educational institutions and connecting these facilities via high-speed networks. One must believe, however, that it could be profound. Many people believe that the university and, indeed, the educational process will be changed in ways quite beyond our ability to predict. Some futurists claim that information technology will radically alter educational methods in institutions, possibly to the point of doing away with schools and universities as we know them (Dunn 1979).

The most profound effects of computers on education may well result not so much from their presence in schools as from their availability elsewhere. At present there are estimated to be more than twenty times as many computers in homes as in schools. To be sure, these home computers are not being used exclusively, probably not even primarily, for educational purposes; and even if they were, it is not clear that what is currently available by way of educational software would produce a large impact. But if we should discover how to exploit effectively the computer's potential as a teaching device—or facilitator of learning—it seems imprudent to assume that the effects will be obtained primarily through those computers that happen to be located within schools.

I have earlier alluded to the much-publicized high rate of illiteracy in this country. The truth is that if literacy is defined as the ability to read *comprehendingly,* we are *all* illiterate with respect to many, if not most, areas of human knowledge. Few people are able to read with deep comprehension in such diverse areas as economics, nuclear physics, paleontology, medicine, and music. This is a sobering but useful fact. First, it reminds us of how very dependent reading comprehension is on knowledge. One simply cannot read comprehendingly in an area about which one knows little or nothing. Second, it is a good antidote to intellectual arrogance. When one begins to feel that one knows quite a lot, one need only browse through a few technical journals outside one's area(s) of interest to be reminded forcefully of the depth of one's ignorance of the world.

The realization that all of us know very little about most areas of human inquiry should prompt us to speculate as to why we do not know more. Is our ignorance a consequence of the limitations of our brains? Or might it be, at least in part, due to the limitations of the ways that we have invented for informing ourselves and, in particular, for representing information for human assimilation? I know of no compelling evidence that we are inherently incapable of assimilating much more information than the most knowledgeable among us now do. Perhaps it is impossible to develop more effective ways of representing information and of making it accessible to people who want it. Intuitively that seems highly unlikely. At least it is not *demonstrably* true that we are doing the best that can be done, or even anything approaching the best.

The challenge is to apply technology to the general problem of finding more effective ways of representing information and of making it accessible to people. Representational schemes are necessarily constrained by the medium that embodies them. The book, that marvelous innovation that democratized knowledge, is an excellent medium for text and static diagrams. It is a poor medium, however, for dynamic representations or for customizing the presentation of information to the knowledge base and cognitive idiosyncrasies of the individual reader. Computer technology presents an opportunity and a challenge to explore new possibilities for conveying information. These possibilities include multimedia systems that can supplement conventional textual representation with voice, film, and animation; they include interactive systems that not only deliver information but respond to questions and permit exploratory learning. Such new approaches could help increase literacy in a general sense. At least that should be a goal in developing them. We should be more bothered than we are, both as individuals and as a society, by how little we know about the world, and we should be motivated to develop the potential that information technology may hold for letting us do a much better job of informing ourselves than we have done in the past.

A view of the future that I find particularly exciting is one put forth by Congressman George Brown, Jr. (1982), in which he envisions the whole community as a learning environment:

Satellite transmission and cable television will bring a wide variety of cultural and educational programs into homes and schools; students themselves will generate much of the programming of interactive TV systems, a prototype of which is already in operation in the Irvine, California, school district. Schools, museums, libraries, and governmental units will be connected through computer and television networks and will have access to a wide variety of data bases. Inexpensive microcomputers in homes will be able to access, through rapid, reliable networks, an almost unlimited range of learning resources. In the workplace, information technology will be used in a wide variety of activities, from routine tasks like electronic mail and electronic funds transfer to sophisticated applications of computer-aided design and robotics. Updating of skills and learning of new skills through satellite transmission and computer-assisted instruction will be a standard feature of industrial and professional training. (52)

Brown notes that the mere existence of networks, home computers, and the gadgetry associated with them will not guarantee that they will be used in socially constructive and equitable ways. To see that they are is a challenge not only to the developers and exploiters of this technology but to society in general. I believe that within information technology there is the potential to transform the educational process—not only to stimulate the cognitive development of children but to enhance the intellectual odysseys of individuals throughout their lives. Whether this potential will be realized remains to be seen. There can be little doubt, however, that for better or worse this technology is going to be responsible for enormous change in educational institutions and processes.

### Information Technology and Disabilities

Although the discoveries and inventions that have resulted from the scientific enterprise have by and large benefited handicapped people as much as they have benefited people without handicaps, there are some technologies that may have served to amplify the effects of specific disabilities. A case in point is the telephone. There can be no doubt that the telephone facilitates communication among hearing people enormously. The existence of this device has had profound effects on our individual and collective lives economically, socially, and psychologically. It is difficult to imagine what the consequences would be if this means of communication were suddenly to cease to exist.

Given that the invention of the telephone was a result, in part at least, of Alexander Graham Bell's lifelong interest in facilitating communication by and with people who are deaf, it is ironic that this technology not only failed to serve the needs of deaf people, at least in the sense of directly assisting their communication, but may indeed have had the effect of increasing their isolation from the hearing world. By providing an extremely effective way for people who can hear to communicate at a distance, this innovation changed profoundly the way in which our society functions and how its business gets done. In particular it assured the emergence of countless jobs in which communication at a distance plays a critical role, jobs that the vast majority of people with severe hearing impairments cannot perform. This is not to say that people with profound hearing impairments have not benefited indirectly from the invention and widespread use of the telephone. Nevertheless, the ubiquity of a device that puts people in immediate touch with other people independently of their location—provided they can hear—is an eminently understandable frustration to people unable to use it.

Among the promises that information technology holds is that of providing the opportunity for people with certain types of communication disorders to communicate with other people, and with various information resources, from a distance. The numerous computer networks, government and commercial, that already exist to serve a variety of purposes should be of great interest to people for whom the conventional telephone is not a suitable vehicle for communication. Although the primary target users of this technology are people who also use telephones, its potential for serving those who do not is slowly being recognized (Cerf 1978; Grignetti, Myer, Nickerson, and Rubenstein 1979; Rubinstein and Goldenberg 1978).

There is some anecdotal evidence that the use of computer-mediated message technology for interperson communication may help to remove some of the psychological barriers that too often inhibit effective communication between people who are handicapped and those who are not. One of these barriers is the difficulty that many "normal" people have in accepting differences. It is a difficulty that few will admit, but the evidence is compelling that people tend to feel awkward and ill at ease when confronted with those whose appearance, speech, or behavior deviates from their stereotype of what is "normal." Such discomfort, an indication of our moral immaturity as a species, should be striven against but nonetheless acknowledged as a fact.

Computer-mediated message systems are opaque to such superficial differences among people who use them and communicate through them. One party to a communication has no way of knowing what the other one looks like, whether he typed in the message with his fingers or a headstick, whether he is six feet tall or two; what is apparent in the message is what the other person thinks, what he feels, what he believes, what he hopes, what he desires, what makes him laugh or cry. Without the distraction of cosmetics, each person can get a better glimpse of the essence of the other.

One might question how satisfying a remote interaction can be from a social or psychological point of view. My guess is that it can be very satisfying, especially for people for whom there is no feasible alternative. Ham radio operators strike up many acquaintances and develop some friendships with people whom they never see nor expect to see. The same is undoubtedly true of people who habitually correspond by letter. I imagine that one could easily document the effectiveness of telephone communication as a means of alleviating the deadly loneliness that can consume people who for one reason or another are cut off from face-to-face interaction with other people.

The possibilities of message technology go beyond the facilitation of person-to-person communication at a distance. It can be the means of bringing work opportunities to people who find it impossible to bring themselves to places where work has conventionally been done. It holds the promise of making various types of information more accessible. It has great potential for bringing educational and recreational resources to people wherever they may happen to be.

Information technology has the potential also to provide challenging job opportunities for people with impairments that limit their ability to perform physically demanding jobs. Realization of this potential may require, in some instances at least, the development of special interfaces and devices to assist interaction with a computer system by individuals with severe motor impairment (Raitzer, Vanderheiden, and Holt 1976; Socha 1984). The prospect of giving a bed-bound person who has very limited motor control both gainful employment and an effective communication link to the world would seem to justify the effort.

Much is happening that should be of interest to people who

would like to see information technology applied to the amelioration of handicapping conditions. Here are a few examples.

- In February 1984, Telesensory Systems, of Mountain View, California, announced the manufacture of TeleBraille, a computer-based device intended to facilitate communication between people who are deaf and blind and people with normal sight and hearing. The deaf-blind user inputs messages to the system by means of a Braille keyboard and receives output through a Braille display that consists of several rows of pins that are mechanically raised to form Braille characters—20 characters at a time. The system also provides access through a conventional typewriter keyboard and visual display for sighted users. Information from both the Braille keyboard and the typewriter can be displayed simultaneously on the Braille and visual displays. The system is battery operated and portable. It can also work with an acoustic coupler over standard telephone lines (*Technology Trends Newsletter* 1984).

- Intex Micro Systems, Birmingham, Michigan, is marketing a portable battery-powered text-to-speech synthesizer for people without speech but with relatively good manual dexterity (*VoiceNews* 1984, April).

- The American Voice Input/Output Society (AVIOS), Palo Alto, California, was formed as a nonprofit organization in 1982 to facilitate the exchange of information about applications of voice input-output technology for disabled persons.

- The National Research and Demonstration Institute of the Association for Retarded Citizens has recently announced the intention of exploring potential applications of computer technology to the provision of increased independence for mentally retarded persons (*Practice Digest* 1983).

- The summer of 1984 saw the first issue of *Computer-Disability News; The Computer Resource Quarterly for People with Disabilities,* published by the National Easter Seal Society. The stated purpose of the quarterly is to present "the most important computer world news for people with disabilities." The lead article in the premier issue was entitled "Computer Programming: New Job Opportunities for People with Disabilities" and it described a program, "Life, Inc.," that trains people with disabilities to be computer programmers and helps place them in jobs.

- Another source of information about applications of computer technology to the problems of handicapped people is *Closing the Gap,* a bimonthly newspaper that carries feature articles, news items, product announcements, and advertisements. The publishers of *Closing the Gap* offer two-day and week-long workshops on several topics relating to the use of computer technology by people with disabilities.

- Still another source is *Bulletins on Science and Technology for the Handicapped,* published quarterly by the Association for the Advancement of Science's Project on the Handicapped in Science. The Fall 1984 issue focuses on computers and disabled people; it contains brief descriptions of several research and development projects in this area and provides numerous pointers to other sources of information.

- The American Association for the Advancement of Science has held several workshops on Science and Technology for the Handicapped and, as a consequence, has published two collections of papers under the title *Technology for Independent Living I* and *II* (Redden and Stern 1983; Stern and Redden 1982).

- The U.S. Department of Education's Division of Educational Services sponsors a Special Education Software Center, in Menlo Park, California, that provides, among other services, information about special education software for persons with various types of handicaps and where to find it.

- The International Council for Computers in Education, located at the University of Oregon, has published two *Resource Guides on Computer Technology for the Handicapped in Special Education and Rehabilitation,* one in 1983 and another in 1985.

- Conferences focusing on the application of computer technology to problems associated with various types of disorders are increasing in number. Calendars of such conferences appear in the AAAS's *Bulletins on Science and Technology for the Handicapped* and in *Closing the Gap.*

One of the more impressive applications of computer technology to problems of handicaps is the Kurzweil reading machine. This device, which was developed and improved over a period of about ten years, makes use of optical character recognition, print-to-grapheme string conversion, and speech synthesis (Kurzweil 1976, 1984; Unger 1976). The machine, which will accept printed material in any typeface, has been widely distributed to colleges and libraries, and efforts are being made to reduce its size and cost to the point where it is affordable and manageable by any blind person who could benefit from its use (Kurzweil 1984).

Efforts have also been made to apply computer technology to the problem of enhancing face-to-face communication by people who are deaf, but considerably less progress has been made here than on the development of reading machines for people who are blind. In the early 1970s an experimental system was built around a PDP-8 computer that would analyze speech in a variety of ways and show selected properties of the speech visually (amplitude, fundamental frequency, degree of nasalization, spectral distribution). The system was used in research at schools for the deaf in Northampton, Massachusetts, and New York City. It proved capable of modifying the speech of participating children in desired ways, but large gains in intelligibility were not realized (Boothroyd, Archambault, Adams, and Storm 1975; Nickerson and Stevens 1972; Nickerson, Kalikow, and Stevens 1976). This problem area remains a challenge; but I believe that computer technology has the potential for contributing to solutions in several ways (Nickerson and Stevens 1980).

Attempts are also being made to use computer technology to facilitate speech reception by people with hearing impairments (Krasner, Huggins, Schwartz, Kimball, Chow, and Makhoul 1984). In this case the ultimate goal is one that was envisioned by Wiener (1948) in his *Cybernetics*—namely, translating the speech signal into text. This goal is still beyond the state of the art; the more immediate objective is to convert the speech into a stream of visually coded information, which includes some phonetic symbols and prosody markers, that will supplement the information that can be acquired from lipreading.

Bowe (1984) is a rich source of information about the use of personal computers by people with various types of disabilities. The book contains numerous accounts of how individuals with special needs have used personal computers to enhance their ability to communicate with other people, to expand their social lives, to increase their job opportunities, and, in general, to attain greater independence and control over their own lives. Effective use of this technology has often required some inventiveness and perseverance on the part of special-needs users, but considerable benefits have attended these efforts. The book points its readers to sources of information relevant to particular disabilities, and should prove a valuable resource.

The application of information technology to the problem of educating children with severe communication disabilities has been examined by Goldenberg (1979). This book was written before personal computers had become commonplace, so it does not focus on their use. But it is a sensitively written account of several efforts to use computer-based tools to give children with severe communication limitations from various causes (cerebral palsy, autism, hearing impairment) more effective means of communication and access to educational resources. It is a forward-looking book that provides exciting glimpses of the possibilities.

Other books dealing with the application of computers, and especially microcomputers, to problems of the disabled are beginning to appear. Among them are Behrmann (1984); McWilliams (1984); Schwartz (1984); Gergen and Hagen (1985); and Hawkridge, Vincent, and Hales (1985).

Perhaps the most compelling evidence of the potential that information technology has for application to disability problems, and the most effective argument for putting considerable energy into the development of that potential, are the testimonies of handicapped people for whom information technology is already providing job tools or employment opportunities that otherwise would not exist (Mallik 1982; Morgan 1984; Wagreich 1982). Further exploration of the possibilities in employment and other areas should include the development and refinement of the following:

- electronic access to libraries by blind people via optical character recognition machines (or reading machines) at the library transmitting Braille (or speech) to the home, either in real-time or store-and-play-later mode;
- devices with high-quality speech-synthesis capability, providing speech as a medium of communication for people who are unable to speak themselves;
- home-based services (shopping, banking) via computer networks;
- home-based employment made possible also by computer networks;
- environmental control devices;
- home security, personal monitoring, emergency alerting systems.

One of the problems that we are all aware of but nobody knows quite how to solve is that some of the most challenging disabilities affect relatively small numbers of people. This has a negative effect on the development of instrumental aids in two ways: first, the cost of producing a device of much complexity is inversely related to the number of devices that are sold; and second, the small size of the potential market means there is little incentive for industry to do the necessary research and development. These facts in turn have several implications. Any research that is done to develop devices for low-incidence problems is typically supported by the government. Because the devices, once they are developed, are not mass-produced, their production costs continue to be high and consequently they are available to most handicapped people only if bought by the government or by "third-party" funds. Because governmental agencies that sponsor research of this kind have limited funds and virtually unlimited demands upon them, they naturally give heavy weight to the question of how many people will be affected by each research proposal under consideration. Because devices that are developed for low-incidence problems are likely to be custom made, maintenance is a continuing problem; it is likely to be expensive and difficult to get. Clearly, from the disabled consumer's point of view, one is better off if one has a problem that many other people also have.

Computer technology might be able to change all this. If, as some are predicting, derivatives of computer technology such as CAD-CAM techniques will make custom production of many things feasible at low cost, people with low-incidence problems could be among the beneficiaries. Personal computers too, because of their versatility, should go some distance toward meeting this same need. What is required to exploit this potential is creative thinking on the part of people who are familiar with both the technology and the needs that have to be served. We have noted some evidence that such thinking is being done, and there have been some successes to report. But the problem deserves far more attention and energy than it has yet received.

A group who share many of the problems experienced by disabled people are the elderly. Since the beginning of this century, the average life expectancy in the United States has nearly doubled. This is not because the ages attained by the oldest people are much greater today than they were in the past; people lived to be 80, 90, or occasionally 100 before this century. The nature of the change is that a much greater percentage of the population now reaches those advanced ages. Many elderly people are in perfectly good health; the incidence of various types of impairments (sensory, motor, and cognitive), however, is naturally greater than in younger groups. Moreover, the elderly may face a variety of problems that are independent of their state of health. These include no longer being gainfully employed and engaged in an absorbing routine, illness or death of a spouse or of friends, possible obsolescence of knowledge and skills, and so on.

How might information technology be used to address these and other problems of the elderly? One can imagine a variety of possibilities, among them the use of computer-based message technology, including electronic bulletin boards, to help elderly people stay in close contact with friends and information resources of various types; computer-based aids to memory; home entertainment; opportunities for engagement in service activities.

There are many indicators of the degree of development of a society. There is none, in my view, more telling than its attitudes and actions toward those of its members who, for whatever reason, face life against unusually unfavorable odds. Information technology has the potential, I believe, for mitigating the effects of many disabilities. The extent to which this potential is realized will depend on the effort we are willing to expend on that goal.

### Toward the Future

The future that we look forward to is richer in possibilities than the future that was seen by any of our predecessors. While opinions may differ as to whether the effects of technology have been beneficial or detrimental on balance, it would be hard to deny that one of those effects has been an increase in the alternatives among which we can choose individually and collectively. Most of us have a far greater range of options regarding where we will live, what kind of education we will get, whom we will associate with, what we will do for work, and how we will spend our leisure time than did our grandparents or even our parents. As a nation we devote only a fraction of our human and capital resources to providing ourselves with food, clothing, and shelter; we have the luxury of deciding how to use the remainder. In short, one of the most obvious and important things that technology has done is to increase our degrees of freedom, as individuals, as nations, and as a species.

Many of the most significant choices that we have regarding the future relate directly to information technology in one way or another. Among the possible future developments are some undesirable ones.

- Serious problems stemming from unemployment due to the automation of both blue- and white-collar jobs, in spite of adequate production of goods.
- Unthinking use of information technology to perform functions more appropriately performed by human beings.
- A widening of the technological and economic gaps between the developed and underdeveloped countries as a consequence of the exploitation of information technology by the former, unaccompanied by serious efforts to share this technology with the latter.
- Increasing use of information technology to support new forms of crime and antisocial behavior.
- The concentration of large amounts of information in tightly controlled repositories and the exploitive use of that information to the benefit of its possessors.

Such possibilities must be recognized and actively resisted if their realization is to be avoided. There are also many desirable possibilities. It is not prudent to assume that they will come about automatically; but they represent both opportunities and challenges that are deserving of our best efforts to assure their realization.

- The emergence of a "learning society" of the type envisioned by G. E. Brown (1982).
- The evolution of democracy into a directly participatory form.
- A great increase in personal freedom and choice.
- Effective application of information technology to problems of worldwide significance, such as hunger and malnutrition.
- A lessening of world tensions as a consequence of increasing possibilities for interperson communication among individuals from different countries.
- Greatly increased accessibility of information to everyone.

The last possibility is, in my view, the most exciting one of all, and the one to which I want to devote the last few pages of this book.

# 18

# A Perspective

Many psychologists these days like to think of human beings as information-processing systems. The metaphor is more apt than most. Among the things we do, there are few that are more important to us than the production, use, and communication of information. To the individual, information is the lifeblood of the intellect. A young mind deprived of it will not mature. A mature mind cut off from it will live off what it has been able to store, and if that proves inadequate, it will shrivel and die.

A business that operates on inaccurate or out-of-date information is heading for serious trouble. Even if the information is accurate, it may not suffice to ensure success if it compares unfavorably with that of the competition in terms of completeness or timeliness.

To a nation, reliable and timely information, and lots of it, is essential to stability and prosperity in peace and to survival in war. Information gathering and interpretation are primary concerns of any government, and history is replete with examples of international disputes and altercations whose outcomes have been determined by differences in the quality of the information the antagonists possessed.

To the species, the ability to accumulate information, represent it in enduring forms, and pass it on from generation to generation has been fundamental to the development of science, technology, and all aspects of civilization. If it is not the single ability that distinguishes humankind as a species, it is certainly among those that do.

In short, whether we think in terms of individuals, organizations, nations, or the species as a whole, information is a necessity, a sine qua non of survival and well-being. The history of humanity could be told in terms of the techniques that have been developed for acquiring information, for storing it, for manipulating it, and for moving it from place to place. Today those techniques are increasingly dependent on computer and communication technologies, so much so in fact that the term "information technology" is sometimes used for computer technology, sometimes for communication technology, and sometimes, more appropriately, for something that encompasses both. As we have noted, the dividing line between computation and communication in modern systems grows ever more difficult to find: large computing systems depend on communications processes of many types, and communication systems make use of computing resources throughout.

One useful way to evaluate what is happening today, and what is likely to happen, in information technology is in terms of the prospects of making information more widely and readily accessible to individuals as well as to corporate and political entities. Indeed, the emergence and development of computer technology provide the prospects of a quantum jump in information accessibility, the full implications of which we can only vaguely foresee.

There have been other quantum jumps in information accessibility. The first, and certainly the most significant, was the acquisition of language. We have not the vaguest idea how this occurred and are not even certain there was a time when humankind existed and language did not. What is clear, however, is that language gives human beings a versatility and potential for information acquisition and use enjoyed by no other species. This is not to overlook the much-debated possibility that other species possess forms of language; but the difference between the language facility enjoyed by humans and that found in any other species is vast indeed. Moreover, language is universal among human beings. Even the most primitive societies have it, and the differences between the languages of primitive and those of "advanced" societies are insignificant compared to the differences between all human languages and the signaling systems of other species.

Individuals acquire information in two ways: by direct observation and by being provided it by other people. It is obvious that the second route is immensely more effective for a species that has language than for one that does not. While perhaps less obvious, it is no less true that language also facilitates the acquisition of information by direct observation, because it helps us both to organize and think about what we observe and to retain information for future reference and use.

A second quantum jump in information accessibility came with the development of written language. The ability to accumulate knowledge, to store it in a tangible form, and to pass it on from generation to generation was immeasurably enhanced by this momentous achievement. It is interesting to speculate about what kind of society an intelligent species might be able to produce in the absence of written language. Certainly technology, as we know it, could not arise. One might argue that it is unthinkable that an intelligent species would fail to develop a means of representing language in some relatively permanent medium; but human beings are believed to have inhabited this planet for perhaps two million years, possibly longer, and to have learned to write less than ten thousand years ago. Why it happened when it did, no one really knows. What life would be like today had it not, is probably beyond our imagination. What we can be sure of is that the information accessible to each of us today is many orders of magnitude greater than would have been the case if the trick of writing still remained to be discovered.

A third quantum jump in information accessibility was caused by the invention of the means of producing books sufficiently inexpensively to make possible their widespread distribution. Although the Chinese had invented a means of printing from blocks as early as 50 B.C., the events that are usually associated with the advent of printing technology are Gutenberg's invention, during the first half of the fifteenth century, of the printing press and movable type and the subsequent discovery of how to make relatively inexpensive and durable paper from linen rags. The long-term consequences of these developments would be difficult to overstate. The effect, in a word, was the democratization of knowledge.

As important as the invention of writing has been to humanity as a whole, until the capacity to mass-produce books was developed, the average person's access to the accumulated knowledge of the species was very limited and indirect. Only a small fraction of the population could read and write, and for the vast majority of people there was neither opportunity nor incentive to learn to do so. Except for a privileged few, including the professional scholars and scribes, literacy had limited utility. What is the point of being able to read if one has no access to material on which to exercise the skill? The laboriousness of producing or copying a lengthy manuscript by hand assured that those that were produced were carefully safeguarded in repositories that were not accessible to the masses. "Library" had a different connotation then; the collections of manuscripts at Alexandria, Pergamum, and elsewhere served a quite different function from that of the public libraries that we take so much for granted today.

It is a significant characteristic of the age in which we live (undoubtedly more true of developed than undeveloped countries) that the average person can become knowledgeable in just about any area he chooses. Information is widely available on any subject in whatever detail one might wish to have. If one wants to become knowledgeable with respect to nuclear physics, nutrition, paleontology, social structures in Polynesia, methods by which coal may be turned into gas, one has access to the necessary information. Effectively availing oneself of such information assumes a certain level of intelligence and literacy; yet the essentially universal opportunity to become literate is also a characteristic of our age. It is too easy for us to forget, or overlook, how very recently this democratization of knowledge occurred.

Another important characteristic of our age is the unprecedented rate at which new information is being acquired. Indeed, perhaps the most obvious and consequential effect of the development of new information-gathering techniques by scientists has been the tremendous acceleration in the accumulation of knowledge. It has been claimed, for example, that whereas determination of the composition of a new carbohydrate would have taken three years in the 1930s, it can now be done in less than three weeks. The difference is due to the development of such analytic techniques as nuclear magnetic resonance, mass spectrometry, and X-ray diffraction analysis (Sharon 1980). Diebold (1969) has pointed out that the amount of scientific literature published in the United States (as indicated by the number of scientific journals) has increased by an order of magnitude every fifty years for the last three centuries. The same writer observed that 90 percent of all the scientists who had ever lived were alive at the time of his writing.

In short, a profoundly important aspect of the history of humanity is the progressive increase in the average individual's access to information about the world. This increase has been greatly accelerated by certain momentous events, such as the development of printing technology a few hundred years ago and the emergence of writing a few thousand years before that. In making information more widely accessible, such events have also ensured the exponential growth of the intellectual treasury of humanity: knowledge builds upon knowledge, and increasing the access to what is already known has had the inevitable effect of equipping more people to participate in the further expansion of the knowledge base. Thus, not only do we today have much greater access to information about the world than did our predecessors of only a few hundred years ago, but there exists much more information to which access is to be had.

Might we now be witnessing developments that will in time produce another quantum jump in information accessibility? The raison d'etre of information systems is to make information more accessible to people who have a use for it. The highest goal of information technologists should be to develop more effective ways of bringing information to people and presenting it to them in usable form. This subsumes a host of subordinate goals having to do with gathering information, evaluating it, organizing it, transforming it, storing it, transmitting it, safeguarding it, updating it, retrieving it, condensing it, and representing it in ways that make it suitable for human use.

Though far more accessible to people today than ever before, information is by no means as accessible as one would like it to be. Information seeking can consume an inordinate amount of time, and this observation can apply equally well to a lawyer searching through legal archives for cases relevant to a specific legal issue, to an historian doing library research on the life of a person or an era, to a physician trying to find out what is known about some rare disease, to a job seeker trying to identify possibilities for employment, to a family trying to determine its options for a vacation. In all these cases one may assume that the needed information exists and that it is available, in the sense that if one knew where to find it one could avail oneself of it. But knowing that information exists and being able to find it when one wants it are quite different things; and finding information can sometimes represent a greater challenge than using it effectively once it has been found, although more attention is given to the problem of using information than to that of finding it. The last point is easily illustrated by reference to recent work on decision making.

Several theoretical treatments of decision making have been published in recent years, as have the results of numerous experimental studies. For the most part what has been written has focused on the question of how to maximize the probability of attaining some decision objective, given the information at one's disposal regarding the decision "space." Some of this work addresses the problem of evaluating and choosing among decision alternatives when those alternatives differ with respect to numerous incommensurate dimensions but the outcome of a given choice is known; and some of it addresses the problem of making rational choices under uncertainty, which is to say when the outcomes of the choices are known only probabilistically, if at all. Prescriptive models have been developed for the purpose of indicating how decisions should be made under specified conditions, and descriptive models have been developed that purport to show how they are made in fact.

This work is of undisputed importance. An aspect of decision making that textbook treatments often overlook, however, is that of information seeking. Whether one is faced with a major decision (selecting a job, purchasing a house, choosing the best response to a military threat) or with one of the many small decisions of daily life (planning a menu, selecting some family entertainment, purchasing a gift), the process of finding out what one's alternatives are is often considerably more difficult than the process of making the choice itself. Moreover, the "worth" of the choice is bounded above by the best of the alternatives of which one is cognizant; the existence of better options (in terms of the decision maker's own value system) does one little good if one is not aware of them at the time the decision must be made.

We should distinguish between access to aggregations of information and access to specific items of information, such as the answers to specific questions. An example of access to aggregations of information is access to a library in which the information one seeks is assumed to be stored, perhaps in several documents. Access to such a repository is not quite the same as actually having in hand the specific information one wants. In the past, information accessibility has been increased more in the first sense than in the second. The quantum jumps in accessibility that we have noted have had the effect of fostering the accumulation of information in various repositories and providing access to these accumulations.

Information technology has the potential to increase accessibility in both senses. By means of computer networks and remote terminals, including wireless terminals, it should be possible in the foreseeable future to provide direct access to major repositories of information from essentially anywhere in the world and to nearly anyone who wants it. Progress on the problem of providing the second type of access is likely to come more slowly, however; consequently, this problem will command more and more attention. Indeed, the need at present is not primarily that of getting more written material to people. Many of us already have in our possession far more than we will ever be able to assimilate in our lifetimes. And, as we have noted, the rate at which written material is increasing is astounding. Licklider (1965) has estimated that given the rate at which information is accumulating, keeping up with new "solid" contributions to a subfield of science or technology would require a peson to read for eight to ten hours a day, every workday, at a speed appropriate for novels. The situation has undoubtedly worsened since Licklider make this observation. So who needs more information, given that most of us are already bombarded with many times as much as we can absorb? In fact, much of what we get is not information that we need or want; much of it is not even information by any reasonable definition of the word. While it is true that few of us need more paper thrown at us, or more words we would like more information—words and images carefully chosen to suit our needs. The problem is to devise ways of getting to people the specific information they want, when they want it, and in a form that facilitates its assimilation and use.

We might refer to this problem as the problem of psychological accessibility, and to the challenge as that of increasing the accessibility of information to the user in a psychological sense. Here the challenge is not only to computer and communication science and technology but to any discipline that includes within its scope an interest in human beings as users, manipulators, processors, storers, and conveyers of information. Among such disciplines are human-factors engineering and cognitive science (the latter term is taken to encompass both cognitive psychology and artificial intelligence). The relevance of human-factors engineering is obvious; that of cognitive science, slightly less apparent but no less real. Many of the questions that must be addressed in the interest of designing information systems that are well matched to their intended users have to do with people's cognitive capabilities and limitations: how to represent information so that it is readily understood and assimilated; how to avoid putting an unrealistic burden on a user's working memory; how to allocate the cognitive aspects of a complex task between people and machines.

Licklider (1965) coined the term "procognitive systems" to denote systems that are intended to promote the advancement and application of knowledge. The purpose of these systems is to facilitate the interaction of people with the stored knowledge of the world in a variety of ways.

A basic part of the overall aim for procognitive systems is to get the user of the fund of knowledge into something more nearly like an executive's or commander's position. He will still read and think and, hopefully, have insights and make discoveries, but he will not have to do all the searching himself, nor all the transforming, nor all the testing for matching or compatibility that is involved in creative use of knowledge. He will say what operations he wants performed on what parts of the body of knowledge, he will see whether the result makes sense, and then he will decide what to have done next. (31).

"Intimate interaction with the fund of knowledge" is the goal. Not many people are interested in this goal, notes Licklider, but then not many people can form a notion of such interaction because there is nothing like it in common experience.

The possibility that access to some information will be increased only for certain segments of the population is a worrisome aspect of the overall picture. Information is power, and concentrated information is concentrated power. The intentional limiting of access to information is typically something that is done by groups of people—governments, industrial power cliques, or whatever—that wish to exercise some measure of control over other people. The best response to this danger, in my view, is the application of the technology itself to the goal of making information, in the most general sense, ever-increasingly accessible to all of us. Widely accessible information is a countermeasure to the development of oppressive control systems. One gains control over other people by having information that they do not have. The better informed a citizenry, the less susceptible it will be to manipulation.

But how much is enough? Suppose it were technically and economically feasible to build a general-purpose question-answering system that had an encyclopedic knowledge base and the necessary search, inference, and production capabilities to extract information that was stored implicitly, as well as what was stored explicitly. Imagine that it could answer questions and requests such as the following:

- What foods are particularly good sources of potassium?
- Who were the most prolific composers during the Italian baroque era? What did they produce?
- What are the major similarities and differences between mitochondria and ribosomes?
- What is the voting record of Senator Jones on issues relating to environmental protection over the past five years?
- Give me a synopsis of the formal education of Winston Churchill.
- What was the closing price of a share of General Motors stock on April 14, 1962?
- Give me a complete list of the symphonic works of Franz Joseph Haydn.
- Who was the U.S. secretary of the treasury during the presidency of James Buchanan?
- How many U.S. secretaries of state have been fluent speakers of some language other than English? And who are they?
- Who was the discoverer of the smallpox vaccine?
- What are the five most abundant elements in the universe?
- How many new corporations came into existence in the United States in 1974?
- How many corporations declared bankruptcy that same year?
- What has been the average annual growth rate of the population over the last ten years in each of the following countries: India, Japan, France, and Egypt?
- What are the main molecular components of the human body and what proportion of the body does each represent?

Would not such a capability stifle curiosity? If questions are too easily answered, would we not lose interest in asking them? My opinion is that the answer to both of these questions is no. This opinion is predicated on the assumption that human curiosity is insatiable, that knowledge fosters curiosity, and that, in general, the more one knows the stronger becomes one's desire to learn still more. Rather than decreasing intellectual activity, the existence of a general fact finder would, in my view, increase it very considerably. How effectively one can think depends to no small degree on the knowledge base one has that can serve as the content of thought.

But one might ask whether having ready access to knowledge is not as good as having the knowledge itself in one's head. Suppose one had at one's command an information system with extensive knowledge on a given topic. Suppose, further, that one's means of interacting with this knowledge store were sufficiently sophisticated and powerful that one could retrieve specific items of information, including those that were stored only implicitly, on demand. Assuming that such a system were completely adequate to meet any practical needs a user might have for information in its field of coverage, would there be any good reason for the human being to store that information in his head?

I believe the answer to this question is unequivocally yes, at least if the human being has any real interest in the topic. One must know something about a subject in order both to frame an intelligent question about it and to understand the answer to the question. In other words, the effective use of an extensive knowledge base requires a knowledgeable user.

I believe that very powerful question-answering systems of the type fantasized above will eventually be a reality. They will not be designed, however, but will gradually evolve. Beginnings will be made (have been made) with databases of modest size and restricted domain, relatively primitive retrieval and inferencing capabilities, and interfaces that are much more highly constrained than natural language. (Interest in question-answering systems goes back at least to the early 1960s [Green, Wolf, Chomsky, and Laughery 1961; Marrill 1963].) Experience gained with these restricted-domain systems will lead to the development of more effective search algorithms, inferencing and production procedures, and natural-language interfaces. New memory designs and multiprocessor architectures will open the way to working effectively with very large databases, and the constraints on both size and subject matter will gradually relax.

As has been generally true, the easier problems will be solved first and the earliest question-answering systems in general use will be much better at answering certain types of questions than others. Our expectations may be guided somewhat by a distinction among several kinds of questions that might be asked of a fact finder with an encyclopedic database. First, there are questions whose answers are stored explicitly in the database and stored in such a way that they are relatively easy to find. The data of a noteworthy historical event, for example, tends to be easy to find in a history book because history books tend to be written with dates in mind. Second, there are questions whose answers may be explicitly represented in the database but are not easy to find. A conventional nonfictionbook, for example, may contain much information that is tangential to its primary themes and that may be difficult to locate by means of its topical organization. Third, there are questions whose answers are not represented explicitly in the database but may be inferred easily from the information that is represented there. Suppose, for example, that one wished to know which of two U.S. presidents was elected first. If the database contained the dates of the terms of office of both, the answer could readily be inferred, provided that the fact-finding system understood the notion of chronological time and how to determine which of two dated events occurred first. Fourth, there are questions whose answers are not explicitly represented in the database and can only be inferred from information that is contained explicitly by means of a relatively sophisticated inferencing capability. Suppose one wanted to know, for example, whether the congressional speeches of a specific legislator were more or less compatible than those of another specific legislator with the stated political objectives of a particular special-interest group. Inasmuch as the speeches are recorded in the *Congressional Record*, the information exists from which to derive an answer. Deriving that answer, however, is clearly a nontrivial task.

Some questions require the making of assumptions or the setting of criteria. For example, in trying to determine how many U.S. secretaries of state have been fluent in some language other than English, one must first decide what it means to be fluent in a foreign language. This example also illustrates the problem of finding information that is scattered. Unless one were fortunate enough to locate a document that was explicitly addressed to the topic of the foreign-language abilities of United States secretaries of state, one would probably have to consult a wide variety of sources to gather this information. The example also illustrates the problem of deciding when the lack of information (or the failure to find information) can be used as the basis for inferring an answer. How is one to decide that a particular individual was *not* fluent in any language other than English? An explicit statement to that effect by a biographer who is considered authoritative might be accepted as adequate evidence. Suppose, however, that the biographer fails to say anything about language capability; should one conclude that if the individual had been fluent in any foreign language, the biographer would have noted the fact? For a person who is sufficiently famous to have been the subject of several biographies, does evidence by omission become increasingly compelling as the number of authors who have omitted that evidence increases?

Clearly, there are some difficult problems to be solved before anything like our fantasized general-purpose question answerer can be developed. But these are the kinds of problems that are currently being studied by researchers on artificial intelligence, and they are slowly yielding. It would be foolhardy to predict when such a system might be a reality; but one can safely say that question-answering systems of ever increasing scope will be appearing on the scene more or less regularly. A truly general-purpose system is an exciting long-range quest.

Of course, a knowledge base that is extensive enough to provide an answer to every question of fact that one might wish to ask, even in a limited domain, is probably not a possibility even conceptually, because one consequence of interacting with a system that held all that was known about a particular topic at a given time would be that the user would learn enough to ask questions that exceeded that knowledge base and could be answered only by further research. This is, after all, the position of the expert who knows, more or less, all that is known in a given domain; he above all others is in a position to ask the really interesting questions about the domain, namely those whose answers require the development of new knowledge.

Surely we know more today than did our predecessors of a few millennia ago. Surely too, we are asking more questions than they did. Many of the questions being asked today could not have been asked in the past. Questioning is a knowledge-based process, and the questions one asks are limited by the extent of the knowledge base from which they derive. There is a very real sense in which the primary effect of increasing our knowledge is an increased awareness of our ignorance. To learn is to get a better glimpse of the enormous disparity between what we know, or what we think we know, and what there is to learn.

# References

Frequently cited associations, conferences, and journals are abbreviated thus:

ACM     Association for Computing Machinery

AFIPS   American Federation of Information Processing Societies

ASIS    American Society for Information Science

CHI     Computer-Human Interaction

IEEE    Institute of Electrical and Electronics Engineers

IJCAI   International Joint Conference on Artificial Intelligence

*IJMMS*  *International Journal of Man-Machine Studies*

Abelson, P. H. (1982). The revolution in computers and electronics. *Science*, *215*, 751–753.

Abraham, E., Seaton, C. T., and Smith, S. D. (1983, February). The optical computer. *Scientific American*, *248*, 85–93.

Abrams, M. H., Goffard, S. J., Kryter, K. D., Miller, G. A., Sanford, J., and Sanford, F. H. (1944). *Speech in noise: A study of factors determining its intelligibility* (OSRD No. 4023, PB 19805). Cambridge, Massachusetts: Harvard University, Psychoacoustics Laboratory.

Abramson, B. (1984). Applied AI. *The Artificial Intelligence Report*, *1*, 3–6.

Adams, J., and Cohen, L. (1969). Time-sharing vs. instant batch processing: An experiment in programming training. *Computers and Automation*, *18*(3), 30–34.

Addis, T. R. (1977). Machine understanding of natural language. *IJMMS*, *9*, 207–222.

Adler, M. J. (1982). *The paideia proposal: An educational manifesto*. New York: Macmillan.

Adult Performance Level Project (1977). *Final report: The adult performance level study*. Austin: University of Texas.

Aikins, J. S., Kunz, J. C., Shortliffe, E. H., and Fallat, R. J. (1982, August). *Puff: An expert system for interpretation of pulmonary function data* (Technical Report STAN-CS-82-931). Stanford, California: Stanford University, Computer Science Department.

Albert, A. F. (1982). The effect of graphic input devices on performance in a cursor positions task. *Proceedings of the Human Factors Society*, 54–58.

Alden, D. G., Daniels, R. W., and Kanarick, A. F. (1972). Keyboard design and operation: A review of the major issues. *Human Factors*, 275–293. [A very similar paper by the same authors: 1970, March (Technical Report 12180-FR1A). St. Paul, Minnesota: Honeywell Systems and Research Center.]

Alexander, T. (1984, August). Why computers can't out-think the experts. *Fortune*, 105–118.

Allik, H., Crowther, W., Goodhue, J., Moore, S., and Thomas, R. (1985, August). Implementation of finite element methods on the Butterfly parallel processor. To be presented at the International Computers in Engineering Conference, American Society of Mechanical Engineers.

Alter, S. (1977). Why is man-computer interaction important for decision support systems? *Interfaces*, 7(2), 109–115.

Ambroggi, R. P. (1980). Water. *Scientific American, 243*(3), 206–231.

Anderson, H. (1981). Homing in on the advanced work-station market. *Mini-Micro Systems*, 117–121.

Anderson, P. (1983, June 27). Computers in the cow barn. *Boston Globe*, 35, 36.

Anderson, R. H., and Gillogy, J. J. (1976). *Rand intelligent terminal agent (RITA): Design philosophy* (Report No. R-1809-ARPA). Santa Monica, California: Rand Corporation.

Anderson, T. H. (1979). Study skills and learning strategies. In H. F. O'Neil, Jr., and C. D. Spielberger (eds.), *Cognitive affective learning strategies*, New York: Academic Press.

Andrews, H. L. (1984). Speech Processing. *Computer, 17*, 315–324.

Antognetta, P., Pederson, D. O., and DeMan, H. (1982). *Computer design age for VLSI circuits*. Alphen aan den Rijn, Netherlands: Sijthoff and Noordhoff.

Armbruster, B., and Anderson, T. (1980). The effect of mapping on the free recall of expository text (Technical Report No. 160). Urbana-Champaign: Center for the Study of Reading, University of Illinois.

Artificial Intelligence Report (1984). DELTA/CATS-1. *The Artificial Intelligence Report, 1*, 7–8.

Assistant Secretary of Defense for Communications, Command, Control, and Intelligence (1981, January 19). *Modernization of the WWMCCS Information System (WIS)*. Washington, D.C.: U. S. Government Printing Office.

Atherton, P. (1971). Bibliographic data bases: Their effect on user interface design in interactive retrieval systems. In D. E. Walker (ed.), *Interactive bibliographic search: The user/computer interface*. Montvale, New Jersey: AFIPS Press.

Atkinson, P., Dalvi, V. S., Drawneek, E. A., Fellgett, P. B., Hovland, H. L., Tring, R. W. H., Walker, B. S., and Whitfield, G. R. (1970). The Picasso low-cost system in relation to graphic communication as a natural language for man-computer interaction. *Proceedings, Conference on Man-Computer Interaction* (Publication No. 68), 172–180. London: Institution of Electrical Engineers.

Atwood, M. E., and Jeffries, J. (1980). *Studies in plan construction I: Analysis of an extended protocol* (Technical Report No. SAI-80-028-DEN). Englewood, Colorado: Science Applications.

Atwood, M. E., and Ramsey, H. R. (1978). *Cognitive structures in the comprehension and memory of computer programs: An investigation of computer program debugging* (Technical Report No. 78-A21). Alexandria, Virginia: U.S. Army Research Institute for the Behavioral and Social Sciences.

Automobile Manufacturers Association (1971). *Automobile facts and figures.* Detroit, Michigan: Automobile Manufacturers Association, Inc.

Bacon, B. (1982). Software, *Science, 215,* 775–779.

Bacon, G. (1984, July). A PC for the teacher. *PC World,* 292–299.

Bahil, A. T. (1983). Computer text processing: New tools for technical writers. *Hardcopy, 12,* 82–83.

Bair, J. H. (1978). Communication in the office of the future: Where the real payoff may be. *Proceedings of the International Computer Communications Conference,* 733–739. Kyoto, Japan.

Baker, J. D. (1984). Dipmeter Advisor: An expert log analysis system at Schlumberger. In P. H. Winston and K. A. Prendergast (eds.), *The AI business.* Cambridge, Massachusetts: MIT Press.

Baker, J. D., and Knerr, B. W. (1981, May). Donovan's brain: An immodest proposal for the year 2000. Paper presented at the National Security Industrial Association First Annual Conference on Personnel and Training Factors in System Effectiveness, San Diego, California.

Baldwin, J. T., and Siklossy, L. (1977). An unobtrusive computer monitor for multi-step problem solving. *IJMMS, 9,* 349–362.

Balzar, R. M., and Shirey, R. W. (1968). *The on-line firing squad simulator* (Technical Report No. RM-5573-ARPA). Santa Monica, California: Rand Corporation (NTIS No. AD 675040).

Bannon, L., Cypher, A., Greenspan, S., and Monty, M. L. (1983, December). Evaluation and analysis of users' activity organization. In A. Janda (ed.), *Proceedings of the CHI'83 Conference on Human Factors in Computing Systems,* 54–57. New York: ACM.

Baran, P. (1964). On distributed communication networks. *IEEE Transactions on Communications Systems, CS-12,* 1–9.

Barmack, J. E., and Sinaiko, H. W. (1966). *Human factors problems in computer-generated graphics displays* (Report No. S-234). Arlington, Virginia: Institute for Defense Analyses (NTIS No. AD 636170).

Barnard, P., Hammond, N., MacLean, A., and Morton, J. (1982). Learning and remembering interactive commands in a text-editing task. *Behavior and Information Technology, 1,* 347–358.

Barnard, P. J., Hammond, N. V., Morton, J., Long, J. B., and Clark, I. A. (1981). Consistency and compatibility in human-computer dialogue. *IJMMS, 15,* 87–134.

Barrett, J. A., and Grems, M. (1960). Abbreviating words systematically. *Communications of the ACM, 3,* 323–324.

Bartee, T. C., Buneman, O. P., Gardner, K. A., and Marcus, M. J. (1979, April). *Computer internetting: $C^3I$ data communications networks.* Institute for Defense Analyses, Science and Technology Division, IDA Paper-1402.

Basil, R., and Edwards, B. M. (1984, February). Medical systems get a second opinion. *Hard Copy*, 32–38.

Bates, M., and Bobrow, R. J. (1984). Natural language interfaces: What's here, what's coming, and who needs it. In W. Reitman (ed.), *Artificial intelligence applications for business*. Norwood, New Jersey: Ablex.

Bavelas, A., Belden, T., Glenn, E., Orlansky, J., Schwartz, J., and Sinaiko, H. W. (1963). *Teleconferencing: Summary of a preliminary research project* (Study S-138). Arlington, Virginia: Institute for Defense Analyses.

Beasley, D. S., Zemlin, W. R., and Silverman, F. H. (1972). Listeners' judgments of sex, intelligibility, and preference for frequency-shifted speech. *Perceptual and Motor Skills*, *34*(3), 782.

Becker, J. (1983). Brain-controlled computers. *Omni*, *5*(10), 30, 31.

Beeler, M. Butterfly parallel processor overview draft. May 3, 1985.

Begg, V. (1984). *Developing expert CAD systems*. London: Kogan Page.

Behrmann, M. (ed.) (1984). *Handbook of microcomputers in special education*. San Diego: College-Hill Press.

Bell, D. (1976, February). Welcome to the post-industrial society. *Physics Today*, 46–49.

Bell, D. (1979, May/June). Thinking ahead. *Harvard Business Review*, 20–22, 26, 28, 32, 36, 40, 42.

Bennet, J. S., and Engelmore, R. S. (1979). SACON: A knowledge-based consultant for structural analysis. *Proceedings of the Sixth IJCAI*, 47–49. Stanford, California: Stanford University, Computer Science Department.

Bennett, E. M., Haines, E. C., and Summers, J. K. (1965). AESOP: A prototype for on-line user control. *AFIPS Conference Proceedings: Fall Joint Computer Conference*, *27*, 435–455.

Bennett, J. (1972). The user interface in interactive systems. *Annual Review of Information Science and Technology*, *7*, 159–196.

Bentley, T. J. (1976). Defining management's information needs. *AFIPS Conference Proceedings*, *45*, 869–876.

Bergman, H., Brinkman, A., and Koelega, H. S. (1981). System response time and problem solving behavior. *Proceedings, Human Factors Society 25th Annual Meeting*, 749–753.

Bertoni, P., and Castleman, P. A. (1976). *BBN and computers: A history*. Cambridge, Massachusetts: Bolt Beranek and Newman Inc.

Birnbaum, J. S. (1982). Computers: A survey of trends and limitations. *Science*, *215*, 760–765.

Black, J., and Moran, T. (1982). Learning and remembering command names. Paper presented at the Conference on Human Factors in Computer Systems, Gaithersburg, Maryland.

Black, J. B., and Sebrechts, M. M. (1981). Facilitating human-computer communication. *Applied Psycholinguistics*, *2*, 149–178.

Blackledge, M. A. (1974). Consideration of additional nuclear effects and fratricide avoidance in QUICK. *NMCSSC, System Planning Manual SPM FD*

*89-74*. Washington, D.C.: National Military Command System Support Center.

Blanc, R. P., and Cotton, I. W. (eds.) (1976). *Computer networking*. New York: IEEE Press.

Blesser, T., and Foley, J. D. (1982). Towards specifying and evaluating the human factors of user-computer interfaces. *Proceedings of Conference, Human Factors in Computer Systems*, Gaithersburg, Maryland, 309–314.

BMDP (1982). *BMDP statistical software*. Los Angeles: BMDP Statistical Software, Inc.

BMDP (undated). *Statcat leads the way*. Los Angeles: BMDP Statistical Software, Inc.

Boehm, B. W. (1973). Software and its impact: A quantitative assessment. *Datamation, 19*(5), 48–59.

Boehm, B. W., and Mobley, R. L. (1969). Adaptive routing techniques for distributed communications systems. *IEEE Transactions on Communication Technology, COM-17*, 340–349.

Boehm, B. W., Brown, J. R., and Lipow, M. (1977). Quantitative evaluation of software quality. *Software phenomenology working papers of the Software Lifecycle Management Workshop*, 81–94.

Boehm, B. W., Seven, M. J., and Watson, R. A. (1971). Interactive problem-solving: An experimental study of "lockout" effects. *AFIPS Conference Proceedings*, 206–210. Montvale, New Jersey: AFIPS Press.

Boies, S. J. (1974). User behavior on an interactive computer system. *IBM Systems Journal, 13*, 2–18.

Boies, S. J., and Gould, J. D. (1974). Syntactic errors in computer programming. *Human Factors, 16*, 253–257.

Bollinger, J. G. (1983, Fall). Computer integrated automation—A quality solution. *CAD/CAM Technology*, 35–36.

Boothroyd, A., Archambault, P., Adams, R. E., and Storm, R. D. (1975). Use of a computer-based system of speech training aids for deaf persons. *Volta Review, 77*(3), 178–193.

Borgatta, L. S. (1983, April). Chips oust clips. *IEEE Spectrum, 20*(4), 42–47.

Borko, H. (1962). *Computer applications in the behavioral sciences*. Englewood Cliffs, New Jersey: Prentice-Hall.

Bott, R. A. (1979). A study of complex learning: Theory and methodology (CHIP Report 82). La Jolla: University of California at San Diego.

Bourne, C. P., and Ford, D. F. (1961). A study of methods for systematically abbreviating English words and names. *Journal of the ACM, 8*, 538–552.

Bourne, C. T. (1961, November). *The world's technical journal literature: An estimate of volume, origin, language, field, indexing, and abstracting*. Menlo Park, California: Stanford Research Institute.

Bowden, V. (1970). The language of computers. *American Scientist, 58*(1), 43–53.

Bowe, F. G. (1984). *Personal computers and special needs*. Berkeley, California: Sybex Computer Books.

Bowen, R. J., Feehrer, C. E., Nickerson, R. S., and Triggs, T. J. (1975, February). *Computer-based displays as aids in the production of Army tactical intelligence* (Technical Paper No. 258). Army Research Institute for the Behavioral and Social Sciences, Alexandria, Virginia.

Bown, H. G., O'Brien, C. D., Lum, Y. F., Sawchuk, W., and Storey, J. R. (1979). Canadian videotex system. *Computer Communications*, 2(2), 65–68.

Bown, H. G., O'Brien, C. D., Sawchuk, W., and Storey, J. R. (1978, December). *A general description of Telidon: A Canadian proposal for videotex systems* (Technical Note No. 697-E). Toronto: Communications Research Centre, Department of Communications.

Branscomb, L. M. (1982). Electronics and computers: An overview. *Sciences*, 215, 755–760.

Branscomb, L. M., and Thomas, J. C. (1983). Ease of use: A system design challenge. In R.E.A. Mason (ed.), *Information processing 83*, 431–438. Amsterdam: Elsevier Science Publishers.

Branscomb, L. M., and Thomas, J. C. (undated). Ease of use: A system design challenge. Unpublished manuscript.

Bridgewater, J. (1954). Human factors in the design of electronic computers. *Computers and Automation*, 3(6), 6–7, 10, 17.

Briggs, G. W., and Schum, D. A. (1965). Automated bayesian hypothesis-selection in a simulated threat-diagnosis system. In J. Spiegel and D. E. Walker (eds.), *Information systems sciences: Proceedings of the Second Congress.* Washington, D.C.: Spartan Books.

Brooks, F. P., Jr. (1975). *The mythical man-month.* Reading, Massachusetts: Addison-Wesley.

Brooks, R. (1982, September). *Towards a theory of the comprehension of computer programs.* Shelton, Connecticut: International Telephone and Telegraph Advanced Technology Center.

Brown, B. S., Dismukes, K., and Rinalducci, E. J. (1982). Video display terminals and vision of workers. Summary and overview of symposium. *Behavior and Information Technology*, 1, 121–140.

Brown, C. M. (1984). Computer vision and natural constraints. *Science*, 224, 1299–1305.

Brown, C. M., Ellis, J. S., Feldman, J. A., LeBlanc, T. J., and Peterson, G. I. (1984–1985). Research with the Butterfly multicomputer. *Computer Science Engineering Research Review*, 3–25.

Brown, G. E., Jr. (1982). A congressional view of the coming information age. In R. A. Kasschau, R. Lachman, and K. R. Laughery (eds.), *Information technology and psychology: Prospects for the future.* New York: Praeger.

Brown, J. S. (1984). The low road, the middle road, and the high road. In P. H. Winston and K. A. Prendergast (eds.), *The AI business.* Cambridge, Massachusetts: MIT Press.

Brown, J. S., Burton, R. R., and de Kleer, J. (1982). Pedagogical natural language and knowledge engineering techniques. In D. Sleman and J. S. Brown (eds.), *Intelligent tutoring systems.* London: Academic Press.

Brown, J. S., Burton, R. R., Bell, A. G., and Bobrow, R. J. (1974). *SOPHIE: A sophisticated instructional environment* (Technical Report No. AFHRL-TR-74-93). Brooks Air Force Base, Texas: Air Force Human Resources Laboratory (NTIS No. AD A010109).

Brown, P. J. (1982). My system gives excellent error messages—or does it? *Software Practice and Experience, 12*(1), 91–94.

Brown, P. J. (1983, April). Error messages: The neglected area of the man/machine interface? *Communications of the ACM, 26*(4), 246–249.

Brown, R. (1977). Use of analogy to achieve new expertise (AI-TR-403). Boston: MIT, Artificial Intelligence Laboratory.

Browndi, T. R. (1983). The coming of age of "knowledge engineering." *Focus*, 56–60.

Browne, J. C. (1984, May). Parallel architectures for computer systems. *Physics Today*, 28–35.

Bryan, G. E. (1967). JOSS: 20,000 hours at a console—A statistical summary. *AFIPS Conference Proceedings*, 769–777. Montvale, New Jersey: AFIPS Press.

Buchanan, B. G., and Feigenbaum, E. A. (1978). DENDRAL and Meta-DENDRAL: Their applications dimension. *Artificial Intelligence, 11*, 5–24.

Buchanan, B. G., Duffield, A. M., and Robertson, A. V. (1971). An application of artificial intelligence to the interpretation of mass spectra. In G. W. A. Milne (ed.), *Mass spectrometry techniques and applications*, 121. New York: Wiley.

Burnette, K. T. (1972). Evaluating the man-display interface. *The Electronic Engineer, 31*(7), 64, 66–67.

Burstyn, H. P. (1983, December). Electronic mail comes of age in the 80's with low-cost computers. *Telecommunication Products and Technology*, 34–37.

Business Outlook (1984, July). Home users drive music software sales. *High Technology*, 64.

Butler, T. W. (1983, December). Computer response time and user performance. In A. Janda (ed.), *Proceedings of the CHI'83 Conference on Human Factors in Computing Systems*, 58–62. New York: ACM.

Button, K. F., and Gambino, S. R. (1973). Laboratory diagnosis by computer. *Computers in Biology and Medicine, 3*, 131–136.

Buzbee, B. L., Ewald, R. H., and Worlton, W. J. (1982, December). Japanese supercomputer technology. *Science*, 1189–1193.

Callahan, N. D., and Grace, G. L. (1967). AUTODOC: Computer-based assistance for document production. *Proceedings of the ACM National Meeting*, 117–185. New York: ACM.

Carbonell, J. R., Elkind, J. I., and Nickerson, R. S. (1968). On the psychological importance of time in a time sharing system. *Human Factors, 10*, 135–142.

Card, S. K., English, W. K., and Burr, B. J. (1978). Evaluation of mouse, rate-controlled isometric joystick, step keys, and text keys for text selection on a crt. *Ergonomics, 21*, 601–613.

Card, S. K., Moran, T. P., and Newell, A. (1980a). Computer text-editing: An information-processing analysis of a routine cognitive skill. *Cognitive Psychology, 12*, 32–74.

Card, S. K., Moran, T. P., and Newell, A. (1980b). The keystroke-level model for user performance time with interactive systems. *Communications of the ACM, 23*(7), 396–410.

Card, S. K., Moran, T. P., and Newell, A. (1983). *The psychology of human-computer interaction.* Hillsdale, New Jersey: Erlbaum Associates.

Card, W. I., Crean, G. P., Evans, C. R., James, B. W., Nicholson, M., Watkinson, G., and Wilson, J. (1970). On-line interrogation of hospital patients by a time-sharing terminal with computer/consultant comparison analysis. *Proceedings, Conference on Man-Computer Interaction* (Conference Publication No. 68), 141–147. London: Institution of Electrical Engineers.

Card, W. I., Nicholson, M., Crean, G. P., Watkinson, G., Evans, C. R., Wilson, J., and Russell, D. (1974). A comparison of doctor and computer interrogation of patients. *International Journal of Bio-Medical Computing, 5,* 175–187.

Carlisle, J. H. (1975). *A selected bibliography on computer-based teleconferencing* (Rep. No. SAI-75-560-WA). Arlington, Virginia: Science Applications, Inc.

Carlisle, J. H. (1976). Evaluating the impact of office automation on top management communication. *AFIPS Conference Proceedings,* 611–616. Montvale, New Jersey: AFIPS Press.

Carpenter, T. P., Corbitt, M. K., Kepner, H., Lindquist, M. M., and Reys, R. W. (1980). Problem solving in mathematics: National assessment results. *Educational Leadership, 37,* 562–563.

Carroll, J. M. (1982a, November). The adventure of getting to know a computer. *Computer Magazine,* 49–58.

Carroll, J. M. (1982b). Creative names for personal files in an interactive computing environment. *IJMMS, 16,* 405.

Carroll, J. M. (1982c). Learning, using, and designing filenames and command paradigms. *Behavior and Information Technology, 1*(4), 327–346.

Carroll, J. M. (1984). *Mental models and software human factors: An overview* (Report No. RC 10616 [47016]). Yorktown Heights, New York: IBM Watson Research Center.

Carroll, J. M., and Carrithers, C. (1984). Training wheels in a user interface. *Communications of the ACM, 27,* 800–806.

Carroll, J. M., and Mack, R. L. (1982a). *Learning to use a word processor: By doing, by thinking, and by knowing* (Research Report 9481). Yorktown Heights, New York: IBM Watson Research Center.

Carroll, J. M., and Mack, R. L. (1982b). *Metaphor, computing systems, and active learning* (Research Report 9636). Yorktown Heights, New York: IBM Watson Research Center.

Carroll, J. M., and Rosson, M. B. (1984). *Usability specifications as a tool in iterative development* (Report No. RC 10437 [46642]). Yorktown Heights, New York: IBM Watson Research Center.

Carroll, J. M., and Thomas, J. C. (1982). Metaphor and the cognitive representation of computing systems. *IEEE Transactions on Systems, Man, and Cybernetics, SMC-12*(2), 107–116.

Carroll, J. M., Thomas, J. C., and Malhotra, A. (1979). A clinical-experimental analysis of design problem solving. *Design Studies*, *12*, 84–92.

Caruso, D. E. (1970). Tutorial programs for operation of on-line retrieval systems. *Journal of Chemical Documentation*, *10*, 98–105.

Castleman, P. A., Russell, C. H., Webb, F. N., Hollister, C. A., Siegel, J. R., Zdonik, S. R., and Fram, D. M. (1974). The implementation of the PROPHET system. *AFIPS Conference Proceedings*, *43*, 457–468.

Castleman, P. A., Whitehead, S. F., Sher, L. D., Hantman, L. M., and Massey, L. D., Jr. (1974). *An assessment of the utility of computer aids in the physician's office* (Report No. 3096). Cambridge, Massachusetts: Bolt Beranek and Newman Inc.

Cavanagh, J. M. A. (1967). Some considerations relating to user-system interaction in information retrieval systems. In A. B. Tonik (ed.), *Information retrieval: The user's viewpoint: An aid to design*. Fourth Annual National Colloquium on Information Retrieval. Philadelphia: International Information, Inc.

CENTACS (1980). *Standardized computer resource interface and management plan*. Ft. Monmouth, New Jersey: Center for Tactical Computer Systems, U.S. Army Communications Research and Development Command.

Cerf, V. (1978). The electronic mailbox: A new communication tool for the hearing impaired. *American Annals of the Deaf*, *123*(6), 768–772.

Chamberlain, R. G. (1975). Conventions for interactive computer programs. *Interfaces* (Pt. 1), *6*(1), 77–82.

Chandrasekaran, B. (1984). Expert systems: Matching techniques to tasks. In W. Reitman (ed.), *Artificial intelligence applications for business*. Norwood, New Jersey: Ablex Publishers.

Chapanis, A. (1965). On the allocation of functions between men and machines. *Occupational Psychology*, *39*, 1–11.

Chapanis, A. (1973). The communication of factual information through various channels. *Information Storage and Retrieval*, *9*, 215–231.

Chapanis, A. (1975). Interactive human communication. *Scientific American*, *232*(3), 36–42.

Chapanis, A. (1981). Interactive human communication: Some lessons learned from laboratory experiments. In B. Shackel (ed.), *Man-computer interaction; Human factors aspects of computers and people*. Rockville, Maryland: Sijthoff and Noordhoff.

Chapanis, A. (1982). Man/computer research at Johns Hopkins. In R. A. Kasschau, R. Lachman, and K. R. Laughery (eds.), *Information technology and psychology: Prospects for the future*. New York: Praeger.

Chapanis, A., and Overbey, C. M. (1974). Studies in interactive communication: III. Effects of similar and dissimilar communication channels and two interchange options on team problem solving [Monograph]. *Perceptual and Motor Skills*, *38*, 343–374.

Chapanis, A., Ochsman, R. B., Parrish, R. N., and Weeks, G. D. (1972). Studies in interactive communication: The effects of four communication modes on the behavior of teams during cooperative problem-solving. *Human Factors, 14*, 487–509.

Chapanis, A., Parrish, R. M., Ochsman, R. B., and Weeks, G. D. (1977). Studies in interactive communication: II. The effects of four communication modes on the linguistic performance of teams during cooperative problem solving. *Human Factors, 19*, 101–126.

Chaudhari, P., Giessen, B. C., and Turnbull, D. (1980). Metallic glasses. *Scientific American, 4*, 55–81.

Cheriton, D. R. (1976). Man-machine interface design for time-sharing systems. *Proceedings of the ACM National Conference*, 362–380. New York: ACM.

Chomsky, C. (1984, January). Finding the best language arts software. *Classroom computer learning*, 61–63.

Christ, R. E. (1975). Review and analysis of color coding research for visual displays. *Human Factors, 17*(6), 542–570.

Chu, W. W. (1974). *Advances in computer communications*. Dedham, Massachusetts: Artech House.

Citrenbaum, R. L. (1972). Planning and artificial intelligence. In H. Sackman and R. L. Citrenbaum (eds.), *Online planning: Towards creative problem solving*. Englewood Cliffs, New Jersey: Prentice-Hall.

Clancey, W. J. (1979). Transfer of rule-based expertise through a tutorial dialogue. Ph.D. Thesis, Department of Computer Science, Stanford University.

Clancey, W. J. (1981). Methodology for building an intelligent tutoring system (Report No. STAN-CS-81-894). Stanford, California: Stanford University.

Clancey, W. J., and Letsinger, R. (1981). Neomycin: Reconfiguring a rule-based expert system for application to teaching. *Proceedings of the Seventh IJCAI*, 829–836.

Clarke, D. C. (1970). Query formulation for on-line reference retrieval: Design considerations from the indexer/searcher viewpoint. *Proceedings of the ASIS, 7*, 83–86.

Clayton, A., and Nisenoff, N. (1976). *A forecast of technology for the scientific and technical information communities*. 4 vols. Arlington, Virginia: Forecasting International, pp. 253, 937.

Clement, J. (1982). Algebra word problem solutions: Thought processes underlying a common misconception. *Journal for Research in Mathematics Education, 13*, 16–30.

Clement, J., Lochhead, J., and Monk, G. S. (1981). Translation difficulties in learning mathematics. *American Mathematical Monthly, 88*(4), 286–290.

Clement, J., Lochhead, J., and Soloway, E. (1979). Translating between symbol systems: Isolating a common difficulty in solving algebra word problems. COINS Technical Report No. 79-19, Department of Computer and Information Science, University of Massachusetts, Amherst.

Codd, E. F. (1974). Seven steps to rendezvous with the casual user. In J. W. Klimbie and K. L. Koffeman (eds.), *Data base management: Proceedings of the*

*International Federation for Information Processing TC-2 Working Conference on Data Base Management Systems*. Amsterdam: North-Holland.

Cohen, D. (1981, October). On holy wars and a plea for peace. *Computer*, 48–54.

Cole, R., Higginson, P., Lloyd, P., and Moulton, R. (1983, June). International net faces problems handling mail and file transfer. *Data Communications*, 175–187.

Coleman, M. L. (1969). Text editing on a graphic display device using hand-drawn proofreader's symbols. In M. Faiman and J. Nievergelt (eds.), *Pertinent concepts in computer graphics: Proceedings of the Second University of Illinois Conference on Computer Graphics*. Urbana: University of Illinois.

The College Board (1983). *Academic preparation for college*. New York: College Entrance Examination Board.

*College Bound Seniors* (1979). New York: College Entrance Examination Board.

Collen, M. F. (ed.) (1970). *Medical information systems*. Springfield, Virginia: National Technical Information Services.

*Computer-Disability News* (1984a). *Vol. 1*. Chicago: National Easter Seal Society.

*Computer-Disability News* (1984b). Featured product: EyeTyper allows eye gaze to run computers. Winter 1984–1985, *1*(3), 5.

*Computer-Disability News* (1984c). Latest in computerized environmental systems: A house that obeys voice commands. Winter 1984–1985, *1*(3), 1.

Conklin, N. F., and Reder, S. M. (1984). The economy of communicative work: a channel analysis of electronic mail usage. *Proceedings of the First IEEE International Conference on Office Automation*.

Coombs, M. J., Gibson, R., and Alty, J. L. (1982). Learning a first computer language: Strategies for making sense. *IJMMS*, *16*, 449–486.

Cortes, C. C. (1983). Business computer graphic systems and users: Who, what, where, why—how much? *Computer Pictures*, *1*(6), 8–17.

Costa, A. L. (1981). Teaching for intelligent behavior. *Educational Leadership*, *39*, 29–32.

Covington, M. V., Crutchfield, R. S., Davies, L., and Olton, R. M. (1974). *The productive thinking program: A course in learning to think*. Columbus, Ohio: Merrill.

Cropper, A. G., and Evans, S. J. W. (1968). Ergonomics and computer display design. *The Computer Bulletin*, *12*(3), 94–98.

Crowther, W., Goodhue, J., Starr, E., Thomas, R., Milliken, W., and Blackadar, T. (undated). *Performance measurements on a 128-node butterfly* (TM parallel processor). Cambridge, Massachusetts: BBN Laboratories.

Cuadra Associates (1984). *Directory of online databases*. Vol. 6. Santa Monica, California: Cuadra Associates.

Cuff, R. N. (1980). On casual users. *IJMMS*, *12*, 163–187.

Cullingford, R. E. (1978). *Script application: Computer understanding of newspaper stories*. Ph.D. Thesis, Department of Computer Science, Yale University, New Haven, Connecticut.

Cumberbatch, J., and Heaps, H. S. (1973). Applications of a non-Bayesian approach to computer-aided diagnosis for upper abdominal pain. *International Journal of Biomedical Computing*, *4*, 105–115.

Cushman, J. R. (1983). Air-land battle mastery and C3 systems for the multinational field commander. *Signal*, *37*, 45–51.

Cushman, R. H. (1972). TOFT: A method for electronic doodling and a first step towards the use of computers on ill-defined problems. *Proceedings of the 1972 International Conference on Cybernetics and Society*, 157–162. New York: IEEE.

Cyert, R. M. (1983). Personal computing in education and research. *Science*, *22*(22), 569.

Dahl, O. J., Dijkstra, E. W., and Hoare, C. A. R. (1972). *Structured programming*. New York: Academic Press.

Dainoff, M. J. (1982). Occupational stress factors in visual display terminal (VDT) operation: A review of empirical research. *Behavior and Information Technology*, *1*(2), 141–176.

Dainoff, J. J., Happ, A., and Crane, P. (1981, August). Visual fatigue and occupational stress in VDT operators. *Human Factors*, *23*(4).

Dalbey, J., and Linn, M. C. (1984, April). Making programming instruction cognitively demanding. In K. Sheingold (chair), *Development studies of computer programming skills*. Symposium conducted at the American Educational Research Association, New Orleans.

Damodaran, L. (1981). The role of user support. In B. Shackel (ed.), *Mancomputer interaction: Human factors aspects of computers and people*. Rockville, Maryland: Sijthoff and Noordhoff.

Danchak, M. M. (1976). CRT displays for power plants. *Instrumentation Technology*, *23*(10), 29–36.

Dansereau, D. F. (1978). The development of a learning strategy curriculum. In H. F. O'Neil, Jr. (ed.) *Learning strategies*. New York: Academic Press.

Dansereau, D. F., and Holley, C. D. (1982). Development and evaluation of a text mapping strategy. In A. Flammer and W. Kintsch (eds.), *Discourse processing*. Amsterdam: North-Holland.

Davies, D. W. (1982). Communication networks to service rapid response computers. *Proceedings of the International Congress for Information Processing: hardware applications*, 650–658

Davies, D. W., and Barber, D. L. A. (1973). *Communications network for computers*. New York: Wiley.

Davis, R. M. (1966). Man-machine communication. In C. A. Cuadra (ed.), *Annual review of information science and technology*, *1*. New York: Wiley.

Davis, R. (1982). Consultation, knowledge acquisition, and instruction: A case study. In P. Szolovits (ed.), *Artificial Intelligence in medicine*. Boulder, Colorado: Westview Press.

Davis, R. (1984). Amplifying expertise with expert systems. In P. H. Winston and K. A. Prendergast (eds.), *The AI business*. Cambridge, Massachusetts: MIT Press.

Davis, R., and King, J. (1977). An overview of production systems. In E. W. Elock and D. Michie (eds.), *Machine intelligence 8*. New York: Wiley.

Davis, R., Buchanan, B. G., and Shortliffe, E. H. (1977). Production rules as a representation for a knowledge-based consultation program. *Artificial Intelligence*, *8*(1), 15–45.

Davis, R., Austin, H., Carlbom, I., Frawley, B., Pruchnik, P., Sneiderman, R., and Gilreath, A. (1981). The dipmeter advisor: Interpretation of geological signals. *Seventh IJCAI*, Vancouver, British Columbia.

Defense Advanced Research Projects Agency (1983). *Strategic computing. New-generation computing technology: A strategic plan for its development and application to critical problems in defense*. U.S. Department of Defense.

DeGreene, K. B. (1970). Man-computer interrelationships. In K. B. De-Greene (ed.), *Systems psychology*. New York: McGraw-Hill.

deHaan, H. J. (1977, June). Speech-rate intelligibility/comprehensibility threshold for speeded and time-compressed connected speech. *Perception and Psychophysics*, *22*, 366–372.

deHaan, H. J., and Schjelderup, J. R. (1978, June). Threshold of intelligibility/comprehensibility of rapid connected speech: Method and instrumentation. *Behavior Research Methods and Instrumentation*, *10*, 841–844.

Denning, P. J. (1985). The science of computing: Computer networks. *American Scientist*, *73*(2), 127–129.

Dertouzos, M. L. (1980, October). Report of the Advisory Committee on Information Network Structure and Functions for the Executive Office of the President. Appendix F in *The future of electronic information handling at the FCC: Blueprint for the 80's*. Washington, D.C.: Federal Communications Commission.

Dertouzos, M. L., and Moses, J. (1980). *The computer age: A twenty-year view*. Cambridge, Massachusetts: MIT Press.

Deutsch, J. T., and Newton, A. R. (1984). A multiprocessor implementation of relaxation-based electrical circuit simulation. *Proceedings of the 21st Design Automation Conference, IEEE*, 350–357.

Dickson, D. (1983). Britain rises to Japan's computer challenge. *Science*, *220*, 799–800.

Dickson, D. (1984a). Europe seeks joint computer research effort. *Science*, *223*, 28–30.

Dickson, D. (1984b). Europeans back computer plan. *Science*, *223*, 1159.

Diebold, J. (1969). *Man and the computer*. New York: Praeger.

Doddington, G. R., and Schalk, T. B. (1981, September). Speech recognition: Turning theory to practice. *IEEE Spectrum*, *18*(9), 26–32.

Dodds, D. W., Jr. (1983). Being most things to most people: Tailorability, predictability, and reliability in the human interface of the Hermes message system. Bolt Beranek and Newman, Inc.

Doherty, W. J., Thompson, C. H., and Boies, S. J. (1972). An analysis of interactive system usage with respect to software, linguistic, and scheduling attributes. *Proceedings of the 1972 International Conference on Cybernetics and*

*Society*, 113–119. New York: IEEE. [Also, 1972 (Technical Report RC-3914). Yorktown Heights, New York: IBM Watson Research Center.]

Dolotta, T. A. (1970). Functional specifications for typewriter-like time-sharing terminals. *Computer Surveys*, *2*, 5–31.

Dooling, D. J., and Klemmer, E. T. (1982). New technology for business telephone users: Some findings from human factors studies. In R. A. Kasschau, R. Lachman, and K. R. Laughery (eds.), *Information technology and psychology: Prospects for the future*. New York: Praeger.

Doughty, J. M. (1967). The AESOP testbed: Test series 1/2. In D. E. Walker (ed.), *Information system science and technology*. Washington, D. C.: Thompson.

Douglas, S. A., and Moran, T. P. (1983, December). Learning text editor semantics by analogy. In A. Janda (ed.), *Proceedings of the CHI'83 Conference on Human Factors in Computing Systems*, 207–211. New York: ACM.

Draper, S. W., and Norman, D. A. (1984, March). Software engineering for user interfaces. *Proceedings of the Seventh International Conference on Software Engineering*, Orlando, Florida.

Dreyfus, H. L. (1972). *What computers can't do: A critique of artificial reason*. New York: Harper and Row.

Duda, R. O., and Shortliffe, E. H. (1983). Expert systems research. *Science*, *220*, 261–268.

Duda, R. O., Gaschnig, J., and Hart, P. E. (1979). Model design in the Prospector consultant system for mineral exploration. In D. Michie (ed.), *Expert system in the microelectronic age*. Edinburgh: Edinburgh University Press.

Duda, R. O., Hart, P. E., Nilsson, N. J., and Sutherland, G. L. (1978). Semantic network representations in rule-based inference systems. In B. A. Waterman and F. Hayes-Roth (eds.), *Pattern-directed inference systems*. New York: Academic Press.

Dumais, S. T., and Landauer, T. K. (1983). Using examples to describe categories. In A. Janda (ed.), *Proceedings of the CHI'83 Conference on Human Factors in Computing Systems*, 112–115. New York: ACM.

Duncan, J., and Ferguson, D. (1974). Keyboard operating posture and symptoms in operating. *Ergonomics*, *17*, 651–662.

Dunn, R. M. (1976). Hardware for wideband interaction for computer graphics. *Proceedings, IEEE International Conference on Cybernetics and Society*, 443–446. New York: IEEE.

Dunn, S. L. (1979). The case of the vanishing colleges. *The Futurist*, *13*(5), 385–393.

Dunsmore, H. E. (1983). Human factors in computer programming. In B. H. Kantowitz and R. D. Sorkin (eds.), *Human factors: Understanding people-system relationships*. New York: Wiley.

Dwyer, B. (1981a). Programming for users: A bit of psychology. *Computers and People*, *30*, 1 and 2, 11–14, 26.

Dwyer, B. (1981b). A user-friendly algorithm. *Communications of the ACM*, *24*(9), 556–561.

Eason, K. D. (1981). A task-tool analysis of manager-computer interaction. In

B. Shackel (ed.), *Man-computer interaction: Human factors aspects of computers and people*, 289–307. Rockville, Maryland: Sijthoff and Noordhoff.

Eason, K. D. (1982). The process of introducing information technology. *Behavior and Information Technology*, *1*(2), 197–213.

Eason, K. D., Damodaran, L., and Stewart, T. F. M. (1975). Interface problems in man-computer interaction. In E. Mumford and H. Sackman (eds.), *Human choice and computers*, 91–105, Amsterdam: North-Holland.

Edwards, B. J. (1965). Probabilistic information processing system for diagnosis and action selection. In J. Spiegel and D. Walker (eds.), *Information system sciences: Proceedings of the Second Congress*. Washington, D.C.: Spartan Books.

Edwards, B. J. (1984). Dynamic alternatives to the keyboard. *Hardcopy*, *13*(10), 34–44.

Edwards, W. (1962). Men and computers. In R. M. Gagne and A. W. Melton (eds.), *Psychological principles in system development*. New York: Holt, Rinehart and Winston.

Ehardt, J. L. (1983). Apple's Lisa: A personal office system. *Seybold Report on Office Systems*, *6*.

Ehrenreich, S. L. (1980). *Design recommendations for query languages* (Report No. TR 484). Alexandria, Virginia: U.S. Army Research Institute for the Behavioral and Social Sciences.

Ehrenreich, S. L. (1981). Query languages: Design recommendations derived from the human factors literature. *Human Factors*, *23*, 709–725.

Ehrlich, K., Soloway, E., and Abbott, V. (1982). *Transfer effects from programming to algebra word problems: A preliminary study* (Technical Report No. 257). Department of Computer Science, Yale University, New Haven, Connecticut.

Ehrlich, K., Abbott, V., Salter, B., and Soloway, E. (1984, April). Issues and problems in studying transfer effects from programming. In K. Sheingold (chair), *Development studies of the computer programming skills*. Symposium conducted at the American Educational Research Association, New Orleans.

Embley, D. W., and Nagy, G. (1981). Behavioral aspects of text editors. *Computer Surveys*, *13*, 33–70.

Engel, S. E., and Granda, R. E. (1975, December). *Guidelines for man/display interfaces* (Technical Report No. TR 00.2720). Poughkeepsie, New York: IBM Poughkeepsie Laboratory.

Engelbart, D. C., and English, W. K. (1968). A research center for augmenting human intellect. *AFIPS Conference Proceedings*, *33*, 395–410.

Engelmore, R., and Nii, H. P. (1977). *A knowledge-based system for the interpretation of protein X-ray crystallographic data* (Memo HPP-77-2). Stanford, California, Department of Computer Science.

English, W. K., Engelbart, D. C., and Berman, M. L. (1967). Display-selection techniques for text manipulation. *Institute of Radio Engineers Transactions on Human Factors in Electronics*, *HFE-8*, 5–15.

English, W. K., Engelbart, D. C., and Huddart, B. (1965, July). *Computer-aided display control* (Final Report, Contract NAS-3988). Menlo Park, California: Stanford Research Institute (NASA No. CR-66111, NTIS No. N66-30204).

Epstein, R. H. (1981). An approach to introducing and evaluating automated office systems. *Electronic Office: Management and Technology*. Auerbach Publishers.

Epstein, R. H. (1982). The wizard of measuring users of computer-based office systems. *AFIPS Office Automation Conference Digest*, 90–95.

Ernst, M. L. (1982). The mechanization of commerce. *Scientific American*, *247*(3), 132–145.

Evanczuk, S., and Manuel, T. (1983, December). Artificial intelligence: Practical systems use natural languages and store human expertise. *Electronics*, 139.

Evans, C. (1979). *The micromillennium*. New York: Washington Square Press.

Fain, J., Hayes-Roth, F., Sowizral, H., and Waterman, D. (1982, February). Programming in ROSIE: An introduction by means of examples. Santa Monica, California: Rand Publications (No. N-1646-ARPA).

Fairbanks, G., Guttman, N., and Miron, M. (1957). Effects of time compression upon the comprehension of connected speech. *Journal of Speech and Hearing Disorders*, *22*, 10–19.

Falk, H. (1975). Technology forecasting I and II (Communication and Computers). *IEEE Spectrum*, *12*(4), 42–49.

Falk, H. (1976). Reaching for a gigaflop. *IEEE Spectrum*, *13*(10), 64–69.

Feigenbaum, E. A. (1983). Knowledge engineering: The applied side. In J. E. Hayes and D. Michie (eds.), *Intelligent systems: The unprecedented opportunity*. New York: Halstead Press.

Feigenbaum, E. A., and McCorduck, P. (1983). *The fifth generation*. Reading, Massachusetts: Addison-Wesley.

Feldman, M. B., and Rogers, G. T. (1982). Toward the design and development of style-independent interactive systems. *Proceedings of the Human Factors in Computer Systems Conference*, 111–113.

Ferguson, D., and Duncan, J. (1974). Keyboard design and operating posture. *Ergonomics*, *17*, 731–744.

Fernbach, S. (1984, December). Applications of supercomputers in the U.S.—today and tomorrow. In K. Hwang (ed.), *Supercomputers: Design and applications—IEEE Computer Society*, 421–428.

Feurzeig, W. (1964, June). Conversational teaching machine. *Datamation*.

Feurzeig, W., Horwitz, P., and Nickerson, R. S. (1981, October). *Microcomputers in education* (Report No. 4798). Cambridge, Massachusetts: Bolt Beranek and Newman Inc.

Feurzeig, W., Munter, P., Swets, J., and Breen, M. (1964). Computer-aided teaching in medical diagnosis. *Journal of Medical Education*, *39*, 746–754.

Feynman, R. P. (1984, June). *Quantum mechanical computers*. Paper presented to IQEC-CLEO Meeting, Anaheim. Pasadena: California Institute of Technology.

Fialka, J. (1980, December). Can the U.S. Army fight? *Washington Star*, Washington, D.C.

Fields, A. F., Maisano, R. E., and Marshall, C. F. (1978). *A comparative analysis of methods for tactical data inputing* (Technical Paper 327). Alexandria, Virginia: U.S. Army Research Institute for the Behavioral and Social Sciences (NTIS No. AD A060 562).

Fields, C., and Negroponte, N. (1977). Using new clues to find data. *Proceedings of the Third International Conference on Very Large Data Bases*. New York: IEEE Press.

Finkelman, J. M. (1976). Information processing load as a human factors criterion for computer systems design. In R. E. Granda and J. M. Finkelman (eds.). *The role of human factors in computers*, 1–6. Proceedings of a symposium cosponsored by the Metropolitan Chapter of the Human Factors Society and Baruch College, City University of New York, New York.

Finman, J., Fram, D. M., Kush, T., and Russell, C. H. (1983, November). An electronic laboratory notebook for VAX, PDP-11, and professional 350. *DEC Professional*, 120–125.

Fischer, M., Fox, R. I., and Newman, A. (1973). A computer diagnosis of acutely ill patient with fever and rash. *International Journal of Dermatology*, *12*, 59–63.

Fitts, P. M. (1962, January). Functions of man in complex systems. *Aerospace Engineering*, 34–39.

Flanagan, J. L. (1972). *Speech analysis, synthesis and perception*. New York: Springer-Verlag.

Flanagan, J. L. (1976). Computers that talk and listen: Man-machine communication by voice. *Proceedings of the IEEE*, *64*, 405–415.

Flanagan, J. L., et al. (Committee on Computerized Speech Recognition Technologies) (1984). *Automatic speech recognition in severe environments*. Washington, D.C.: National Academy Press.

Fleiss, J. L., Spitzer, R. L., Cohen, J., and Endicott, J. (1972). Three computer diagnosis methods compared. *Archives of General Psychiatry*, *27*, 643–649.

Fletcher, H. (1929). *Speech and hearing*. New York: Van Nostrand.

Foley, J. D., and van Dam, A. (1982). *Fundamentals of interactive computer graphics*. Reading, Massachusetts: Addison-Wesley.

Foley, J. D., and Wallace, V. L. (1974). The art of natural graphic man-machine conversation. *Proceedings of the IEEE*, *62*, 462–471.

Folley, L., and Williges, R. (1982, March). User models of text editing command languages. In *Proceedings of Human Factors in Computer Systems Conference*. Gaithersburg, Maryland: National Bureau of Standards.

Forgie, J. W. (1975). Speech transmission in packet-switched store-and-forward networks. *Proceedings of the National Computer Conference*, 137–141.

Forgie, J. W., Feehrer, C. E., and Weene, P. L. (1979, March). *Voice conferencing technology program* (Technical Report No. ESD-TR-79-78). Lexington, Massachusetts: Lincoln Laboratory, Massachusetts Institute of Technology.

Forsdick, H. C., and Thomas, R. H. (1982, October). *The design of Diamond— A distributed multimedia document system* (BBN Report No. 5204). Cambridge, Massachusetts: Bolt Beranek and Newman Inc.

Foss, D., Rosson, M. B., and Smith, P. (1982). Reducing manual labor: An experimental analysis of learning aids for a text editor. *Proceedings of Human Factors in Computer Systems Conference.* Gaithersburg, Maryland: National Bureau of Standards.

Foulke, E. (1971). The perception of time-compressed speech. In D. Horton and J. Jenkins (eds.), *The perception of language.* Columbus, Ohio: Merrill.

Fox, J (1977). Medical computing and the user. *IJMMS, 9,* 669–686.

Frank, H., Kahn, R. E., and Kleinrock, L. (1972). Computer communications network design—experience with theory and practice. *AFIPS Conference Proceedings, 40,* 255–270.

Frankhuizen, J. L., and Vrins, T. G. M. (1980). Human factors studies with Viewdata. *Proceedings of the Ninth International Symposium on Human Factors in Telecommunications,* Holmdel, New Jersey.

Franklin, J., and Dean, E. (1974, May–June). Some expected and not so expected reactions to a computer-aided design with interactive graphics (candid) system. *Society for Information Display Journal,* 5–13.

Frase, L. T. (1983). The UNIX writer's workbench software: Philosophy. *Bell System Technical Journal, 62*(6), 1883–1890.

Frederiksen, J. R. (1975, July). *Survey of the state of the art in human factors in computers* (Technical Report No. SAI-75-533-WA). Arlington, Virginia: Science Applications.

Freedy, A., and Weltman, G. (1974). Adaptive computer aiding in dynamic decision processes. In K. S. Fu and J. T. Tou (eds.), *Learning systems and intelligent robots.* New York: Plenum.

Freedy, A., Davis, K. B., Steeb, R., Samet, M. G., and Gardiner, P. C. (1976, August). *Adaptive computer aiding in dynamic decision processes: Methodology, evaluation, and applications* (Technical Report No. PFTR-1016-76-8/30). Woodland Hills, California: Perceptronics, Inc.

Furnas, G. W., Landauer, T. K., Gomez, L. M., and Dumais, S. T. (1983). Statistical semantics: Analysis of the potential performance of keyword information systems. In J. C. Thomas and M. Schneider (eds.), *Human factors and computer systems.* Norwood, New Jersey: Ablex Press.

Gade, P. A., and Gertman, D. (1979, August). *Listening to compressed speech: The effects of instructions, experience, and preference (Technical Paper 369).* Alexandria, Virginia: Army Research Institute for the Behavioral and Social Sciences.

Gaines, B. R., and Facey, P. V. (1975, June). Some experience in interactive system development and application. *Proceedings of the IEEE, 63*(6), 894–911.

Galambos, J. A., Wikler, E. S., Black, J. B., and Sebrechts, M. M. (1983, December). How you tell your computer what you mean: Ostension in interactive systems. In A. Janda (ed.), *Proceedings of the CHI'83 Conference on Human Factors in Computing Systems,* 182–185. New York: ACM.

Galitz, W. O. (1981). *Handbook of screen for format design.* Wellesley, Massachusetts: Q.E.D. Information Science, Inc.

Gannon, J. D. (1976). An experiment for the evaluation of language features. *IJMMS, 8,* 61–73.

Gardner, D. P., et al. (National Commission on Excellence in Education) (1983). *A nation at risk: The imperative for educational reform.* Washington, D.C.: U.S. Department of Education.

Gentner, D. (1980). The structure of analogical models in science (BBN Technical Report No. 4451). Cambridge, Massachusetts: Bolt Beranek and Newman Inc.

Gergen, M., and Hagen, D. (eds.) (1985). Computer technology for the handicapped. *Proceedings of the 1984 Closing the Gap Conference.* Henderson, Minnesota: Closing the Gap.

Gershman, A. (1982, July). Building a geological expert system for dipmeter interpretation. *Proceedings of the European Conference on Artificial Intelligence.*

Getty, D. J. (ed.) (1982). *Three-dimensional displays: Perceptual research and applications to military systems.* Washington, D.C.: U.S. Naval Air Systems Command.

Gibbons, J. H. (Director, Office of Technology Assessment) (1981, October). *Computer-based national information systems: Technology and public policy issues summary* (Report No. OTA-CIT-147). Washington, D.C.: Office of Technology Assessment.

Gilb, T. (1977). *Software metrics.* Cambridge, Massachusetts: Winthrop Publishers.

Gingrich, P. S. (1983). The UNIX Writer's Workbench software: Results of a field study. *Bell System Technical Journal, 62*(6), 1909–1921.

Ginzburg, E. (1982). The mechanization of work. *Scientific American, 247*(3), 66–75.

Giuliano, V. E. (1974). In defense of natural language. *Proceedings of the ACM National Conference.* New York: ACM.

Giuliano, V. E. (1982). The mechanization of office work. *Scientific American, 247*(3), 148–164.

Glass, A. M. (1984). Materials for optical information processing. *Science, 226,* 657–662.

Glazebrook, R. R. (1984). "Saving" literary classics with software. *Hardcopy, 13*(7), 59–64.

Gledhill, V. X., Mathews, J. D., and Mackay, J. R. (1972). Computer-aided diagnosis: A study of bronchitis. *Methods of Information in Medicine, II,* 228–232.

Gold, M. M. (1969). Time-sharing and batch-processing: An experimental comparison of their values in a problem-solving situation. *Communications of the ACM, 12,* 249–259.

Golden, D. (1980). A plea for friendly software. *Software Engineering Notes, 4* and *5.*

Goldenberg, E. P. (1979). *Special technology for special children.* Baltimore: University Park Press.

Goldstine, H. H. (1972). *The computer from Pascal to von Neumann.* Princeton, New Jersey: Princeton University Press.

Goodlad, J. I. (1983). *A place called school: Prospects for the future.* Heightstown, New Jersey: McGraw-Hill.

Goodman, T. J., and Spence, R. (1981). The effect of computer system response time variability on interactive graphical problem solving. *IEEE Transactions on Systems, Man, and Cybernetics, SMC-11,* 207–216.

Goodwin, N. C. (1974, March). *Intro: In which a smart terminal teaches its own use* (Technical Paper MTP-150). Bedford, Massachusetts: Mitre Corporation.

Goodwin, N. C. (1975). Cursor positioning on an electronic display using lightpen, lightgun, or keyboard for three basic tasks. *Human Factors, 17,* 289–295.

Gould, J. D. (1968). Visual factors in the design of computer-controlled CRT displays. *Human Factors, 10,* 359–375.

Gould, J. D. (1975). Some psychological evidence on how people debug computer programs. *IJMMS, 7,* 151–182.

Gould, J. D., and Ascher, R. N. (1975, February 20). Use of an IQF-like query language by non-programmers (IBM Research Report RC 5279).

Gould, J. D., and Boies, S. J. (1983). Human factors challenges in creating a principal support office system—The speech filing system approach. *ACM Transactions on Office Information Systems.*

Gould, J. D., and Drongowski, P. (1974). An exploratory study of computer program debugging. *Human Factors, 16,* 258–277.

Gould, J. D., and Lewis, C. (1983, December). Designing for usability—key principles and what designers think. In A. Janda (ed.), *Proceedings of the CHI'83 Conference on Human Factors in Computing Systems,* 50–53. New York: ACM.

Gould, J. D., Conti, J., and Hovanyecz, T. (1983, April). Composing letters with a simulated listening typewriter. *Communications of the ACM, 26*(4), 295–308.

Granda, R. E. (1977). Some considerations in defining the role of human factors in computers. In R. E. Granda and J. M. Finkelman (eds.), *The role of human factors in computers: Proceedings of a symposium co-sponsored by the Metropolitan Chapter of the Human Factors Society and Baruch College, City University of New York.* New York: Human Factors Society, Metropolitan Chapter.

Grant, E. E. (1966, May). *An empirical comparison of on-line and off-line debugging* (Report No. SP-2441). Santa Monica, California: System Development Corporation (NTIS No. AD 633907).

Grant, E. E., and Sackman, H. (1967). An exploratory investigation of programmer performance under on-line and off-line conditions. *IEEE Transactions on Human Factors in Electronics, HFE-8,* 33–48.

Green, B. F., Jr. (1963). *Digital computers in research.* New York: McGraw-Hill.

Green, B. F., Wolf, A. K., Chomsky, C., and Laughery, K. (1961). Baseball: An automatic question-answer. *Proceedings of the Western Joint Computer Conference, 19,* 219–224.

Green, T. R. G., Payne, S. J., and Van der Veer (eds.) (1983). *The psychology of computer use.* London: Academic Press.

Greenblatt, D., and Waxman, J. (1978). A study of three database query languages. In B. Schneiderman (ed.), *Databases: Improving usability and responsiveness.* New York: Academic Press.

Greene, R. A. (1969). *A computer-based system for medical record keeping by physicians.* Harvard University.

Gregory, D. S. (1969). Compressed speech—the state of the art. *IEEE Transactions on Engineering, Writing, and Speech, EWS-12,* 12–17.

Greist, J. H., Klein, M. H., and Van Cura, L. J. (1973). A computer interview for psychiatric patients. *Archives of General Psychiatry, 29,* 247–253.

Grether, W. F., and Baker, C. A. (1972). Visual presentation of information. In H. P. Van Cott and R. G. Kinkade (eds.), *Human engineering guide to equipment design,* revised edition. Washington D.C.: U.S. Government Printing Office.

Grignetti, M. C., and Miller, D. C. (1970). Modifying computer response characteristics to influence command choice. *Proceedings, Conference on Man-Computer Interaction,* 201–205 (Publication No. 68). London: Institution of Electrical Engineers.

Grignetti, M. C., Miller, D. C., Nickerson, R. S., and Pew, R. W. (1971). *Information processing models and computer aids for human performance: Task 2: Human-computer interaction models* (Technical Report No. AFOSR-TR-71-2845) (NTIS No. AD732913).

Grignetti, M., Myer, T., Nickerson, R. S., and Rubenstein, R. (1977, December). *Computer aided communications for the deaf* (Report No. 3738). Cambridge, Massachusetts: Bolt Beranek and Newman Inc.

Grossberg, M., Wiesen, R. A., and Yntema, D. B. (1976, March). An experiment on problem solving with delayed computer responses. *IEEE Transactions, Man and Cybernetics,* 219–222.

Grupp, P. (1984, March). Microsoft's word. *Business Computing, 2,* 58–60.

GTE Telenet (1984). *Directory of computer-based services.* Vienna, Virginia: GTE Telenet Communications.

Gunn, T. G. (1982). The mechanization of design and manufacturing. *Scientific American, 247*(3), 114–130.

Haider, M., Kundi, M., and Weisenbock, M. (1980). Worker strain related to VDT with differently colored characters. In E. Grandjean and E. Vigliani (eds.), *Ergonomic aspects of visual display terminals.* London: Taylor and Francis.

Halasz, F., and Moran, T. P. (1982, March). Analogy considered harmful. *Proceedings of the Human Factors in Computer Systems Conference,* 383–386. Gaithersburg, Maryland: National Bureau of Standards.

Haley, P, Kowalski, J., McDermott, J., and McWhorter, R. (1983). PTRANS: A rule-based management assistant (Technical Report). Pittsburgh, Pennsylvania: Carnegie Mellon University, Department of Computer Science.

Halpern, M. (1967, March). Foundations of the case for natural-language programming. *IEEE Spectrum, 4*(3), 140–149.

Hammond, N. V., Long, J. B., Clark, I. A., Barnard P. J., and Morton, J. (1980). Documenting human-computer mismatch in the interactive system.

*Proceedings of the Ninth International Symposium on Human Factors in Telecommunications*, Holmdel, New Jersey, 17–24.

Hammond, N., Jorgensen, A., MacLean, A., Barnard, P., and Long, J. (1983, December). Design practice and interface usability: Evidence from interviews with designers. In A. Janda (ed.), *Proceedings of the CHI'83 Conference on Human Factors in Computing Systems*, 40–44. New York: ACM.

Hanes, L. F. (1975). Human factors in international keyboard arrangement. In A. Chapanis (ed.), *Ethnic variables in human factors engineering*. Baltimore: Johns Hopkins University Press.

Hanes, R. M., and Gebhard, J. W. (1966, September). The computer's role in command decision. *U.S. Naval Institute Proceedings, 92*(9), 60–68.

Hansen, W. J. (1971). User engineering principles for interactive systems. *Proceedings of the Fall Joint Computer Conference, 39*, 523–532. Montvale, New Jersey: AFIPS Press.

Haralambopoulos, G., and Nagy, G. (1977). Profile of a university computer user community. *IJMMS, 9*, 287–313.

Harlan, J. R. (1976, September). The plants and animals that nourish man. *Scientific American*, 89–97.

Hart, P. E., Duda, R. O., and Einaudi, M. T. (1978). PROSPECTOR—A computer-based consultation system for mineral exploration. *Mathematical Geology, 10*(5).

Harwood, K. A. (1955). Listenability and rate of presentation. *Speech Monographs, 22*, 57–59.

Hawkridge, D. (1983). *New information technology in education*. Baltimore: Johns Hopkins University Press.

Hawkridge, D., Vincent, T., and Hales, G. (eds.) (1985). *Handbook of microcomputers in special education*. San Diego: College-Hill.

Hayes, J. R. (1981). *The complete problem solver*. Philadelphia: Franklin Institute Press.

Hayes-Roth, F. (1984, October). Knowledge based expert systems. *Computer, 17*(10), 263–273.

Heady, E. O. (1976, September). The agriculture of the U.S. *Scientific American*, 107–123.

Heart, F. E. (1975, September). The ARPANET network. In R. L. Grimsdale and F. F. Kuo (eds.), *Computer communication networks: 1973 Proceedings of the NATO Advanced Study Institute*. Leyden, The Netherlands: Noordhoff International Publishing.

Heart, F., Kahn, R. E., and Kleinrock, L. (1972). Network design—experience with theory and practice. *Proceedings of the Spring Joint Computer Conference*, 255–270. Montvale, New Jersey: AFIPS Press.

Heart, F., Kahn, R. E., Ornstein, S., Crowther, W., and Walden, D. (1970). The interface message processor for the ARPA computer network. *Proceedings of the Spring Joint Computer Conference*, 551–567. Montvale, New Jersey: AFIPS Press.

Heart, F., McKenzie, A., McQuillan, J., and Walden, D. (1978, January). *ARPANET completion report*. Cambridge, Massachusetts: Bolt Beranek and Newman Inc.

Hedberg, B. (1970). *On man-computer interaction in organizational decision making: A behavioral approach*. Gothenburg University.

Heglin, H. J., Saben, R., and Driver, L. L. (1972). Digital message entry device (DMED): A human factors analysis. In *Proceedings of the 16th Annual Meeting of the Human Factors Society*, 403–409. Santa Monica, California: Human Factors Society.

Heidorn, G. E. (1972). Natural language inputs to a simulation programming system. Monterey, California: Naval Postgraduate School (Technical Report No. NPS-55HD72101A).

Heidorn, G. E. (1975). Augmented phrase structure grammars. In B. L. Nash-Webber and R. C. Schank (eds.), *Theoretical issues in natural language processing*. Association for Computational Linguistics.

Heidorn, G. E., Jensen, K., Miller, L. A., Byrd, R. J., and Chodorow, M. S. (1982). The EPISTLE text-critiquing system. *IBM Systems Journal, 21*(3), 305–326.

Helps, F. G. (1970). Minimizing human problems when introducing automation. *Applied Ergonomics, 1*, 130–133.

Hemingway, P. W., Kubala, A. L., and Chastain, G. D. (1979, May). *Study of symbology for automated graphic displays* (Technical Report No. 79-A18). Alexandria, Virginia: Army Research Institute.

*High Technology* (1980). Videodisc battle coming for market that may top $3 billion by 1985. *High Technology, 1*(5), 31–39.

Hill, I. D. (1972). Wouldn't it be nice if we could write computer programs in ordinary English—or would it? *Honeywell Computer Journal, 6*(2), 76–83.

Hillis, W. D. (1981a). *The connection machine* (Report AIM-646). Cambridge, Massachusetts: Artificial Intelligence Laboratory, Massachusetts Institute of Technology.

Hillis, W. D. (1981b). A high resolution imaging touch sensor. *International Journal of Robotics Research 1*(2). Based on a Master's Thesis, Massachusetts Institute of Technology, Cambridge, Massachusetts.

Hiltz, S. R., and Turoff, M. (1978). *The network nation: Human communication through computer*. Reading, Massachusetts: Addison-Wesley.

Hiltz, S. R., Johnson, K., Aronovitch, C., and Turoff, M. (1980, August). *Face-to-face versus computerized conferences: A controlled experiment* (Report 12). New Jersey Institute of Technology Computerized Conferencing and Communications Center.

Hirsch, R. S. (1981). Procedures of the human factors center at San Jose. *IBM Systems Journal, 20*, 123–171.

Hirsh, A. T. (1984, January). The future of integrated software. *Business Computing*, 42–45.

Hirst, E. (1973). Energy intensiveness in passenger and freight transport modes, 1950–1970. Oakridge, Tennessee: Oakridge National Laboratory.

Hodge, M. H., and Pennington, F. M. (1973). Some studies of word abbreviation behavior. *Journal of Experimental Psychology, 98,* 350–361.

Holden, C. (1984). Will home computers transform school? *Science, 225,* 296.

Hollander, C. R., Iwasaki, Y., Courteille, J. M., and Fabre, M. (undated). *The drilling advisor,* unpublished.

Holley, C. D., Dansereau, D. F., McDonald, B. A., Garland, J. C., and Collins, K. W. (1979). Evaluation of a hierarchical mapping technique as an aid to prose processing. *Contemporary Educational Psychology, 4,* 227–237.

Holt, D. A. (1985). Computers in production agriculture. *Science, 228,* 422-427.

Hopper, D. (1976). The development of agriculture in developing countries. *Scientific American, 235*(3), 196–205.

Hormann, A. M. (1965, October). *Designing a machine partner—prospects and problems* (Technical Report SP-2169/000/01). Santa Monica, California: System Development Corporation (NTIS No. AD 626173).

Hormann, A. M. (1971a). A man-machine synergistic approach to planning and creative problem solving: Part I. *IJMMS, 3,* 167–184.

Hormann, A. M. (1971b). A man-machine synergistic approach to planning and creative problem solving: Part II. *IJMMS, 3,* 241–267.

Hormann, A. M. (1972). Planning by man-machine synergism. In H. Sackman & R. L. Citrenbaum (eds.), *Online planning: Toward creative problem-solving.* Englewood Cliffs, New Jersey: Prentice-Hall.

Horning, J. J. (1974). What the compiler should tell the user. In Bauer and Eickel (eds.), *Compiler construction.* Berlin: Springer-Verlag.

Hornsby, M. E. (1981). A comparison of full- and reduced-alpha keyboards for aircraft data entry. *Proceedings of the Human Factors Society, 25th Annual Meeting,* 257.

Horrocks, J. C., and de Dombal, F. T. (1973). Human and computer-aided diagnosis of "dyspepsia." *British Journal of Surgery, 60,* 910.

Houghton, B., and Wisdom, J. C. (1980). Non-bibliographic online databases: An investigation into their uses within the fields of economics and business studies. In J. Wanger et al. (eds.), *Proceedings of the Third International Online Information Meeting,* London, 4–6 December 1979. Oxford: Learned Information.

Howard, J. (1981). What is good documentation? *Byte, 6,* 132–150.

Howe, J. A. M., O'Shea, T., and Pane, J. (1979). *Teaching mathematics through logo programming* (Technical Report No. 115). Artificial Intelligence Laboratory, University of Edinburgh.

Howell, W. C. (1967, September). *Some principles for the design of decision systems: A review of six years of research on a command-control system simulation* (Technical Report AMRL-TR-67-136). Wright-Patterson Air Force Base, Ohio: Aerospace Medical Research Laboratories (NTIS No. AD 665469).

Howell, W. C., and Getty, C. F. (1968, November). *Some principles for design of decision systems: A review of the final phase of research on a command-control system simulation* (Technical Report No. AMRL-TR-68-158). Wright-Patterson Air

Force Base, Ohio: Aerospace Medical Research Laboratories (NTIS No. AD 684548).

Hudson, C. A. (1982). Computers in manufacturing. *Science, 215,* 818–825.

Hunting, W., Laubli, T. H., and Grandjean, E. (1981, December). Postural and visual loads at VDT workplaces: Constrained postures. *Ergonomics, 24*(12), 917–931.

Igersheim, R. H. (1976). Managerial response to an information system. *Proceedings of the National Computer Conference, 45,* 877–882. Montvale, New Jersey: AFIPS Press.

Ingersoll, A. P. (1983). The atmospheres. *Scientific American, 249*(3), 162–174.

Isa, B. S., Boyle, J. M., Neal, A. S., and Simons, R. M. (1983, December). A methodology for objectively evaluating error messages. In A. Janda (ed.), *Proceedings of the CHI'83 Conference on Human Factors in Computing Systems,* 68–71. New York: ACM.

Ivergard, T. B. K. (1976, September). Man-computer interaction in public systems. Paper presented at the Advanced Study Institute on Man-Computer Interaction, Mati, Greece.

Jackson, C. L. (1980). The allocation of the radio spectrum. *Scientific American, 242,* 34–39.

Jacob, R. J. K. (1983a, December). Executable specifications for a human-computer interface. In A. Janda (ed.), *Proceedings of the CHI'83 Conference on Human Factors in Computing Systems,* 28–34. New York: ACM.

Jacob, R. J. K. (1983b, April). Using formal specifications in the design of a human-computer interface. *Communications of the ACM, 26*(4), 259–264.

Jacques, J. A. (ed.) (1972). *Computer diagnosis and diagnostic methods.* Springfield, Illinois: Charles C. Thomas.

Janick, J., Noller, C. H., and Rhykerd, C. L. (1976, September). The cycles of plant and animal nutrition. *Scientific American,* 75–86.

Japan Information-Processing Development Center (1981, October). *Proceedings of International Conference on Fifth Generation Computer Systems.*

Jarke, M., and Vassiliou, Y. (1984). Coupling expert systems with database management systems. In W. Reitman (ed.), *Artificial intelligence applications for business.* Norwood, New Jersey: Ablex Publishers.

Jeffries, R., Turner, A. A., Polson, P. G., and Atwood, M. E. (1981). The processes involved in designing software. In J. Anderson (ed.), *Cognition and problem solving.* Hillsdale, New Jersey: Lawrence Erlbaum Associates.

Johansen, R., Vallee, J., and Collins, K. (1978). Learning the limits of teleconferencing: Design of a teleconference tutorial. In M. C. J. Elton, W. A. Lucas, and D. W. Conrath (eds.), *Evaluating new telecommunication systems.* New York: Plenum.

Johansen, R., Vallee, J., and Spangler, K. (1979). *Electronic meetings: Technical alternatives and social choices.* Menlo Park, California: Addison-Wesley.

Johnson, R. C. (1981, February). Thirty-two bit microprocessors inherit mainframe features. *Electronics,* 138–141.

Jones, V. E. (1980, July). *Final report of the software acquisition and development working group*. Prepared for the Assistant Secretary of Defense for Communications, Command, Control, and Intelligence.

Jordan, N. (1963). Allocation of functions between man and machines in automated systems. *Journal of Applied Psychology, 47*, 161–165.

Joyce, J. D., and Cianciolo, M. J. (1967). Reactive displays: Improving manmachine graphical communication. *AFIPS Conference Proceedings, 31*, 713–721.

Jutila, S. T., and Baram, G. (1971). A user-oriented evaluation of a timeshared computer system. *IEEE Transactions on Systems, Man, and Cybernetics, SMC-1*, 344–349.

Kahn, R. E. (1972a). Resource-sharing computer communications networks. *Proceedings of the IEEE, 60*(11), 1397–1407.

Kahn, R. E. (1972b). Terminal access to the ARPA computer network. In R. Rustin (ed.), *Computer networks*. Englewood Cliffs, New Jersey: Prentice-Hall.

Kahn, R. E., and Crowther, W. R. (1972, June). Flow control in a resource sharing computer network. *IEEE Transactions on Communication Technology of the ACM, 15*(4), 221–230.

Kahneman, D., and Tversky, A. (1974). Judgment under uncertainty: Heuristics and biases. *Science, 185*, 1124–1131.

Karlin, J. E., and Alexander, S. N. (1962, May). Communication between man and machine. *Proceedings of the Institute of Radio Engineers, 50*(5), 1124–1128.

Karplus, R. (1974). *Science curriculum improvement study: Teacher's handbook*. Berkeley: University of California.

Kay, A. C. (1977). Microelectronics and the personal computer. *Scientific American, 237*(3), 230–244.

Kay, A. (1984). Comment in Today and tomorrow: A discussion. In P. H. Winston and K. A. Prendergast (eds.), *The AI business*, Cambridge, Massachusetts: MIT Press.

Kelly, M. J. (1975, December). *Studies in interactive communication: Limited vocabulary natural language dialogue* (Technical Report No. 3). Baltimore: Johns Hopkins University (NTIS No. AD A019198).

Kelly, M. J., and Chapanis, A. (1977). Limited vocabulary natural language dialogue. *IJMMS, 9*, 479–501.

Kemeny, J. G. (1972). *Man and the computer*. New York: Charles Scribner's Sons.

Kempton, W. (in press). Two theories used for home heat control. In N. Quinn and D. Holland (eds.), *Cultural models in language and thought*. Cambridge: Cambridge University Press.

Kennedy, T. C. S. (1974). The design of interactive procedures for manmachine communication. *IJMMS, 6*, 309–334.

Kennedy, T. C. S. (1975). Some behavioral factors affecting the training of naive users of an interactive computer system. *IJMMS, 7*, 817–834.

Kerr, E. B., and Hiltz, S. R. (1982). *Computer-mediated communication systems: Status and evaluation.* New York: Academic Press.

Kerr, R. A. (1985). Fifteen years of African drought. *Science, 227,* 1453–1454.

Kibler, A. W., Watson, S. R., Kelly, C. S., III, and Phelps, R. H. (1978, November). *A prototype aid for evaluating alternative courses of action for tactical engagement* (Technical Report TR-78-A38) (AD A064 275).

King, W. R., and Cleland, D. I. (1975). The design of management information systems: An information analysis approach. *Management Science, 22,* 286–297.

Klatt, D. H. (1977). Review of the ARPA speech understanding project. *Journal of the Acoustical Society of America, 62,* 1345–1366.

Klein, S. (1983, July). Picture this . . . the new business graphics. *Computer Decisions,* 182–196.

Kleine, H., and Citrenbaum, R. L. (1972). Interactive management planning. In H. Sackman and R. L. Citrenbaum (eds.), *Online planning: Towards creative problem solving.* Englewood Cliffs, New Jersey: Prentice-Hall.

Kleinmutz, B. (1968). The processing of clinical information by men and machines. In B. Kleinmutz (ed.), *Formal representation of human judgment.* New York: Wiley.

Klemmer, E. T. (1971). Keyboard entry. *Applied Ergonomics, 2,* 2–6.

Klimbie, J. W. (1982). Digital optical recording: Principle, and possible applications. In L. Tedd et al. (eds.), *Proceedings of the Fifth International Online Information Meeting,* London, 8–10 December 1981. Oxford: Learned Information.

Klinger, A. (1973, December). *Natural language, linguistic processing, and speech understanding: Recent research and future goals* (Report No. R-1377-ARPA). Santa Monica, California: Rand Corporation.

Knapp, B. G., Moses, F. M., and Gellman, L. H. (1982). Information highlighting on complex displays. In A. Badre and B. Schneiderman (eds.), *Directions in human-computer interactions.* Norwood, New Jersey: Ablex Publishers.

Knowles, A. C. (1982, December). Third-generation personal computers as professional workstations. Presentation at Strategic Issues Conference, Monterey, California.

Knuth, D. E. (1972). An empirical study of FORTRAN programs. *Software—Practice and Experience, 1,* 105–133.

Kolata, G. (1982). How can computers get common sense? *Science, 217,* 1237–1238.

Kowalski, R. (1979). *Logic for problem solving.* New York: Elsevier North-Holland.

Kozdrowicki, E. W., and Theis, D. J. (1980, November). Second generation of vector supercomputers. *Computer, 13*(11), 9–21.

Kraft, A. (1984). XCON: An expert configuration system at Digital Equipment Corporation. In P. H. Winston and K. A. Prendergast (eds.), *The AI business.* Cambridge, Massachusetts: MIT Press.

Krasner, M. A., Huggins, A. W. F., Schwartz, R. M., Kimball, O. A., Chow, Y., and Makhoul, J. J. (1984, September). Development of a VIDVOX speech communication aid. Presented at the American Voice Input-Output Systems Conference, Arlington, Virginia.

Kriebel, C. H. (1970, September). *The evaluation of management information systems* (Report No. RR-226). Pittsburgh: Carnegie-Mellon University, Management Sciences Research Group (NTIS No. AD 723083).

Kriloff, H. A. (1976, October). Human factor considerations for interactive display systems. Position paper, ACM/Special Interest Group in Graphics Workshop on User-Oriented Design of Interactive Graphics Systems.

Kristal, J. (1983, November/December). The new wave board room. *Restaurant and Hotel Design*, 93–97.

Kroemer, K. H. Eberhard (1972). Human engineering the keyboard. *Human Factors, 14*(1), 51–63.

Kubitz, W. J. (1980). Computer technology: A forecast for the future. In F. W. Lancaster (ed.), *The role of the library in an electronic society. Proceedings of the 16th Annual Clinic on Library Applications of Data Processing,* 135–161. Urbana: University of Illinois, Graduate School of Library Science.

Kulikowski, C. A., and Weiss, S. M. (1982). Representation of expert knowledge for consultation: The CASNET and EXPERT projects. In P. Szolovits (ed.), *Artificial intelligence in medicine.* Boulder, Colorado: Westview Press.

Kull, D. (1983). Electronic mail: Should computers carry your mail? *Computer Decisions, 15,* 164–184.

Kurland, D. M., Mawby, R., and Cahir, N. (1984, April). The development of programming expertise. In K. Sheingold (chair), *Developmental Studies of Computer Programming Skills.* Symposium conducted by the American Educational Research Association, New Orleans.

Kurzweil, R. (1976). The Kurzweil reading machine: A technical overview. In M. R. Redden and W. Schwandt (eds.), *Science, Technology and the Handicapped* (AAAS Report No. 76-R-11, pp. 3–7). Washington, D.C.: American Association for the Advancement of Science.

Kurzweil, R. (1984, October). The coming age of intelligent machines or "what is 'AI' anyway?" *Proceedings, IEEE International Conference on Computer Design,* 1–14.

Kush, T. (1981). Designing decision support systems for the health sciences. *Proceedings, International Conference on Cybernetics and Society,* 168–172. New York: IEEE Systems, Man, and Cybernetics Society.

Lambert, J. V., Shields, J. L., Gade, P. A., and Dressel, J. D. (1978, June). *Comprehension of time-compressed speech as a function of training* (Technical Paper 295). Alexandria, Virginia: Army Research Institute for the Behavioral and Social Sciences.

Lampson, B. W. (1967). A critique of "an exploratory investigation of programmer performance under on-line and off-line conditions." *IEEE Transactions on Human Factors in Electronics, HFE-8,* 48–51.

Lancaster, F. W. (1982). *Libraries and librarians in an age of electronics.* Arlington, Virginia: Information Resources Press.

Landauer, T. K., Galotti, K. M., and Hartwell, S. (1983). Natural command names and initial learning: A study of text-editing terms. *Communications of the ACM, 26*(7), 495–503.

Landauer, T. K., Dumais, S. T., Gomez, L. M., and Furnas, G. W. (1982). Human factors in data access. *Bell System Technical Journal, 61,* 2487–2509.

Langridge, R., Ferrin, T. E., Kuntz, I. D., and Connolly, M. L. (1981). Real-time color graphics in studies of molecular interactions. *Science, 211,* 661–666.

Latermore, G. B. (1984, February). The business of writing for profit. *Business Computing,* 38–42.

Laubli, T. H., Hunting, W., and Grandjean, E. (1981, December). Postural and visual loads at VDT workplaces: Lighting conditions and visual impairments. *Ergonomics, 24*(12), 933–944.

Lea, W. A. (1968). Establishing the value of voice communication with computers. *IEEE Transactions on Audio and Electroacoustics, AU-16,* 184–197.

Lea, W. A. (ed.) (1980). *Trends in speech recognition.* Englewood Cliffs, New Jersey: Prentice-Hall.

Lea, W. A., and Shoup, J. E. (1980). Specific contributions to the ARPA SUR project. In W. A. Lea (ed.), *Trends in speech recognition.* Englewood Cliffs, New Jersey: Prentice-Hall.

Lederberg, J. (1978). Digital communications and the conduct of science. The new literacy. *Proceedings of the IEEE, 66,* 1314–1319.

Ledgard, H. F. (1971). Ten mini-languages: A study of topical issues in programming languages. *Computing Surveys, 3,* 115–146.

Ledgard, H., Whiteside, J. A., Singer, A., and Seymour, W. (1980). The natural language of interactive systems. *Communications of the ACM, 23*(10).

Lenat, D. B. (1984). Computer software for intelligent systems. *Scientific American, 251*(3), 204–213.

Leontief, W. W. (1980). The world economy of the year 2000. *Scientific American, 243*(3), 206–231.

Leontief, W., Dunchin, F., and Szyld, D. B. (1985). New approaches in economic analysis: *Science, 228,* 419–422.

Lesgold, A., and Reif, F. (1983). *Computers in education: Realizing the potential.* Chairmen's report of a research conference. Washington, D.C.: U.S. Department of Education and National Institute of Education.

Levine, R. D. (1982). Supercomputers. *Scientific American, 246*(1), 118–135.

Levit, R. A., Heaton, B. J., and Alden, D. G. (1975, November). *Development and application of decision aids for tactical control of battlefield operations: Decision support in a simulated tactical operations system (simtos).* Minneapolis: Honeywell Systems and Research Center.

Lewell, J. (1983, January/February). The pioneers—David Evans, Evans and Sutherland. *Computer Pictures,* 16–22.

Lewin, R. (1984). National networks for molecular biologists. *Science, 223,* 1379–1380.

Licklider, J. C. R. (1960). Man-computer symbiosis. *Institute of Radio Engineers Transactions on Human Factors Electronics, HFE-1,* 4–11.

Licklider, J. C. R. (1961). Preliminary experiments in computer-aided teaching. In J. E. Coulson (ed.), *Programmed learning and computer-based instruction.* New York: Wiley.

Licklider, J. C. R. (1962). One-line man-computer communication. *AFIPS Proceedings, 21,* 113–128.

Licklider, J. C. R. (1965, May). Man-computer partnership. *International Science and Technology, 41,* 18–26.

Licklider, J. C. R. (1967). Graphic input: A survey of techniques. In F. Gruenberger (ed.), *Computer graphics.* Washington, D.C.: Thompson.

Licklider, J. C. R. (1968). Man-computer communication. In C. A. Cuadra (ed.), *Annual review of information science and technology, vol. 3.* Chicago: Encyclopedia Britannica.

Licklider, J. C. R., and Clark, W. E. (1962). On-line man-computer communication. *AFIPS Proceedings, 21,* 113–128.

Lindsay, R. K., Buchanan, B. G., Feigenbaum, E. A., and Lederberg, J. (1980). *Applications of artificial intelligence for organic chemistry: The DENDRAL Project.* New York: McGraw-Hill.

Linn, M. C., and Fisher, C. W. (1983). Computer education: The gap between promise and reality. *Proceedings of making our schools more effective: A conference for educators.* San Francisco: Farwest Laboratory.

Linstone, H. A. (1975). Technology forecasting IV (where to look ahead). *IEEE Spectrum, 12*(4), 60–61.

Lipman, A., and Negroponte, N. (1979, September). *Graphical input techniques* (Technical Report 409). Cambridge, Massachusetts: MIT Architecture Machine Group.

Lodwick, G. S. (1965). A probabilistic approach to diagnosis of bone tumors. *Radiologic Clinics of North America, 3.*

Loeb, K. M. C. (1983). Membrane keyboards and human performance. *Bell System Technical Journal, 62*(6), 1733–1749.

Long, G., Hein, R., Coggiola, D., and Pizzente, M. (1978). *Networking: A technique for understanding and remembering instructional material.* Rochester, New York: Department of Research and Development, National Technical Institute for the Deaf, Rochester Institute of Technology.

Loomis, R. S. (1976, September). Agricultural systems. *Scientific American,* 99–105.

Lowe, T. C. (1966, December). *Design principles for an on-line information retrieval system* (Technical Report No. 67–14). Philadelphia: University of Pennsylvania, Moore School of Electrical Engineering (NTIS No. AD 647196).

Lusted, L. B. (1965). Computer techniques in medical diagnosis. In R. W. Stacy and B. Waxman (eds.), *Computers in biomedical research, 1.* New York: Academic Press.

McAllister, C., and Bell, J. M. (1971). Human factors in the design of an interactive library system. *ASIS Journal, 22,* 96–104.

McCall, J. A., Richards, P. K., and Walters, G. F. (1977). *Factors in software quality* (Report 77C1502). Sunnyvale, California: General Electric Company.

McCarthy, J. (1966). Information. *Scientific American, 215*(3), 64–73.

MacClay, H., and Osgood, C. E. (1959). Hesitation phenomena in spontaneous English speech. *Word, 15,* 19.

McCloskey, M, Caramazza, A., and Green, B. (1980). Curvilinear motion in the absence of external forces: native beliefs about the motion of objects. *Science, 210,* 1139–1141.

McCorduck, P. (1979). *Machines who think: A personal inquiry into the history and prospects of artificial intelligence.* San Francisco: Freeman.

McCorduck, P. (1984, November). The conquering machine. *Science, 5*(9), 131–138.

McCormick, E. J., and Sanders, M. S. (1982). *Human factors in engineering and design,* 5th edition. New York: McGraw-Hill.

McCosh, A. M., and Scott-Morton, M. S. (1978). *Management decision support systems.* New York: Wiley.

McDermott, J. (1982a). R1: A ruled-based configurer of computer systems. *Artificial Intelligence, 19*(1), 39–88.

McDermott, J. (1982b). XSEL: A computer salesperson's assistant. In J. E. Hayes, D. Michie, and Y.-H. Pao (eds.), *Machine Intelligence 10.* New York: Wiley.

McDermott, J. (1984). Building expert systems. In W. Reitman (ed.), *Artificial intelligence applications for business.* Norwood, New Jersey: Ablex Publishers.

McDermott, J., and Steele, B. (1981). Extending a knowledge-based system to deal with ad hoc constraints. *Proceedings of the Seventh IJCAI,* 824–828.

MacDonald, N. (1965, September). A time-shared computer system: The disadvantages. *Computers and Automation, 14*(9), 21–22.

MacDonald, N. H. (1983). The UNIX writer's workbench software: Rationale and design. *Bell System Technical Journal, 62*(6), 1891–1908.

MacDonald, N. H., Frase, L. T., Gingrich, P. S., and Keenan, S. A. (1982). The writer's workbench: Computer aids for text analysis. *IEEE Transactions on Communication, Con-30*(1), 105–110.

Mace, D. J., Harrison, P. C., Jr., and Seguin, E. L. (1979, August). *Prevention for and remediation of human input errors in ADP operations* (Technical Report No. 395). Alexandria, Virginia: U.S. Army Research Institute for the Behavioral and Social Sciences.

McEwen, S. A. (1981, May). An investigation of user search performance on a Telidon information retrieval system. *Telidon Behavioral Research 2: The Design of Videotex Tree Indices.*

McGirr, E. M. (1969). Computers in clinical diagnosis. In J. Rose (ed.), *Computers in medicine,* 19–29. London: J. A. Churchill.

Machover, C. (1972, November/December). Interactive CRT terminal selection. *Society for Information Display Journal, 1*(4), 10–17.

Mack, R., Lewis, C. H., and Carroll, J. (1983). Learning to use word processors: Problems and prospects. *ACM Transactions on Office Information Systems, 1*(3), 254–271.

McKeithen, K. B., Reitman, J. S., Rueter, H. H., and Hirtle, S. C. (1981). Knowledge organization and skill differences in computer programmers. *Cognitive Psychology, 13,* 307–325.

McQuillan, J. M., and Walden, D. C. (1977). The ARPA network design decisions. *Computer Networks, 1.*

McWilliams, P. A. (1984). *Personal computers and the disabled.* New York: Doubleday.

Madarasz, T. (1983, March/April). Television news by the keyboard. *Computer Pictures, 1*(2), 6–14.

Maddock, I. (1983). Technology and the future of work. In J. E. Hayes and D. Michie (eds.), *Intelligent systems: The unprecedented opportunity.* New York: Wiley.

Magnuson, R. A. (1966). Computer assisted writing. *Datamation, 12,* 49–52, 57, 59.

Mahler, H. (1980). People. *Scientific American, 243*(3), 66–77.

Malhotra, A., Thomas, J. C., Carroll, J. M., and Miller, L. A. (1980). Cognitive processes in design. *IJMMS, 14,* 269–282.

Mallik, K. (1982). Rehabilitation engineering: Making the disabled person able. In V. W. Stern and M. R. Redden (eds.), *Proceedings of the 1980 Workshops on Science and Technology for the Handicapped,* 188–193. Washington, D.C.: Project on the Handicapped in Science, American Association for the Advancement of Science.

Manual, T., and Evanczuk, S. (1983, November). Artificial intelligence: Commercial products begin to emerge in decades of research. *Electronics,* 127–129.

Marrill, T. (1963, November). *Libraries and question-answering systems* (Report No. 1071). Cambridge, Massachusetts: Bolt Beranek and Newman Inc.

Marrill, T., and Roberts, L. A. (1966). Cooperative network of timesharing computers. *Proceedings of the AFIPS 1966 Spring Joint Computer Conference,* 425–431.

Martin, A. (1972). A new keyboard layout. *Applied Ergonomics, 3*(1).

Martin, J. (1973). *Design of man-computer dialogues.* Englewood Cliffs, New Jersey: Prentice-Hall.

Martin, J. (1984, March). Softwood Systems' Multimate. *Business Computing, 2,* 62–65.

Martin, N., Friedland, P., King, J., and Stefik, M. (1977). Knowledge base management for experiment planning in molecular genetics. *Proceedings of the Fifth International Joint Conference on Artificial Intelligence.* Massachusetts Institute of Technology.

Martin, T. H., and Parker, E. B. (1971). Designing for user acceptance of an interactive bibliographic search facility. In D. E. Walker (ed.), *Interactive bibliographic search: The user/computer interface.* Montvale, New Jersey: AFIPS Press.

Martin, T. H., Carlisle, J., and Treu, S. (1973). The user interface for interactive bibliographic searching: An analysis of the attitudes of nineteen information scientists. *ASIS Journal, 24*, 142–147.

Martins, G. R. (1984). The overselling of expert systems. *Datamation, 30*, 76–80.

Matisoo, J. (1980). The superconducting computer. *Scientific American, 242*(5), 50–65.

Matula, R. A. (1981). Effects of visual display units on the eyes: A bibliography (1972–1980). *Human Factors, 23*(5), 581.

Mayer, J. (1976, September). The dimensions of human hunger. *Scientific American,* 40–49.

Mayer, R. E. (1981). The psychology of how novices learn computer programming. *Computing Surveys, 13*, 121–141.

Mayer, S. R. (1967). Computer-based subsystems for training the users of computer systems. *IEEE Transactions on Human Factors in Electronics, HFE-8*, 70–75.

Mayo, J. S. (1977). The role of microelectronics in communication. *Scientific American, 237*(3), 192–209.

Mayzner, M. S., and Dolan, T. R. (eds.) (1978). *Minicomputers in sensory and information-processing research.* Hillsdale, New Jersey: Lawrence Erlbaum Associates.

Mead, C., and Conway, L. (1980). *Introduction to VLSI systems.* Reading, Massachusetts: Addison-Wesley.

Meadow, C. T. (1970). *Man-machine communication.* New York: Wiley.

Meindl, J. D. (1982). Microelectronics and computers in medicine. *Science, 215*, 792–797.

Meister, D., and Sullivan, D. J. (1969, August). *Guide to human engineering design for visual displays.* Canoga Park, California: Defense Systems Division, Bunker-Ramo Corporation.

Metz, W. (1983, October 18). Forgotten Zenith scores big with its personal computer. *Boston Globe.*

Michie, D., and Buchanan, B. G. (1974). Current status of the heuristic Dendral program for applying artificial intelligence to the interpretation of mass spectra. In R. A. G. Carrington (ed.), *Computers for Spectroscopy.* London: Adam Hilger.

Miller, A. R. (1971). *The assault on privacy: Computers, data banks, and dossiers.* Ann Arbor: University of Michigan Press.

Miller, L. A. (1974). Programming by non-programmers. *IJMMS, 6*, 237–260. [Also: 1973 (Research Report RC-4280). Yorktown Heights, New York: IBM Watson Research Center.]

Miller, L. A. (1978). *Behavioral studies of the programming process* (Research Report). IBM, Yorktown Heights, New York.

Miller, L. A. (1981). Natural language programming: Styles, strategies, and contrasts. *IBM Systems Journal, 21*(2), 184–215.

Miller, L. A. (1982, June). Natural language texts are not necessarily grammatical and unambiguous. Or even complete. Paper presented as part of a panel, "Building non-normative systems—the search for robustness," Toronto, Canada. IBM Research Report RC 9441 (41672), Yorktown Heights, New York.

Miller, L. A., and Becker, C. A. (1974, November). *Programming in natural English* (Technical Report No. RC-5137). Yorktown Heights, New York: IBM: Watson Research Center (NTIS No. AD A003923).

Miller, L. A., and Thomas, J. C., Jr. (1976, December). *Behavioral issues in the use of interactive systems* (Technical Report No. RC-6326). Yorktown Heights, New York: IBM Watson Research Center.

Miller, L. A., Heidorn, G. E., and Jensen, K. (1981). Text-critiquing with the EPISTLE system: An author's aid to better syntax. *AFIPS Conference Proceedings, 50,* 649–655.

Miller, R. A., Pople, H. E., Jr., and Myers, J. D. (1982, August). Internist-1, an experimental computer-based diagnostic consultant for general internal medicine. *New England Journal of Medicine,* 468–476.

Miller, R. B. (1965, February). *Psychology for a man-machine problem-solving system* (Technical Report TROO. 1246). Poughkeepsie, New York: IBM Corporation [Also: 1967. In L. Thayer ed.), *Communication theory and research: Proceedings of the First International Symposium,* 310–347. Springfield, Illinois: Charles C. Thomas (NTIS No. AD 640283)].

Miller, R. B. (1968). Response time in man-computer conversational transactions. *Proceedings of the Spring Joint Computer Conference, 33,* 267–277. Montvale, New Jersey: AFIPS Press.

Miller, R. B. (1969). Archetypes in man-computer problem solving. *Ergonomics, 12,* 559–581.

Minsky, M. (1984). The problems and the promise. In P. H. Winston and K. A. Prendergast (eds.), *The AI business.* Cambridge, Massachusetts: MIT Press.

Mintzberg, H. (1973). *The nature of managerial work.* New York: Harper and Row.

Mjosund, A. (1975). Toward a strategy for information needs analysis. *Computers and Operations Research, 2,* 39–47.

Molzberger, P. (1983, December). Aesthetics and programming. In A. Janda (ed.), *Proceedings of the CHI'83 Conference on Human Factors in Computing Systems,* 247–250. New York: ACM.

Montgomery, C. A. (1972). Is natural language an unnatural query language? *Proceedings of the ACM National Conference,* 1075–1078. New York: ACM.

Monty, R. A., Geller, E. S., Savage, R. E., and Perlmuter, L. C. (1979). The freedom to choose is not always so choice. *Journal of Experimental Psychology: Human Learning and Memory, 5,* 170–178.

Mooers, C. D. (1982, August). *The Hermes guide* (Report No. 4995). Cambridge, Massachusetts: Bolt Beranek and Newman Inc.

Mooers, C. D. (1983). Changes that users demanded in the human interface to the Hermes message system. In A. Janda (ed.), *Proceedings of the CHI'83 Conference on Human Factors in Computing Systems*, 88–92. New York: ACM.

Moore, E. (1755, September). Advantages of labor. *World.* Reprinted in J. Ferguson (ed.), *The British essayists, 28.* London: T. C. Hansard (1819).

Moore, R. C. (1982). The role of logic in knowledge representation and common sense reasoning. *Proceedings of the National Conference on Artificial Intelligence*, 428–433.

Moran, T. P. (1983, December). Getting into a system: External-internal task mapping analysis. In A. Janda (ed.), *Proceedings of the CHI'83 Conference on Human Factors in Computing Systems*, 45–49. New York: ACM.

Morefield, M. A., Wiesen, R. A., Grossberg, M., and Yntema, D. B. (1969, June). *Initial experiments on the effects of system delay on on-line problem-solving* (Technical Note ESD-TR-69-158). Lexington, Massachusetts: Lincoln Laboratory, Massachusetts Institute of Technology.

Morgan, C. T., Chapanis, A., Cook, J. S., III, and Lund, M. W. (eds.) (1963). *Human engineering guide to equipment design.* New York: McGraw-Hill.

Morgan, K. (1984). Contract work by computer—at home. *Bulletins on Science and Technology for the Handicapped, 4*(1), 2.

Morrill, C. S. (1967). Computer-aided instruction as part of a management information system. *Human Factors, 9,* 251–256.

Morrill, C. S., Goodwin, N. C., and Smith, S. L. (1968). User input mode and computer-aided instruction. *Human Factors, 10,* 225–232.

Morton, M. S. S. (1967). Computer-driven visual display devices—their impact on the management decision-making process. Doctoral Dissertation, Harvard University, Graduate School of Business Administration.

Moses, F. L., and Ehrenreich, S. L. (1981). Abbreviations for automated systems. Paper presented at the Human Factors Society Meeting, Rochester, New York.

Moses, F. L., and Potash, L. M. (1979). *Assessment of abbreviation methods for automated tactical systems* (Technical Report No. 398). Alexandria, Virginia: U.S. Army Research Institute for the Behavioral and Social Sciences (NTIS No. AD A077 840).

Moshier, S. L., Osborn, R. R., Baker, J. M., and Baker, J. K. (1980). Dialog systems' automated speech recognition capabilities—Present and future. In S. Harris (ed.), *Proceedings of Symposium on Voice-Interactive Systems: Applications and Payoffs*, 163–187, Dallas, Texas.

Mosteller, F., and Wallace, D. L. (1964). *Inference and disputed authorship: The Federalist.* Reading, Massachusetts: Addison-Wesley.

Mosteller, W. (1981). Job entry control language errors. *Proceedings, SHARE 57,* 149–155. Chicago: SHARE, Inc.

Moto-Oka, T. (1983). Overview to the fifth-generation computer system project. *Proceedings of the Fifth-Generation Computer System Project*, 596–601. New York: ACM.

Moto-Oka, T. (ed.) (1982). *Proceedings of the international conference on the fifth generation computer systems.* Amsterdam: Elsevier-North Holland.

Mourant, R., Lakshman, R., and Chantadisai, R. (1981, October). Visual fatigue and CRT display terminals. *Human Factors, 23*(5), 529.

Murphy, J. A. (1983, May). Integrated office-automation systems. *Mini-Micro Systems,* 181–188.

Murray, W. E., Moss, C. E., Parr, W. H., Cox, C., Smith, M. J., Cohen, B. F. G., Stammerjohn, L. W., and Happ, A. (1981). *Potential health hazards of video display terminals.* U.S. Department of Health and Human Services, Public Health Service, Center for Disease Control, National Institute of Occupational Safety and Health, Division of Biomedical and Behavioral Science, Division of Surveillance, Hazard Evaluations and Field Studies. Washington, D.C.: U.S. Government Printing Office.

Murrell, S. (1983, December). Computer communication system design affects group decision making. In A. Janda (ed.), *Proceedings of the CHI'83 Conference on Human Factors in Computing Systems,* 63–67. New York: ACM.

Muter, P., Latremouille, S. A., Treurniet, W. C., and Beam, P. (1982). Extended reading of continuous text on television screens. *Human Factors, 24,* 501–508.

Myer, T. H. (1980, September). Future message system design: Lessons from the Hermes experience. *Proceedings of Computer Conference Fall '80,* IEEE Computer Society.

Myer, T. H., and Barnaby, J. R. (1971, January). *TENEX executive language: Manual for users.* (Revised by W. W. Plummer, April 1973.) Cambridge, Massachusetts: Bolt Beranek and Newman Inc.

Myer, T. H., and Mooers, C. C. (1976). *Hermes users' guide.* Cambridge, Massachusetts: Bolt Beranek and Newman Inc.

Naisbitt, J. (1984). *Megatrends.* New York: Warner Brothers.

Nakatani, L. H., and Rohrlich, J. A. (1983, December). Soft machines: A philosophy of user-computer interface design. In A. Janda (ed.), *Proceedings of the CHI'83 Conference on Human Factors in Computing Systems,* 19–23. New York: ACM.

National Assessment of Educational Progress (1981). *Reading, thinking, writing: A report on the 1979–1980 assessment,* NAEP Publications C.M. 6710, Princeton, New Jersey.

National Center for Education Statistics (1980, Fall). *Student use of computers in school* (FRSS Report No. 12). Washington, D.C.: U.S. Government Printing Office (NCES 81-243).

National Center for Education Statistics (1982, September). Early release, fast response survey system: Instructional use of computers in public schools. Washington, D.C.: U.S. Government Printing Office (NCES 82-245).

The National Science Board Commission on Precollege Education in Mathematics, Science, and Technology (1983, September). *Educating Americans for the 21st century: A plan of action for improving mathematics, science, and technology education for all American elementary and secondary students so that their achievement is the best in the world by 1995.* Washington, D.C.: National Science Foundation.

Nawrocki, L. H., Strub, M. H., and Cecil, R. M. (1973). Error categorization and analysis in man-computer communication systems. *IEEE Transactions on Reliability, R-22,* 135–140.

Neal, A. S., and Simons, R. M. (1983, December). Playback: A method for evaluating the usability of software and its documentation. *CHI'83 Proceedings,* 78–82. In A. Janda (ed.), *Proceedings of the CHI'83 Conference on Human Factors in Computing Systems,* 78–82. New York: ACM.

Negroponte, N. (1975, June). Idiosyncratic systems: Toward personal computers and understanding context. Cambridge, Massachusetts: Massachusetts Institute of Technology, Architecture Machine Group.

Negroponte, N., Herot, C., and Weinzapfel G. (1978, October). *One-point touch input of vector information for computer displays* (Technical Report TR-78-TH3). Cambridge, Massachusetts: Massachusetts Institute of Technology, Architecture Machine Group (AD A064 278).

Newell, A. (1965). The possibility of planning languages in man-computer communication. In F. A. Geldard (ed.), *Communication processes.* New York: Macmillan.

Newell, A., and Simon, H. A. (1963). GPS, a program that stimulates human thought. In E. A. Feigenbaum and J. Feldman (eds.), *Computers and thought.* New York: McGraw-Hill.

Newell, A., and Simon, H. A. (1972). *Human problem solving.* Englewood Cliffs, New Jersey: Prentice-Hall.

Newell, A., and Sproull, R. F. (1982). Computer networks: Prospects for scientists. *Science, 215,* 843–852.

Newell, A., Barnett, J., Forgie, J., Green, C., Klatt, D., Licklider, J. C. R., Munson, J., Reddy, R., and Woods, W. (1971, May). Speech-understanding systems: Final report of a study group. Pittsburgh, Pennsylvania: Computer Science Department, Carnegie-Mellon University.

Newman, M., and Sproull, F. (1979). *Principles of interactive computer graphics: Second edition.* New York: McGraw-Hill.

Newsted, P. R., and Wynne, B. E. (1976). Augmenting man's judgment with interactive computer systems. *IJMMS, 8,* 29–59.

Nicholson, R. M., Wiggins, B. D., and Silver, C. A. (1972, February). *An investigation into software structures for man/machine interactions.* Arlington, Virginia: Analytics, Inc. (NTIS No. AD 737266).

Nickerson, R. S. (1969). Man-computer interaction: A challenge for human factors research. *Ergonomics, 12,* 501–517. (Reprinted 1969. *IEEE Transactions, Man-Machine Systems, MMS-10,* 164–180.)

Nickerson, R. S. (1976, October). On conversational interaction with computers. In S. Treu (ed.), *User-oriented design of interactive graphics systems.* Proceedings of ACM/SIGGRAPH Workshop, Pittsburgh, Pennsylvania.

Nickerson, R. S. (1980, September/October). *Some human factors implications of the blurring of the line between communications and computation* (Report No. 4577). Cambridge, Massachusetts: Bolt Beranek and Newman Inc. Paper based on an invited talk given at the 9th International Symposium on Human Factors in Telecommunications, Red Bank, New Jersey.

Nickerson, R. S. (1981). Some characteristics of conversations. In B. Shackel (ed.), *Man-computer interaction: Human factors aspects of computers and people.* Rockville, Maryland: Sijthoff and Noordhoff. (Also, Report No. 3498. Cambridge, Massachusetts: Bolt Beranek and Newman Inc.)

Nickerson, R. S. (1982). Foreword: Information technology and psychology—An invitation for speculation. Conclusion: Information technology and psychology—A retrospective look at some views of the future. In R. A. Kasschau, R. Lachman, and K. R. Laughery (eds.), *Information technology and psychology: Prospects for the future.* New York: Praeger.

Nickerson, R. S. (1983). Computer programming as a vehicle for teaching thinking skills. *Thinking, The Journal of Philosophy for Children, 4*(3 and 4), 42–48.

Nickerson, R. S. (1985). Understanding understanding. *American Journal of Education, 93*(2), 201–239.

Nickerson, R. S., and Feehrer, C. E. (1975, August). *Decision making and training: A review of theoretical and empirical studies of decision making and their implications for the training of decision makers* (Technical Report: NAVTRA-EQUIPCEN 73-C-0128-1). Cambridge, Massachusetts: Bolt Beranek and Newman Inc.

Nickerson, R. S., and Huggins, A. W. F. (1977). *The assessment of speech quality* (Report No. 3486). Cambridge, Massachusetts: Bolt Beranek and Newman Inc.

Nickerson, R. S., and Pew, R. W. (1977, February). Chapter 6, Person-computer interaction, in *The C³-system user: Vol. 1. A review of research on human performance as it relates to the design and operation of command, control, and communication systems* (Report No. 3459). Cambridge, Massachusetts: Bolt Beranek and Newman Inc.

Nickerson, R. S., and Stevens, K. S. (1972). Teaching speech to the deaf: Can a computer help? *Proceedings of the ACM,* 25th Anniversary Conference, Boston, August 1972, 240–251.

Nickerson, R. S., and Stevens, K. S. (1980). Approaches to the study of the relationship between intelligibility and physical properties of speech. In J. Subtly (ed.), *Speech assessment and speech improvement for the hearing impaired.* Washington, D.C.: A. G. Bell Association for the Deaf.

Nickerson, R. S., Kalikow, D. N., and Stevens, K. N. (1976). Computer-aided speech training for the deaf. *Journal of Speech and Hearing Disorders, 41,* 120–132.

Nickerson, R. S., Perkins, D. M., and Smith, E. E. (1985). *The teaching of thinking.* Hillsdale, New Jersey: Lawrence Erlbaum Associates.

Nickerson, R. S., Myer, T. H., Miller, D. C., and Pew, R. W. (1981, September). *User-computer interaction: Some problems for human factors research* (Report No. 4719). Cambridge, Massachusetts: Bolt Beranek and Newman Inc.

Nilles, J. M., Carlson, F. R., Gray, P., and Hanneman, G. (1976). Telecommuting—An alternative to urban transportation congestion. *IEEE Transactions on Systems, Man, and Cybernetics, SMC-6,* 77–84.

Nisbett, R., and Ross, L. (1980). *Human inference: Strategies and shortcomings of social judgment.* Englewood Cliffs, New Jersey: Prentice-Hall.

Nora, S., and Minc, A. (1980). *The computerization of society.* Cambridge, Massachusetts: MIT Press.

Norman, D. A. (1983a). Design principles for human-computer interfaces. In A. Janda (ed.), *Proceedings of the CHI'83 Conference on Human Factors in Computing Systems,* 1–10. New York: ACM.

Norman, D. A. (1983b). Design rules based on analyses of human error. *Communications of the ACM, 26*(4), 254–258.

Norrie, C. (1984, March). Supercomputers for super problems: An architectural introduction. *IEEE Computer,* 62–74.

Novell, M. (1967). An information retrieval system for the inexperienced-experienced user: How a user would view the system. In A. B. Tonik (ed.), *Information retrieval: The user's viewpoint: An aid to design.* Fourth Annual National Colloquium on Information Retrieval. Philadelphia: International Information, Inc.

Noyce, R. S. (1977, September). Microelectronics. *Scientific American, 237,* 63–69.

NRC Committee on Computerized Speech Recognition Technologies (1984). *Automatic speech recognition in severe environments.* Washington, D.C.: National Research Council.

Obermayer, R. W. (1977). Accuracy and timeliness in large-scale data-entry subsystems. *Proceedings of the 21st Annual Meeting of the Human Factors Society,* 173–177. Santa Monica, California: Human Factors Society.

Oestberg, O. (1975, November/December). CRTs pose health problems for operators. *International Journal of Occupational Health and Safety, 44*(6), 24–26, 46, 50, 52.

Oestberg, O. (1976, September). Office computerization in Sweden: Worker participation, workplace design considerations, and the reduction of visual strain. Paper presented at NATO Advanced Study Institute on Man-Computer Interaction, Mati, Greece. (Reprinted by Department of Human Work Sciences, University of Lulea, Lulea, Sweden.)

"The Office of the Future" (1975, June 30). *Business Week,* 48–84.

Office of Technology Assessment (1981, October). *Computer-based national information systems: Technology and public policy issues.* Washington, D.C.: U.S. Government Printing Office.

Office of the Undersecretary of Defense for Research and Engineering (1982, April). *Report of the Defense Science Board, 1981 summer study panel on operational readiness with high performance systems.* Washington, D.C.: U.S. Government Printing Office.

Oldham, W. G. (1977). The fabrication of microelectronic circuits. *Scientific American, 237*(3), 110–128.

O'Neal, J. D. (1976). We increased typing productivity 340%. *The Office,* 95–97.

Orlansky, J., and String, J. (1979). *Cost-effectiveness of computer-based instruction in military training* (IDA Paper P-1375). Arlington, Virginia: Institute for Defense Analysis.

Ornstein, S. M., Heart, F. E., Crowther, W. R., Russell, S. B., Rising, H. K., and Mitchell, A. (1972). The terminal IMP for the ARPA computer network. *AFIPS Conference Proceedings, 40,* 243–254. Montvale, New Jersey: AFIPS Press.

Ortony, A. (1979). The role of similarity in similes and metaphors. In A. Ortony (ed.), *Metaphor and thought.* New York: Cambridge University Press.

Osborn, J., Fagan, L., Fallat, R., McClung, D., and Mitchell, R. (1979). Measuring the data from respiratory measurements. *Medical Instrumentation, 13*(6).

Parnas, D. L. (1969). On the use of transition diagrams in the design of a user interface for an interactive computer system. *Proceedings, 24th National ACM Conference,* 379–385.

Parsons, H. M. (1970). The scope of human factors in computer-based data processing systems. *Human Factors, 12,* 165–175.

Pauker, S. G., and Kassirer, J. P. (1975). Therapeutic decision making: A cost benefit analysis. *New England Journal of Medicine, 293,* 229–234.

Pauker, S. G., Gorry, G. A., Kassirer, J. P., and Schwartz, W. B. (1976). Toward the simulation of clinical cognition: Taking a present illness by computer. *American Journal of Medicine, 60,* 981–995.

Pavelle, R., Rothstein, M., and Fitch, J. (1981). Computer algebra. *Scientific American, 245*(6), 136–152.

Pea, R. D., and Kurland, D. M. (1984a). *Logo programming and the development of planning skills* (Technical Report 16). New York: Center for Children and Technology, Bank Street College of Education.

Pea, R. D., and Kurland, D. M. (1984b). On the cognitive effects of learning computer programming. *New Ideas in Psychology, 2*(1).

Peace, D. M. S., and Easterby, R. S. (1973). The evaluation of user interaction with computer-based management information systems. *Human Factors, 15,* 163–177.

Pennington, N. (1982, July). *Cognitive components of expertise in computer programming: A review of the literature* (Technical Report No. 46). Ann Arbor: University of Michigan, Center for Cognitive Science.

Perlis, A. J., Sayward, F. G., and Shaw, M. (eds.) (1984). *Software metrics: An analysis and evaluation.* Cambridge, Massachusetts: MIT Press.

Perlman, G. (1982, December). *Natural artificial languages: Low level processes* (ONR Report 8202). San Diego: University of California, Center for Human Information Processing.

Perlmuter, L. C., Scharff, K., Karsh, R., and Monty, R. A. (1980). Perceived control: A generalized state of motivation. *Motivation and Emotion, 4*(1), 35–45.

Perrone, G. (1984, February). Review: Einstein letter series. *Infor World, 6,* 53–54.

Petrie, H. G. (1979). Metaphor and learning. In A. Ortony (ed.), *Metaphor and thought.* New York: Cambridge University Press.

Pew, R. W., and Rollins, A. M. (1975). *Dialog specification procedure* (Report No.

3129, Revised Edition). Cambridge, Massachusetts: Bolt Beranek and Newman Inc.

Pew, R. W., Baron, S., Feehrer, C. E., and Miller, D. C. (1977, March). *Critical review and analysis of performance models applicable to man-machine systems evaluation* (Report No. 3446). Cambridge, Massachusetts: Bolt Beranek and Newman Inc. (NTIS No. AD A038597).

Phipps, C. H. (1982). Implementation of computer technology in the 1980s: A semiconductor perspective. In R. A. Kasschau, R. Lachman, and K. R. Laughery (eds.), *Information technology and psychology: Prospects for the future.* New York: Praeger.

Plath, W. J. (1972, November). *Restricted English as a user language.* Yorktown Heights, New York: IBM Watson Center.

Plucknett, D. L., and Smith, N. J. H. (1984). Networking in international agricultural research. *Science, 225,* 989–993.

Pohm, A. V. (1984). High-speed memory systems. *Computer, 17,* 162–171.

Pollack, A. (1984, March 1). A document that can talk. *New York Times,* Technology column.

Pollack, L., and Weiss, H. (1984). Communications satellites: Countdown for INTLSAT VI. *Science, 223,* 553–559.

Pomerantz, D. (1982, August). Electronic mail: A new medium for the message. *Today's Office,* 41–47.

Pooch, U. W. (1976, August). Computer graphics, interactive techniques, and image processing 1970–1975: A bibliography. *Computer,* 46–64.

Pool, I. S. (1983). Tracking the flow of information. *Science, 221,* 609–613.

Pople, H. E., Jr. (1977). The formation of composite hypotheses in diagnostic problem solving: An exercise in synthetic reasoning. *Proceedings of the Fifth International Joint Conference on Artificial Intelligence.* Pittsburgh, Pennsylvania: Carnegie-Mellon University.

Pople, H. E., Jr. (1982). Heuristic methods for imposing structure on ill-structured problems: The structuring of medical diagnostics. In P. Szolovits (ed.), *Artificial Intelligence in medicine,* 119–190. Boulder, Colorado: Westview Press.

Pople, H. E. (1984a). Knowledge-based expert systems: The buy or build decision. In W. Reitman (ed.), *Artificial intelligence applications for business,* 23–40. Norwood, New Jersey: Ablex Publishers.

Pople, H. E. (1984b). Caduceus: An experimental expert system for medical diagnosis. In P. H. Winston and K. A. Prendergast (eds.), *The AI business,* 67–80. Cambridge, Massachusetts: MIT Press.

Poppel, H. L. (1982). Who needs the office of the future? *Harvard Business Review, 60*(11), 739–751.

Porat, M. (1977). *Information economy: Definition and measurement.* U.S. Department of Commerce, Office of Telecommunications. Washington, D.C.: U.S. Government Printing Office.

*Practice Digest* (1983). Projects and resources, a sampling: Computer to enhance retarded clients' lives. *Practice Digest, 6*(3), 31.

Predicasts (1984). *Prediction from Predicasts, 2*(7).

Preston, G. W. (1983). The very large scale integrated circuit. *American Scientist, 71*(5), 466–472.

Prince, M. D. (1972). *Interactive graphics for computer-aided design.* Reading, Massachusetts: Addison-Wesley.

Prince, S. D. (1983, March/April). PROJECTS: Inside "The Works." *Computer Pictures, 1*(2), 16–20.

Proctor, J. H. (1963). Normative exercising: An analytic and evaluative aid in system design. *IEEE Transactions on Engineering Management, EM-10,* 183–192.

Project Athena (1984, May). Faculty/Student Projects, Massachusetts Institute of Technology.

Quillian, M. R. (1968). Semantic memory. In M. Minsky (ed.), *Semantic information processing.* Cambridge, Massachusetts: MIT Press.

Quillian, M. R. (1969). The teachable language comprehender: A simulation program and theory of language. *Communications of the ACM, 12,* 459–476.

Raben, T. (1985). Computer applications in the humanities: *Science, 228,* 434–438.

Rabiner, L. R., and Levison, S. E. (1981). Isolated and connected word recognition—Theory and selected applications. *IEEE Transactions on Communications, 29*(5), 621–659.

Rabiner, L. R., and Schafer, R. W. (1976). Digital techniques for computer voice responses: Implementation of applications. *Proceedings of the IEEE, 64,* 416–432.

Raitzer, G. A., Vanderheiden, G. C., and Holt, C. S. (1976). Interfacing computers for the physically handicapped: A review of international approaches. *AFIPS Conference Proceedings, 45,* 209–216.

Ramsey, H. R., and Atwood, M. E. (1979, September). *Human factors in computer systems: A review of the literature* (Technical Report SAI-79-111-DEN). Englewood, Colorado: Science Applications, Inc.

Ramsey, H. R., Atwood, M. E., and Kirshbaum, P. J. (1978, May). *A critically annotated bibliography of the literature of human factors in computer systems* (Technical Report No. SAI-78-070-DEN). Englewood, Colorado: Science Applications, Inc.

Rapp, M. H. (1972). Man-machine interactive transit system planning. *Socio-Economic Planning Sciences, 6,* 95–123.

Rasmussen, W. D. (1982). The mechanization of agriculture. *Scientific American, 247*(3), 76–89.

Rauch-Hindin, W. (1983). Artificial intelligence: A solution whose time has come—part 1. *Systems and Software,* 150–177.

Rauch-Hindin, W. (1984, March). Communication standards: ISO poised to make its mark. *Systems and Software,* 104–126.

Redden, M. R., and Stern, V. M. (1983). *Technology for independent living II.* Washington, D.C.: American Association for the Advancement of Science.

Reder, S. M., and Conklin, N. F. (1984). Some issues in electronically mediated communication. Paper presented at the Fifth Annual Ethnography in Education Research Forum at the University of Pennsylvania, March 30–April 1.

Reid, B. K., and Walker, J. H. (1979, July). Scribe: Introductory user's manual (Second Edition). Cambridge, Massachusetts: Bolt Beranek and Newman Inc.

Reisner, P. (1977). Use of psychological experimentation as an aid to development of a query language. *IEEE Transactions on Software Engineering, SE-3*(3), 218–229.

Reisner, P. (1981). Formal grammar and human factors design of an interactive graphics system. *IEEE Transactions on Software Engineering, SE-7*, 229–240.

Reisner, P. (1982, March). Further developments toward using formal grammar as a design tool. *Proceedings of the Conference on Human Factors in Computer Systems*, 304–308. Gaithersburg, Maryland: National Bureau of Standards.

Reisner, P., Boyce, R. F., and Chamberlain, D. D. (1975). Human factors evaluation of two data base query languages: Square and sequel. *AFIPS Conference Proceedings*, 447–452. Montvale, New Jersey: AFIPS Press.

Reitman, W. (1984a). Managing the acquisition of an AI capability: Some observations, suggestions, and conclusions. In W. Reitman (ed.), *Artificial intelligence applications for business*. Norwood, New Jersey: Ablex Publishers.

Reitman, W. (1984b). Artificial intelligence applications for business: Getting acquainted. In W. Reitman (ed.), *Artificial intelligence applications for business*. Norwood, New Jersey: Ablex Publishers.

Reitman, W. (1984c). Managing the acquisition of an AI capability: Some observations and conclusions. In W. Reitman (ed.), *Artificial intelligence applications for business*. Norwood, New Jersey: Ablex Publishers.

Remington, R. J., and Rogers, M. (1969, February). *Keyboard literature survey: Phase I: Bibliography* (Technical Report No. TR29.0042). Research Triangle Park, North Carolina: IBM Systems Development Division.

Renner, J. W., and Lawson, A. E. (1973). Promoting intellectual development through science teaching. *The Physics Teacher, 11.*

Revelle, R. (1976). The resources available for agriculture. *Scientific American, 235*(3), 165–178.

Rich, C. (1984). The programmer's apprentice. In P. H. Winston and K. A. Prendergast (eds.), *The AI business*. Massachusetts: MIT Press.

Richards, J. T., and Boies, S. J. (1981). The IBM audio distribution system. *Proceedings of the IEEE MIDCON Conference.*

Reiger, C. A., and Greenstein, J. S. (1982). The allocation of tasks between the human and computer in automated systems. *Proceedings of the International Conference on Cybernetics and Society*, 204–208. New York: IEEE Systems, Man, and Cybernetics Society.

Riganati, J. P., and Schneck, P. B. (1984, October). Supercomputing. *Computer*, 97–111.

Ritchie, G. J., and Turner, J. A. (1975). Input devices for interactive graphics. *IJMMS*, 7, 639–660.

Roach, J. W., and Nickson, M. (1983, December). Formal specifications for modeling and developing human/computer interfaces. In A. Janda (ed.), *Proceedings of the CHI'83 Conference on Human Factors in Computing Systems*, 35–39. New York: ACM.

Roberts, L. G. (1973). The ARPANET. In N. Abramson and F. Kuo (eds.), *Computer-communication networks*. Englewood Cliffs, New Jersey: Prentice-Hall.

Roberts, L. G., and Wessler, B. D. (1970). Computer network development to achieve resource sharing. *Proceedings of the Spring Joint Computer Conference*, 543–549. Montvale, New Jersey: AFIPS Press.

Roberts, T. L., and Moran, T. P. (1983). The evaluation of text editors: Methodology and empirical results. *Communications of the ACM, 26*(4), 265–283.

Robinson, A. L. (1984a). Computing without dissipating energy. *Science, 223*, 1164–1166.

Robinson, A. L. (1984b). Experimental memory chips reach one megabit. *Science, 224*, 590–592.

Robinson, A. L. (1984c). One billion transistors on a chip? *Science, 223*, 267–278.

Robinson, J. A. (1983). Logical reasoning in Machines. In J. E. Hayes and D. Michie (eds.), *Intelligent systems*. Chichester, England: Ellis Horwood Limited.

Robson, D. (1981, August). Object-oriented software systems. *Byte, 6*, 74–86.

Rockart, J. F., and Treacy, M. E. (1982). The COE goes on line. *Harvard Business Review, 60*, 82–88.

Rodriguez, H., Jr. (1977, August). *Measuring user characteristics on the multics system* (Technical Report No. MIT/LCS/TM-89). Cambridge, Massachusetts: Massachusetts Institute of Technology.

Rosati, R. D., McNeer, J. F., and Stead, E. A. (1975). A new information system for medical practices. *Archives of Internal Medicine, 135*, 1017–1024.

Rosenberg, J. (1982). Evaluating the suggestiveness of command names. *Behavior and Information Technology, 1*, 371–400.

Rosenfeld, A. (1969). *Picture processing by computer*. San Francisco: Academic Press.

Rosnick, P., and Clement, J. (1980). Learning without understanding: The effect of tutoring strategies on algebra misconceptions. *The Journal of Mathematical Behavior, 3*, 3–27.

Rouse, W. B. (1975). Design of man-computer interfaces for on-line interactive systems. *Proceedings of the IEEE*, 847–857. New York: IEEE.

Rubenstein, M. F. (1975). *Patterns of problem solving*. Englewood Cliffs, New Jersey: Prentice-Hall.

Rubenstein, M. F. (1980). A decade of experience in teaching an interdisciplinary problem-solving course. In D. T. Tuma and F. Reif (eds.), *Problem solving and education: Issues in teaching and research.* Hillsdale, New Jersey: Lawrence Erlbaum Associates.

Rubin, A., and Bruce, B. (1983, September). *QUILL: Reading and writing with a microcomputer* (BBN Report No. 5410). To appear in B. A. Huston (ed.), *Advances in reading/language research 3.* Greenwich, Connecticut: JAI Press.

Rubinstein, R., and Goldenberg, E. P. (1978, June). Using a computer message system for promoting reading and writing in a school for the deaf. *Proceedings of the 5th Annual Conference on Systems and Devices for the Disabled,* 135–138.

Rupp, B. (1984). *Human factors of work stations with visual displays.* Yorktown Heights, New York: IBM Corporation.

Saal, H. J., and Weiss, Z. (1977). An empirical study of APL programs. *Computer Languages, 2*(3), 47–60.

Sackman, H. (1968, October). Time-sharing versus batch processing: The experimental evidence. *AFIPS Conference Proceedings,* 1–10. Montvale, New Jersey: AFIPS Press. (Also published with additional summary section: [1967] Technical Report SP2975. Santa Monica, California: System Development Corporation [NTIS No. AD 661665].)

Sackman, H. (1970a). Experimental analysis of man-computer problem-solving. *Human Factors, 12,* 187–201.

Sackman, H. (1970b). *Man-computer problem solving.* Princeton, New Jersey: Auerbach.

Sackman, H. (1972). Advanced research in online planning. In H. Sackman and R. L. Citrenbaum (eds.), *Online planning: Towards creative problem-solving.* Englewood Cliffs, New Jersey: Prentice-Hall.

Sackman, H. (1981). Outlook for man-computer symbiosis: Toward a general theory of man-computer problem solving. In B. Shackel (ed.), *Man-computer interaction: Human factors aspects of computers and people.* Rockville, Maryland: Sijthoff and Noordhoff.

Sackman, H., and Citrenbaum, R. L. (eds.) (1972). *Online planning: Towards creative problem-solving.* Englewood Cliffs, New Jersey: Prentice-Hall.

Sammet, J. E. (1966). The use of English as a programming language. *Communications of the ACM, 9,* 228–230.

Sammet, J. E. (1969). *Programming languages: History and fundamentals.* Englewood Cliffs, New Jersey: Prentice-Hall.

Samuel, A. L. (1963). Some studies in machine learning using the game of checkers. In E. Feigenbaum and J. Feldman (eds.), *Computers and thought.* New York: McGraw-Hill.

Sandberg-Diment, E. (1984, June 26). Do word processors spoil writers? *The New York Times,* 19.

Sauter, S. L., Gottlieb, M. S., Jones, K. C., Dodson, V. N., and Rohrer, K. M. (1983, April). Job and health implications of VDT use: Initial results of the Wisconsin-NIOSH study. *Communications of the ACM, 26*(4), 264–294.

Scapin, D. (1981). Computer commands in restricted natural language: Some aspects of memory and experience. *Human Factors, 23,* 365–375.

Schank, R., and Abelson, R. (1977). *Scripts, plans, goals, and understanding: An inquiry into human knowledge structures.* Hillsdale, New Jersey: Lawrence Erlbaum Associates.

Schatzoff, M., Tsao, R., and Wiig, R. (1967). An experimental comparison of time sharing and batch processing. *Communications of the ACM, 10,* 261–265.

Scherr, A. L. (1965). An analysis of time-shared computer systems. Ph.D. Thesis, Massachusetts Institute of Technology, Cambridge, Massachusetts (MAC-TR-18).

Schneider, M. L., Wexelblat, R. L., and Jende, M. S. (1980). Designing control languages from the user's perspective. In D. Beech (ed.), *Command language directions: Proceedings of the International Federation for Information Processing Working Conference on Command Languages.* New York: North-Holland.

Schoenfeld, A. H. (1980). Teaching problem-solving skills. *American Mathematical Monthly, 87*(10), 794–805.

Schoichet, S. (1981, April). Personal workstations: A concept evolves into a blooming industry. *Mini-Micro Systems,* 98–113.

Schure, A. (1983). What lies in the future for the management of computer graphics? *Computer Pictures, 1*(6), 36–42.

Schwartz, A. (ed.) (1984). *Handbook of microcomputer applications in communication disorders.* San Diego: College-Hill Press.

Science and the Citizen (1984). And the poor get sicker. *Scientific American, 251*(3), 82–91.

Scott, R., and Simmons, D. (1974). Programmer productivity and the Delphi technique. *Datamation, 20*(5), 71–73.

Scrimshaw, N. S., and Taylor, L. (1980). Food. *Scientific American, 243*(3), 78–88.

Seaman, I. (1983). Voice mail: Should computers carry your mail? *Computer Decisions, 15,* 188–200.

Sedelow, S. Y. (1970). The computer in the humanities and fine arts. *Computing Surveys, 2,* 89–110.

Seibel, R. (1972). Data entry devices and procedures. In H. P. Van Cott and R. G. Kinkade (eds.), *Human engineering guide to equipment design* (Revised Edition). Washington, D.C.: U.S. Government Printing Office.

Senders, J. W. (1963). Information storage requirements for the contents of the world's libraries. *Science, 141,* 1067–1068.

Severance, L. S., and Granato, D. J. (1983, March). C3I, silent partner. *Signal,* 37–40.

Seybold, J. W. (1981, May). The Xerox star, a "professional" workstation. *The Seybold Report on Word Processing, 4*(5), 1–19.

Seybold, J. W. (1983, March 7). *The Seybold report on professional computing, 1*(2).

Shackel, B. (1969). Man-computer interaction: The contribution of the human sciences. *Ergonomics, 12,* 485–499.

Shackel, B. (1981, September). The concept of usability. *Proceedings of IBM Software and Information Usability Symposium.*

Shackel, B. (ed.). (1981). *Man-computer interaction: Human factors aspects of computers and people.* Rockville, Maryland: Sijthoff and Noordhoff.

Shackel, B., and Shipley, P. (1970, February). *Man-computer interaction: A review of ergonomics literature and related research* (Report No. DMP-3472). Hayes, Middlesex, England: EMI Electronics Ltd.

Shank, C. V., and Auston, D. H. (1982). Ultrafast phenomena in semiconductor devices. *Science, 215,* 797–801.

Sharon, N. (1980). Carbohydrates. *Scientific American, 2*(5), 90–116.

Shaw, J. C. (1965, May). *Joss: Experience with an experimental computing service for users at remote typewriter consoles* (Report No. P-3149). Santa Monica, California: The Rand Corporation.

Sheil, B. A. (1982). Coping with complexity. In R. A. Kasschau, R. Lachman, and K. R. Laughery (eds.), *Information technology and psychology: Prospects for the future.* New York: Praeger.

Sheil, B. (1983). Power tools for programmers. *Datamation, 29*(2), 131–144.

Sherman, H. (1981, January). *A comparative study of computer-aided clinical diagnosis of birth defects.* S.M. Thesis, Department of Electrical Engineering and Computer Science, Massachusetts Institute of Technology, Cambridge, Massachusetts.

Sherman, R. H., and Gable, M. G. (1983, June). Considerations in interconnecting diverse local nets. *Data Communications,* 145–154.

Sherwood, G. A. (1979, August). The computer speaks. *IEEE Spectrum,* 18–25.

Shneiderman, B. (1977). Measuring computer program quality and comprehension. *IJMMS, 9,* 465–478.

Shneiderman, B. (1980a). *Software psychology: Human factors in computer and information systems.* Cambridge, Massachusetts: Winthrop Publishers.

Shneiderman, B. (1980b). System message design: Guidelines and experimental results. In A. Badre and B. Shneiderman (eds.), *Directions in human-computer interaction.* Norwood, New Jersey: Ablex Publishers.

Shneiderman, B. (1982, September). Designing computer system messages. *Communications of the ACM, 25*(9), 604–605.

Shneiderman, B., and McKay, D. (1976). Experimental investigations of computer program debugging and modification. *Proceedings of the 6th International Congress of the International Ergonomics Association.*

Shneiderman, B., and Mayer, R. (1979). Syntactic/semantic interactions in programmer behavior: A model and experimental results. *International Journal of Computer and Information Sciences, 7,* 219–238.

Shneiderman, B., Mayer, R., McKay, D., and Heller, P. (1977). Experimental investigations of the utility of detailed flowcharts in programming. *Communications of the ACM, 20,* 373–381.

Short, J., Williams, E., and Christie, B. (1976). *The social psychology of telecommunications.* New York: Wiley.

Shortliffe, E. H. (1976). *Computer based medical consultations: MYCIN.* New York: American Elsevier.

Shortliffe, E. H., and Buchanan, B. G. (1975). A model of inexact reasoning in medicine. *Mathematica Biosciences, 23,* 351.

Shuford, E. H. (1965). A computer-based system for aiding decision making. In J. Spiegel and D. Walker (eds.), *Information system sciences, Proceedings of the Second Congress,* 157–168. Washington, D.C.: Spartan Books.

Shurtleff, D. A. (1980). *How to make displays legible.* La Mirada, California: Human Interface Design.

Siegel, J. H., and Farrell, E. J. (1973). A computer simulation model to study the clinical observability of ventilation and perfusion abnormalities in human stock states. *Surgery.*

Siegel, J. H., and Strom, B. L. (1972). The computer as a "living textbook" applied to the care of the critically injured patient. *Journal of Trauma, 12,* 739–755.

Sime, M. E. (1981). The empirical study of computer language: "So I said in the most natural way, *if x* = 0, then begin . . . ." In B. Shackel (ed.), *Man-computer interaction: Human factors aspects of computers and people.* Rockville, Maryland: Sijthoff and Noordhoff.

Sime, M. E., Green, T. R. C., and Guest, D. J. (1973). Psychological evaluation of two conditional constructions used in computer languages. *IJMMS, 5,* 105–113.

Sizer, T. R. (1983, June). High school reform: The need for engineering. *Phi Delta Kappan, 64*(10).

Small, D. W. (1983). An experimental comparison of natural and structured query languages. *Human Factors, 25,* 253–263.

Small, D. W., and Weldon, L. J. (1977, September). The efficiency of retrieving information from computers using natural and structured query languages (Report SAI-78-655 WA). Arlington, Virginia: Science Applications, Inc.

Smarr, L. L. (1985). An approach to complexity: numerical computations. *Science, 228,* 403–408.

Smith, D. C., Harslem, E., Irby, C., and Kimball, R. (1982, June). Star user interface: An overview. *Proceedings of the AFIPS 1982 National Computer Conference, 50,* 515–528.

Smith, D. C., Irby, C., Kimball, R., Verplank, W., and Harslem, E. (1982, April). Designing the star user interface. *Byte, 74,* 242–282.

Smith, L. B. (1967). A comparison of batch processing and instant turnaround. *Communications of the ACM, 10*(8), 495–500.

Smith, M. J., Cohen, B. G. F., Stammerjohn, I. W., and Happ, A. (1981, August). An investigation of health complaints and job stress in video display operations. *Human Factors, 23*(4), 387–399.

Smith, S. L. (1974). *An on-line model of traffic control in a communication network* (Technical Report MTR-2813). Bedford, Massachusetts: MITRE Corporation.

Smith, S. L., and Goodwin, N. C. (1970). Computer-generated speech and man-computer interaction. *Human Factors, 12,* 215–223. (Also, 1969. *Proceed-*

*ings of the International Symposium on Man-Machine Systems: Vol. 1*, 8–12 (IEEE Conference Record No. 69C58-MMS). New York: IEEE.)

Socha, J. (1984, January). For the handicapped. Toggeling shift keys. *PC*, 199–204.

Soloway, E., Ehrlich, K., Bonar, J., and Greenspan, J. (1982). What do novices know about programming? In B. Shneiderman and A. Badre (eds.), *Directions in human-computer interactions*. Norwood, New Jersey: Ablex Publishers.

Spinrad, R. J. (1982). Office automation. *Science, 215*, 808–813.

Splinter, W. E. (1976). Center-pivot irrigation. *Scientific American, 234*(6), 90–99.

Sridharan, N. S. (1985, Fall). Evolving systems of knowledge. *AI Magazine*, 108–120.

Stallman, R., and Sussman, G. J. (1976). *Forward reasoning and dependency-directed backtracking in a system for computer-aided circuit analysis* (Memo 380). Massachusetts Institute of Technology, AI Laboratory.

Stanford Research Institute (1972). *Patterns of energy consumption in the United States*. Washington, D.C.: U.S. Government Printing Office.

Stefick, M. (1981a). Planning and meta-planning (MOLGEN: Part I). *Artificial Intelligence, 16*, 11–140.

Stefick, M. (1981b). Planning and meta-planning (MOLGEN: Part II). *Artificial Intelligence, 16*, 141–170.

Steier, R. (ed.) (1983, June). The STARS program. *Communications of the ACM, 26*(6), 399, 460.

Sterling, T. D. (1974). Guidelines for humanizing computerized information systems: A report from Stanley House. *Communications of the ACM, 17*, 609–613.

Sterling, T. D. (1975). Humanizing computerized information systems. *Science, 190*, 1168–1172.

Stern, M. R., and Redden, V. W. (1982). *Technology for independent living I*. Washington, D.C.: American Association for the Advancement of Science.

Stevens, A. L., Roberts, B. R., Stead, L. S., Forbus, K. D., Steinberg, C., and Smith, B. C. (1981). *Steamer: Advanced computer aided instruction in propulsion engineering* (Report No. 4702). Cambridge, Massachusetts: Bolt Beranek and Newman Inc.

Stewart, T. F. M. (1976). Displays and the software interface. *Applied Ergonomics, 7*, 137–146.

Stewart, T. F. M. (1981). The specialist user. In B. Shackel (ed.), *Man-computer interaction: Human factors aspects of computers and people*. Rockville, Maryland: Sijthoff and Noordhoff.

Streeter, L. A., Ackroff, J. M., and Taylor, G. A. (1983). On abbreviating command names. *Bell System Technical Journal, 62*(6), 1807–1826.

Streeter L. A., Ackroff, J. M., Taylor, G. A., and Galotti, K. M. (1979). Cited in Landauer, Galotti, and Hartwell (1983).

Strehlo, K. (1984, January). When the objective is efficient project management. *Personal Computing*, 132–141, 194.

Summers, J. K., and Bennett, E. (1967). AESOP: A final report. A prototype on-line interactive information control system. In D. E. Walker (ed.), *Information system science and technology*. Washington, D.C.: Thompson (papers prepared for the Third Congress).

Sunn, M. (1983). The Pentagon's ambitious computer plan. *Science, 222,* 1213–1215.

Sussman, G. J., and Stallman, R. M. (1975). Heuristic techniques in computer aided circuit analysis. *IEEE Transactions on Circuits and Systems, CAS-22*(11).

Sutherland, I. E., and Mead, C. A. (1977). Microelectronics and computer science. *Scientific American, 237*(3), 210–228.

Swets, J. A. (1963). Information retrieval systems. *Science, 141,* 245–250.

Swets, J. A., Green, D. M., and Winter, E. F. (1961). Learning to identify nonverbal sounds. *Journal of the Acoustical Society of America, 33,* 855(A).

Swets, J. A., Harris, J. R., McElroy, L. S., and Rudloe, H. (1964). *Further experiments on computer-aided learning of sound identification* (Report: NAV-TRADEVCEN 789-2). Port Washington, New York: U.S. Naval Training Device Center.

Swets, J. A., Harris, J. R., McElroy, L. S., and Rudloe, H. (1966). Computer-aided instruction in perceptual identification. *Behavioral Science 11*(2), 98–104.

Swets, J. A., Millman, S. H., Fletcher, W. E., and Green, D. M. (1962). Learning to identify nonverbal sounds: An application of a computer as a teaching machine. *Journal of the Acoustical Society of America, 34*(7), 928–935.

Synectics Corporation (1980). Development of design guidelines and criteria for user/operator transactions with battlefield automated systems Fairfax, Virginia: author.

Szolovits, P. (ed.) (1982). *Artificial intelligence in medicine*. Boulder, Colorado: Westview Press.

Szolovits, P., and Pauker, S. G. (1978). Categorical and probabilistic reasoning in medical diagnosis. *Artificial Intelligence, U*(11), 115–144.

Tanenbaum, A. S. (1981). *Computer networks*. Englewood Cliffs, New Jersey: Prentice-Hall.

Tarasoff, B. J. (1978, October). *Changing communications technology and its impact on data processing markets*. New York: Blyth, Eastman, Dillon Industry Report.

Tate, P. (1984, November). The blossoming of European AI. *Datamation,* 86–88.

*Technology Trends Newsletter,* March 1984 Telesensory Systems News Release (1984). Revolutionary deaf-blind communication aid announced. Mountain View, California: Telesensory Systems, Inc.

Teger, S. L. (1983). Impacting the evolution of office automation. *Proceedings of the IEEE, 71*(4), 503–511.

Teitelman, W. (1972, April). "Do what I mean": The programmer's assistant. *Computers and Automation, 21*(4), 8–11.

Teitelman, W., and Masinter, L. (1981). The Interlisp programming environment. *Computer, 14*(4), 25–33.

Tesler, L. G. (1984, September). Programming languages. *Scientific American, 251*(3), 70–93.

Thomas, J. C. (1976). *Quantifiers and question-asking* (Research Report RC 5866). New York: IBM Watson Research Center.

Thomas, J. C. (1977). Psychological issues in database management. *Proceedings of the Third International Conference on Very Large Data Bases*, Tokyo.

Thomas, J. C., and Carroll, J. M. (1981). Human factors in communication. *IBM Systems Journal 20*(2), 236–263.

Thomas, J. C., and Carroll, J. M. (1982). Metaphor and the cognitive representation of computing systems. *IEEE Transactions on Systems, Man, and Cybernetics, SMC-12*(2).

Thomas, J. C., and Gould, J. D. (1975). A psychological study of query-by-example. *Proceedings of the National Computer Conference*, 439–445. Montvale, New Jersey: AFIPS Press.

Thomas, R. H., Forsdick, H. C., Crowley, T. R., Robertson, G. G., Schaaf, R. W., Tomlinson, R. S., and Travers, V. M. (1985, in press). Diamond: A multimedia message system built upon a distributed architecture. *Computer Magazine*.

Thompson, D. A. (1971). Interface for an interactive information retrieval system: A literature survey and a research system description. *ASIS Journal, 22*, 361–373.

Toffler, A. (1970). *Future shock.* New York: Random House.

Toffler, A. (1980). *The third wave.* New York: Bantam.

Tolnay, T. (1983, March/April). Weathering the computer graphics storm. *Computer Pictures, 1*(2), 22–27.

Tomeski, E. A., and Lazarus, H. (1975). *People-oriented computer systems.* New York: Van Nostrand Reinhold.

Toong, H. D. (1977). Microprocessors. *Scientific American, 237*(3), 146–161.

Toong, H. D., and Gupta, A. (1982). Personal computers. *Scientific American, 247*(6), 86–107.

Treu, S. (1975). Interactive command language design based on required mental work. *IJMMS, 7*, 135–149.

Turn, R. (1974, January). *Speech as a man-computer communication channel* (Report No. P-5120). Santa Monica, California: Rand Corporation.

Turoff, M., and Hiltz, R. (1977, May). Meeting through your computer. *IEEE Spectrum*, 58–67.

Turoff, M. W., Whitescarver, J., and Hiltz, S. R. (1978). The human-machine interface in a computerized conferencing environment. *Proceedings of the IEEE Conference on Interactive Systems, Man, and Cybernetics*, 145–157. New York: IEEE.

Tversky, A., and Kahneman, D. (1974). Judgment under uncertainty: Heuristics and biases. *Science, 185*, 1124–1131.

Twentieth Century Fund (1983). *Making the grade.* New York: Twentieth Century Fund.

Uber, G. T., Williams, P. E., Hisey, B. L., and Siekert, R. G. (1968). The organization and formatting of hierarchical displays for the on-line input of data. *AFIPS Conference Proceedings,* 219–226. Montvale, New Jersey: AFIPS Press.

Uhlig, R. P. (1977, May). Human factors in computer message systems. *Datamation,* 120–126.

Unger, B. (1976). Bringing the disabled user into the design process: Improving new technology for the disabled. In M. R. Redden and W. Schwandt (eds.), *Science, technology, and the handicapped* (AAAS Report No. 76-R-11, pp. 7–17). Washington, D.C.: American Association for the Advancement of Science.

U.S. Army Materiel and Readiness Command (1980, October). *The future of electronic information handling at the FCC: Blueprint for the 80s.* Washington, D.C.: Federal Communications Commission.

Uttal, W. R. (1967). *Real-time computers: Technique and applications in the psychological sciences.* New York: Harper and Row.

Vallee, J., Johansen, R., Randolph, R. H., and Hastings, A. C. (1974, November). *Group communication through computers, vol. 2: A study of social effects* (Report R-33). Menlo Park, California: Institute for the Future.

Vance, D. (1969). A data book of museum holdings. *ICRH Newsletter.* New York University, Institute for Computer Research in the Humanities.

van Dam, A. (1984, September). Computer software for graphics. *Scientific American, 251*(3), 146–161.

van Dam, A., and Rice, D. E. (1971). On-line text editing: A survey. *Computing Surveys, 3,* 93–114.

Van Melle, W. (1979). A domain-independent production rule system for consultation programs. *Proceedings of the Sixth IJCAI,* 923–925. Stanford: Department of Computer Science, Stanford University.

Van Melle, W., Scott, A. C., Bennett, J. S., and Peairs, M. (1981). *The EMYCIN manual* (Technical Report STAN-CS-81-885). Stanford, California: Stanford University, Computer Science Department.

Voekler, C. H. (1938). An experimental study of the comparative rate of utterance of deaf and normal hearing speakers. *American Annals of the Deaf, 83,* 274–284.

*VoiceNews* (1984, April). *4*(5). Rockville, Maryland: Stoneridge Technical Services.

*VoiceNews* (1984, May). *4*(5). Rockville, Maryland: Stoneridge Technical Services.

Waern, Y., and Rollenhagen, C. (1983). Reading text from visual display units. *IJMMS, 18,* 441–465.

Wagreich, B. (1982). Technology for the sensory impaired: Present and future. In V. W. Stern and M. R. Redden (eds.), *Proceedings of the 1980 Workshops on Science and Technology for the Handicapped,* 194–200. Project on the Handicapped in Science—American Association for the Advancement of Science, Washington, D.C.

Walden, D. (1972, April). A system for interprocess communication in a resource sharing computer network. *Communications of the ACM, 15*(4), 221–230.

Waldrop, M. M. (1984a). Artificial intelligence (I): Into the world. *Science, 223*, 802–805.

Waldrop, M. M. (1984b). Artificial intelligence in parallel. *Science, 225*, 608–610.

Waldrop, M. M. (1985). NSF commits to supercomputers. *Science, 228*, 568–571.

Walker, A. (1984). SYLLOG: An Approach to PROLOG for nonprogrammers. In M. van Caneghem and D. H. D. Warren (eds.), *Logic programming and its applications*. Norwood, New Jersey: Ablex Publishers.

Walker, A., and Porto, A. (1983). *Kbol: A knowledge-based garden store assistant* (Technical Report). San Jose, California: IBM Research Laboratory.

Walsh, J. (1983). Super competing over super computers. *Science, 220*, 581–584.

Walsh, J. (1984). NSF plans help with big computer problems. *Science, 223*, 797–798.

Walston, C. E., and Felix, C. P. (1977). A method of programming measurement and estimation. *IBM Systems Journal, 16*(1), 54–73.

Walther, G. H., and O'Neil, H. F., Jr. (1974). On-line user-computer interface: The effects of interface flexibility, terminal type, and experience on performance. *AFIPS Conference Proceedings*, 379–384. Montvale, New Jersey: AFIPS Press.

Waltz, D. L. (1982). Artificial intelligence. *Scientific American, 247*(4), 118–133.

Waltz, D. L. (1983, November). Helping computers understand natural language. *IEEE Spectrum*, 81–84.

Wason, P. C., and Johnson-Laird, P. (1972). *Psychology of reasoning: Structure and content*. London: Batsford.

Wasserman, A. I., and Shewmake, D. T. (1982). Rapid prototyping of interactive information systems. *ACM SIGSOFT Software Engineering Notes, 7*, 171–180.

Wasserman, T. (1973). The design of idiot-proof interactive systems. *Proceedings of the National Computer Conference, 42*. Montvale, New Jersey: AFIPS Press.

Waterman, D. A., and Jenkins, B. M. (1977, March). *Heuristic modeling using rule-based computer systems* (Report No. P-5811). Santa Monica, California: Rand Corporation.

Waterman, D. A., Anderson, R. H., Hayes-Roth, F., Klahr, P., Martin, G., and Rosenschein, S. J. (1979, May). *Design of a rule-oriented system for implementing expertise* (Report No. N-1158-1-ARPA). Santa Monica, California: Rand Corporation.

Waters, R. C. (1982). The programmer's apprentice: Knowledge based program editing. *IEEE Transactions on Software Engineering, SE-8, 1.*

Watson, R. W. (1976). User interface design issues for a large interactive system. *AFIPS Conference Proceedings,* 357–364. Montvale, New Jersey: AFIPS Press.

Watt, D. (1984a, June). The feds are coming! *Popular Computing,* 91–94.

Watt, D. (1984b, January). Tools for writing. *Popular Computing,* 75–78.

Weinberg, G. M. (1971). *The psychology of computer programming.* New York: Van Nostrand Reinhold.

Weiss, S., Kulikowski, C. A., and Safir, A. (1978). Glaucoma consultation by computer. *Computers in Biomedical Research, 8,* 24–40.

Weiss, S., Kern, K., Kulikowski, C. A., and Safir, A. (1976). System for interactive analysis of a time-sequenced optimal logical data base. *Proceedings of the Third Illinois Conference on Medical Information Systems.* Chicago: University of Illinois Medical Center.

Weiss, S., Kulikowski, C. A., Amarel, S., and Safir, A. (1978). A model-based method for computer-aided medical decision-making. *Artificial Intelligence, 11,* 145–172.

Weizenbaum, J. (1966). Eliza—A computer program for the study of natural language communications between man and machine. *Communications of the ACM, 9,* 36–45.

Weizenbaum, J. (1967). Contextual understanding by computers. *Communications of the ACM, 10*(8), 474–480.

Weizenbaum, J. (1976). *Computer power and human reason: From judgment to calculation.* San Francisco: Freeman.

Welty, C. (1979). A comparison of a procedural and a nonprocedural query language: Syntactic metrics and human factors. Ph.D. Dissertation, Computer and Information Science Department, University of Massachusetts, Amherst, Massachusetts.

Westin, A. F., and Baker, M. A. (1972). *Databanks in a free society: Computers, record-keeping and privacy.* New York: Quadrangle.

Weyls, S., Fries, J., Weiderhood, G., and Germano, F. (1975). A modular self-describing clinical data bank system. *Computers in Biomedical Research, 8,* 279–293.

Whalen, T., and Latremouille, S. (1981, May). The effectiveness of a tree-structured index when the existence of information is uncertain. *Telidon behavioral research 2: The design of Videotex tree indices.*

White, B. W. (1962). Studies of perception. In H. Borko (ed.), *Computer applications in the behavioral sciences.* Englewood Cliffs, New Jersey: Prentice-Hall.

Whitehead, S. F., and Castleman, P. A. (1974). Evaluation of an automated medical history in office practice. Fourth Annual Conference of the Society for Computer Medicine, New Orleans, Louisiana. Abstract in *Clinical Medicine and the Computer: Proceedings of the Conference.*

Whitfield, D. (1976, July). Human factors aspects of interactive conflict resolution in air traffic control. Paper presented at Sixth Congress of the International Ergonomics Association, University of Maryland. Symposium on Computer Aiding in Cognitive Tasks.

Whitfield, D., and Stammers, R. (1976, May). Human factors aspects of computer aiding for air traffic controllers. Paper presented at NATO Training Course on Human Engineering, Utrecht.

Whitted, T. (1982). Some recent advances in computer graphics. *Science, 215,* 767–774.

Wiener, N. (1948). *Cybernetics.* New York: Wiley.

Wiener, N. (1950). *The human use of human beings: Cybernetics and society.* Boston: Houghton Mifflin.

Wiener, N. (1960). Some moral and technical consequences of automation. *Science, 131,* 1355–1358.

Wiener, N. (1964). *God & Golem, Inc.* Cambridge, Massachusetts: MIT Press.

Wiesner, J. B. (1971, December). Science, technology, and the quality of life. *Technology Review,* 15–18.

Williams, M. D., Hollan, J., and Stevens, A. L. (1981). An overview of STEAMER: An advanced computer-assisted instructional system for propulsion engineering. *Behavioral Methods and Instrumentation, 2*(13).

Williams, M. E. (1985). Electronic databases. *Science, 228,* 445–456.

Williams, M. E., Lannom, L., and Robins, C. G. (eds.) (in press). *Computer-readable database: A directory and data sourcebook.* 2 vols. Chicago: American Library Association.

Williges, R. C., and Williges, B. H. (1981, September). *Users' considerations in computer based information systems* (Technical Report CSIE-81-2). Blacksburg, Virginia: Virginia Polytechnic Institute and State University (NTIS No. AD A106194).

Winograd, T. (1984, September). Computer software for working with language. *Scientific American, 251*(3), 130–145.

Winston, P. (1981). Learning and reasoning by analogy. *Communications of the ACM, 23,* 689–703.

Winston, P. H. (1977). *Artificial intelligence.* Reading, Massachusetts: Addison-Wesley.

Winston, P. H. (1984). Perspective. In P. H. Winston and K. A. Prendergast (eds.), *The AI business.* Cambridge, Massachusetts: MIT Press.

Winston, P. H., and Prendergast, K. A. (eds.) (1984). *The AI business: The commercial uses of Artificial Intelligence.* Cambridge, Massachusetts: MIT Press.

Winterbotham, F. W. (1974). *The ultra secret.* New York: Harper and Row.

Wirth, N. (1984, September). Data structures and algorithms. *Scientific American, 251*(3), 60–69.

Witten, I. H., and Madams, P. H. C. (1977). The telephone enquiry service: A man-machine system using synthetic speech. *IJMMS, 9,* 449–464.

Wixon, D., Whiteside, J., Good, M., and Jones, S. (1983, December). Building a user-defined interface. In A. Janda (ed.), *Proceedings of the CHI'83 Conference on Human Factors in Computing Systems,* 24–27. New York: ACM.

Wolfram, S. (1984, September). Computer software in science and mathematics. *Scientific American, 251*(3), 188–203.

Woods, W. A. (1973). Progress in natural language understanding: An application to lunar geology. *AFIPS Conference Proceedings, 42*, 441–450. Montvale, New Jersey: AFIPS Press.

Woods, W. A. (1977). A personal view of natural language understanding. *Special Interest Group in Artificial Intelligence, Newsletter, 61*, 17–20.

Woods, W. A. (1984). Natural language communication with machines: An ongoing goal. In W. Reitman (ed.), *Artificial intelligence applications for business.* Norwood, New Jersey: Ablex Publishers.

Woodson, W. E., and Conover, D. W. (1964). *Human engineering guide for equipment designers.* Second Edition. Berkeley: University of California Press.

Wright, P. (1983). Manual dexterity: A user-oriented approach to creating computer documentation. In A. Janda (ed.), *Proceedings of the CHI'83 Conference on Human Factors in Computing Systems,* 11–18. New York: ACM.

Wright, P., and Bason, G. (1982). Detour routes to usability: A comparison of alternative approaches to multipurpose software design. *IJMMS, 18*, 391–400.

Wright, P., and Lickorish, A. (1983). Proof-reading texts on screen and paper. *Behavior and Information Technology, 2*(3), 227–235.

Yeh, S. Y., Betyar, L., and Hon, E. H. (1972). Computer diagnosis of fetal heart rate patterns. *American Journal of Obstetrics and Gynecology, 114*, 890–897.

Yntema, D. B., and Klem, L. (1965). Telling a computer how to evaluate multidimensional situations. *IEEE Transactions on Human Factors Engineering,* 3–13.

Yntema, D. B., and Torgerson, W. S. (1961). Man-computer cooperation in decisions requiring common sense. *Institute of Radio Engineers Transactions on Human Factors in Electronics,* 20–26.

Young, R. M. (1983). Surrogates and mappings: Two kinds of conceptual models for interactive devices. In A. L. Stevens and D. Gentner (eds.), *Mental models.* Hillsdale, New Jersey: Lawrence Erlbaum Associates.

Youngs, E. A. (1974). Human errors in programming. *IJMMS, 6*, 361–376.

Yourdon, E. (1971). Maybe the computers can save us after all. *Computers and Automation, 20*(5), 21–26.

Yu, V. L., Buchanan, B. G., Shortliffe, E. H., Wraith, S. M., Davis, R., Scott, A. C., and Cohen, S. N. (1979). Evaluating the performance of a computer-based consultant. *Computer Programs in Biomedicine, 9*, 95–102.

Yu, V. L., Fagan, L. M., Wraith, S. M., Clancey, W. J., Scott, A. C., Hannigan, J. F., Blum, R. L., Buchanan, B. G., and Cohen, S. N. (1979). Antimicrobial selection by a computer—a blinded evaluation by infectious disease experts. *Journal of the American Medical Association, 242*, 1279–1282.

Zeigler, J. F., and Lanford, W. A. (1979). Effect of cost rays on computer memories. *Science, 206*, 776–788.

Zelmin, W. R., Daniloff, R. G., and Skinner, T. H. (1968). The difficulty of listening to time-compressed speech. *Journal of Speech and Hearing Research, 11,* 875–881.

Zloof, M. M. (1977). Query-by-Example—A data base language. *IBM Systems Journal, 4,* 324–343.

Zoltan, E., and Chapanis, A. (1982). What do professional persons think about computers? *Behavior and Information Technology, 1,* 55–68.

Zraket, C. A. (1984). Strategic command, control, communications, and intelligence. *Science, 224,* 1306–1311.

# Name Index

Abbott, V., 267, 268
Abelson, P. H., 5, 286
Abraham, E., 58
Abrams, M. H., 142
Abramson, B., 289
Ackroff, J. M., 124, 127
Adams, R. E., 343
Addis, T. R., 131
Alexander, S. N., 74, 309
Amarel, S., 294
Anderson, R. H., 36, 115, 167, 216
Archambault, P., 343
Aronovitch, C., 181
Ascher, R. N., 191
Atwood, M. F., 50, 74, 226, 260, 261, 262
Austin, H., 289

Babbage, Charles, 30–31
Bacon, B., 260, 270
Bacon, G., 335–336
Bahil, A. T., 161
Baker, C. A., 99
Baker, J. D., 46
Baker, M. A., 194, 289
Baldwin, J. T., 253
Balzar, R. M., 86
Bannon, L., 77, 225
Baram, G., 79
Baran, P., 21
Barber, D. L. A., 21
Barmack, J. E., 99
Barnaby, J. R., 150
Barnard, P. J., 124, 223, 245
Baron, S., 79
Barrett, J. A., 127
Barter, T. C., 40
Basil, R., 292, 301
Bason, G., 222
Bates, M., 131
Bavelas, A., 181

Beasley, D. S., 142
Becker, C. A., 260
Becker, J., 108
Beeler, M., 65
Begg, V., 11, 169
Behrmann, M., 344
Belden, T., 181
Bell, A. G., 295
Bell, D., 5
Bell, J. M., 191, 253
Bennett, E., 74
Bennett, J. S., 167, 225
Bentley, T. J., 221
Bergman, H., 81
Berman, M. L., 90
Bertoni, P., 20
Betyar, L., 296
Birnbaum, J. S., 110, 258, 269
Black, J. B., 124, 245
Blackadar, T., 65
Blackledge, M. A., 168
Blanc, R. P., 22
Blesser, T., 230
Blum, R. L., 292
Bobrow, R. J., 131, 253, 295
Boehm, B. W., 21, 81, 97, 263, 277
Boies, S. J., 79, 152, 165, 225, 254, 262
Bollinger, J. G., 36
Boothroyd, A., 343
Borgatta, L. S., 10, 217
Borko, H., 49
Bott, R. A., 250
Bowden, Lord Vivian, 4
Bowe, F. G., 343
Bowen, R. J., 168
Bown, H. G., 188, 234–235
Boyce, R. F., 191
Boyle, J. M., 252
Branscomb, L. M., 9, 10, 217, 254
Breen, M., 296
Bridgewater, J., 74

Heglin, H. J., 90
Heidorn, G. E., 129, 163, 164
Hein, R., 115
Heller, P., 263
Helps, F. G., 324
Hemingway, P. W., 105
Herot, C., 107
Higginson, P., 237
Hill, I. D., 133
Hillis, W. D., 65, 90
Hiltz, S. R., 174, 181, 183, 225
Hirsch, R. S., 90
Hirsh, A. T., 102
Hirst, E., 209
Hirtle, S. C., 261
Hisey, B. L., 106
Hoare, C. A. R., 262
Hodge, M. H., 127
Holden, C., 334
Hollan, J., 295
Hollander, C. R., 290
Hollerith, Herman, 30
Holley, C. D., 115
Holt, D. A., 36
Hon, E. A., 296
Hopper, D., 327
Hormann, A. M., 75, 86
Horning, J. J., 252
Hornsby, M. E., 90
Horrocks, J. C., 296
Horwitz, P., 334
Houghton, B., 189
Hovanyecz, T., 145, 224
Hovland, H. L., 97
Howard, J., 250
Howe, J. A., 267
Howell, W. C., 225
Huddart, B., 90
Hudson, C. A., 169
Huggins, A. W. F., 142, 343
Hunting, W., 100

Igersheim, R. H., 243, 324
Irby, C., 18
Isa, B. S., 252
Ivergard, T. B., 190
Iwasaki, Y., 290

Jackson, C. L., 26
Jacob, R. J. K., 230, 231
Jacquard, Joseph, 30
Jacques, J. A., 296
Janick, J., 325

Jarke, M., 297
Jeffries, R., 260, 261
Jende, M. S., 229
Jenkins, B. M., 167
Jensen, K., 163, 164
Johansen, R., 181, 183
Johnson, R. C., 10
Johnson-Laird, P., 72
Jones, K. C., 100
Jones, S., 233
Jones, V. E., 40
Jordan, N., 86
Jorgensen, A., 223
Joyce, J. D., 97
Jutila, S. T., 79

Kahn, R. E., 21, 22
Kahneman, D., 72, 168
Kalikow, D. N., 343
Kanarick, A. F., 90
Karlin, J. F., 74
Karplus, R., 313, 331
Karsh, R., 243
Kassirer, J. P., 286, 295
Kay, A. C., 248, 264, 276
Keenan, S. A., 162
Kelly, C. S., 168
Kelly, M. J., 145
Kemeny, J. G., 74
Kennedy, T. C. S., 225, 251
Kepner, H., 267, 331
Kern, K., 294
Kerr, E. B., 181
Kerr, R. A., 328
Kibler, A. W., 168
Kimball, O. A., 343
Kimball, R., 18
King, J., 291
King, W. R., 243
Kirshbaum, P. J., 74
Klein, S., 97, 215
Kleine, H., 75
Kleinmutz, B., 296
Kleinrock, L., 22
Klem, L., 167
Klemmer, E. T., 90, 256
Klimbie, J. W., 12
Klinger, A., 131
Knapp, B. G., 105
Knerr, B. W., 46
Knowles, A. C., 9
Knuth, D. E., 262
Koelega, H. S., 81

Whalen, T., 119
White, B. W., 48
Whitehead, S. F., 243, 296
Whitescarven, J., 225
Whiteside, J. A., 127, 233
Whitfield, D., 77, 168
Whitfield, G. R., 97
Whitted, T., 169
Wiener, Norbert, 32–33, 198, 205,
   323, 343
Wiesen, R. A., 81, 82
Wiesner, Jerome, 322
Wiggins, B. D., 74
Wiig, R., 80
Wikler, E. S., 245
Williams, E., 106, 181, 295
Williges, R. C., 225, 248
Winograd, T., 130, 134
Winston, P. H., 247, 275, 276, 287
Winter, E. F., 47
Winterbotham, F. W., 31
Wirth, N., 120
Wisdom, J. C., 189
Witten, I. H., 91
Wixon, D., 233
Wolf, A. K., 357
Wolfram, S., 44–45
Woods, W. A., 130, 131, 224
Woodson, W. E., 99
Worlton, W. J., 16, 68
Wraith, S. M., 292
Wright, P., 100, 222, 251
Wynne, B. E., 168

Yasukama, 263
Yeh, S. Y., 296
Yntema, D. B., 81, 82, 167
Youngs, E. A., 262
Yourdon, E., 184
Yu, V. L., 292

Zdonik, S. R., 296
Zeigler, J. F., 14
Zemlin, W. R., 142
Zloof, M. M., 224
Zoltan, E., 240
Zraket, C. A., 39

# Subject Index

# ꓐ *Bradford Books*